CHINESE OUTBOUND TOURISM 2.0

Advances in Hospitality and Tourism

CHINESE OUTBOUND TOURISM 2.0

Edited by
Xiang (Robert) Li, PhD

Apple Academic Press Inc.
3333 Mistwell Crescent
Oakville, ON L6L 0A2
Canada

Apple Academic Press Inc.
9 Spinnaker Way
Waretown, NJ 08758
USA

Exclusive worldwide distribution by CRC Press, a member of Taylor & Francis Group

Printed in the United States of America on acid-free paper

International Standard Book Number-13: 978-1-77188-180-7 (Hardcover)

International Standard Book Number-13: 978-1-77188-181-4 (eBook)

Library and Archives Canada Cataloguing in Publication

Chinese outbound tourism 2.0 / edited by Xiang (Robert) Li, PhD.

(Advances in hospitality and tourism book series)
Includes bibliographical references and index.
Issued in print and electronic formats.
ISBN ISBN 978-1-77188-180-7 (hardcover).--ISBN 978-1-77188-181-4 (pdf)
1. Chinese--Travel. 2. Tourism. 3. International travel--China. 4. Consumer behavior--China. I. Li, Xiang (Professor of tourism), author, editor II. Series: Advances in hospitality and tourism book series

G155.C6C45 2015 910.89'951 C2015-906150-4 C2015-906151-2

Library of Congress Cataloging-in-Publication Data

Chinese outbound tourism 2.0 / edited by Xiang (Robert) Li, PhD.

pages cm
Includes bibliographical references and index.
ISBN 978-1-77188-180-7 (alk. paper)
1. Tourism--China. 2. Chinese--Travel--Foreign countries. 3. Tourism--Management.
4. Tourism--Marketing. I. Li, Xiang (professor of tourism management) editor of compilation.

G155.C55C4827 2015 910.89'951--dc23 2015034298

Apple Academic Press also publishes its books in a variety of electronic formats. Some content that appears in print may not be available in electronic format. For information about Apple Academic Press products, visit our website at **www.appleacademicpress.com** and the CRC Press website at **www.crcpress.com**

ADVANCES IN HOSPITALITY AND TOURISM BOOK SERIES

Editor-in-Chief:
Mahmood A. Khan, PhD
Professor, Department of Hospitality and Tourism Management,
Pamplin College of Business,
Virginia Polytechnic Institute and State University,
Falls Church, Virginia, USA
email: mahmood@vt.edu

BOOKS IN THE SERIES

Food Safety: Researching the Hazard in Hazardous Foods
Editors: Barbara Almanza, PhD, RD, and Richard Ghiselli, PhD

Strategic Winery Tourism and Management: Building Competitive Winery Tourism and Winery Management Strategy
Editor: Kyuho Lee, PhD

Sustainability, Social Responsibility and Innovations in the Hospitality Industry
Editor: H. G. Parsa, PhD

Consulting Editor: Vivaja "Vi" Naraparreddy, PhD
Associate Editors: SooCheong (Shawn) Jang, PhD, Marival Segarra-Oña, PhD, and Rachel J. C. Chen, PhD, CHE

Managing Sustainability in the Hospitality and Tourism Industry: Paradigms and Directions for the Future
Editor: Vinnie Jauhari, PhD

Management Science in Hospitality and Tourism: Theory, Practice, and Applications
Editors: Muzaffer Uysal, PhD, Zvi Schwartz, PhD, and Ercan Sirakaya-Turk, PhD

Tourism in Central Asia: Issues and Challenges
Editors: Kemal Kantarci, PhD, Muzaffer Uysal, PhD, and Vincent Magnini, PhD

Poverty Alleviation through Tourism Development: A Comprehensive and Integrated Approach
Robertico Croes, PhD, and Manuel Rivera, PhD

Chinese Outbound Tourism 2.0
Editor: Xiang (Robert) Li, PhD

Hospitality Marketing and Consumer Behavior: Creating Memorable Experiences
Editor: Vinnie Jauhari, PhD

ABOUT THE EDITOR

Xiang (Robert) Li, PhD

Dr. Xiang (Robert) Li is currently a Professor and Washburn Senior Research Fellow at the School of Tourism and Hospitality Management, Temple University. Previously, he was a professor at the School of Hotel, Restaurant, and Tourism Management, University of South Carolina. Dr. Li's research mainly focuses on destination marketing and tourist behavior, with special emphasis on international destination branding, customer loyalty, and tourism in Asia. He has authored over 100 scientific publications and is serving on the boards of over ten journals and book series. He is currently guest-editing a special series on "Emerging Market Tourist Behavior" for the *Journal of Travel Research* and guest co-editing a special issue on "Transcultural experiences within and beyond home" for the *Journal of Business Research.* Dr. Li's academic work has been recognized by a number of awards, including the 2013 "Emerging Scholars of Distinction" Award by the International Academy for the Study of Tourism and the first ever "Best Emerging Scholar in Tourism (B.E.S.T.)" Award by the International Tourism Studies Association.

Dr. Li has conducted extensive tourism research in various areas and has been awarded over $1.25 million research funding. His clients include the United States Department of Commerce/Office of Travel and Tourism Industries (OTTI), China National Tourism Administration (CNTA), the United States Travel Association (UST, formerly TIA), and National Tour Association (NTA), as well as numerous destination marketing organizations and companies. His research on China and Japan outbound tourism and destination marketing was cited by numerous American and international media, such as *USA Today*, *Time*, *Las Vegas Sun Times*, *The Globe and Mail* (Canada), and *China Daily*. During 2012–13, Dr. Li served on the Global Insights Advisory Council for Brand USA. He holds/held adjunct or visiting faculty appointments in universities in Australia, Hong Kong, and Mainland China. He is also a frequent speaker at numerous international and industry conferences.

CONTENTS

Dedicated To

My Beloved Parents

FOREWORD BY DAVID SCOWSILL

Any future historian looking back on the 21st century will undoubtedly be struck by the potency of travel and tourism to enrich people's lives and spur economic growth. Currently, around 260 million people worldwide are employed directly or indirectly in the travel and tourism sector – that's 1 in 11 jobs on the planet. Ours is a sector whose GDP growth, at 3.5% per annum, outstrips that of the wider economy. More than a billion international trips are taken every year and countless domestic ones.

The outlook for the next 10 years is equally robust as our appetite for exploring countries and cultures seems insatiable. By 2024, travel and tourism's total economic contribution is forecast to account for 10% of global GDP, US$ 11 trillion, and 1 in 10 jobs. By 2024, our industry is forecast to support 81 million new jobs.

Much of this growth will be driven by increasing wealth among the Asia region's middle classes, especially in China where it is estimated that there will be 1 billion middle-class consumers by 2030 – 70% of the population.

China's travel and tourism businesses already support 65 million jobs and the sector contributes 9% of China's GDP. While some may worry about shrinking or static growth within the Chinese economy, China's travel and tourism is forecast to keep growing at over 7% each year for the next 10 years.

Enormous numbers of Chinese people are not only traveling domestically, but also venturing overseas. Almost 100 million trips were made internationally by the Chinese in 2013, and Chinese international tourists account for 8% of global international travel expenditure, second only to the United States at 11%, according to WTTC economic impact statistics. By 2024, WTTC forecasts that China will overtake the United States as the world's largest travel and tourism economy, measured in total GDP terms and the size of the outbound market.

But will we also see China follow – and potentially overtake – another trend which the Americans started during their initial boom in overseas travel? Back in the 1960s, the growth of USA outbound travelers rapidly brought with it the establishment of global brands that are well known today: Hilton, Marriott, Holiday Inn, Hertz and American Airlines – to name a few. But no other major traveling nation since has had the same influence on the global brands, which tourists of all nationalities choose. Will the situation be different for China?

Home Inns & Hotels Management Shanghai and Shanghai Jin Jiang International Hotel Group Co. are the ninth and tenth largest hotel companies in the world, respectively, by room count, even though almost 100% of their room stock is in China – compared with other leading chains whose inventory is spread across 50 to 100 countries. These brands could ride the wave of Chinese international tourism and expand outside of China. In fact, Dalian Wanda Group is already building its first 5-star hotel outside China in central London.

As millions more Chinese look to explore the world, hugely successful domestic travel brands in China could set their sights on worldwide operations and easily overtake their more established international rivals.

Any future historian looking back at when China began to dominate travel and tourism at home and abroad may well pinpoint 2014–2015 as a turning point. China will become increasingly central to our sector's growth over the next decade, but our sector will also become increasingly crucial to the world's fastest-growing economy. No other sector besides travel and tourism can offer such a mutual advantage. It's an interesting and reassuring thought, and for that, we applaud Dr. Xiang (Robert) Li and his colleagues' insightful discussion on the latest development and trends of Chinese Outbound Tourism.

David Scowsill
President & CEO
World Travel & Tourism Council

[The World Travel & Tourism Council is the global authority on the economic and social contribution of Travel & Tourism. It promotes sustainable growth for the sector, working with governments and international institutions to create jobs, to drive exports, and to generate prosperity. Its Members are the Chairs, Presidents, and Chief Executives of the world's leading, private sector Travel & Tourism businesses. These Members bring specialist knowledge to guide government policy and decision-making, raising awareness of the importance of Travel & Tourism as an economic generator of prosperity. For more information on WTTC: www.wttc.org.]

LIST OF CONTRIBUTORS

Wolfgang Georg Alrt

Prof. Dr. Wolfgang Georg Arlt, FRGS, is a professor in the International Tourism Management at West Coast University of Applied Sciences Germany, Heide/Germany, founder and director of COTRI China Outbound Tourism Research Institute, and fellow of the Royal Geographical Society (London). His main field of interest for the past three decades has been China's outbound tourism, especially the market development, Chinese travel motivations and behavior and implications of the globalization of travel source markets for tourism theory development.

Stanley Chan

After obtaining his Master Degree of Philosophy (majoring in Sociology) in the Chinese University of Hong Kong, Stanley started his career as a research practitioner in the field of social and marketing research. Stanley worked in different multinational research firms in Hong Kong and China. In 1998, Stanley established his own research company – MIC Consultancy in China. In these 23 years of research work, Stanley's research focus falls mainly on tourism, B2B corporate imagery and market entry consultancy.

Yun Chen

Yun Chen, PhD candidate of Fudan University, is a lecturer at History, Culture and Tourism Institute of Jiangsu Normal University in Xuzhou city, China. Her research interests are in tourism economics, tourism geography, and tourism adventure.

Li Jason Chen

Dr. Jason Li Chen is a lecturer in Tourism and Events Management in the School of Hospitality and Tourism Management, University of Surrey, UK. His background is in economics with research interests in tourism forecasting, tourist behavior, and applications of quantitative research methods.

Chia-Kuen Cheng

Dr. Chia-Kuen Cheng is an associate professor at the Department of Horticulture and Landscape Architecture, National Taiwan University. His research interest includes human–place relationships, landscape perception, and destination.

Margaret Deery

Prof. Margaret Deery is at the University of Surrey, UK, visiting professor at the Leeds Beckett University, UK, Adjunct professor at Curtin University, Australia, and editor-in-chief of the Journal of Hospitality and Tourism Management. Her research interests cover a number of tourism areas including sport tourism, visitor information centers, business events, volunteers in heritage attractions, and the social impacts of tourism on communities.

Lei Fang

Dr. Lei Fang received his PhD from Zhejiang University of China and joined the School of Hotel and Tourism Management as Research Assistant in 2013. He is specialized in spatial analysis, GIS, tourism and hospitality marketing, and tourism planning. He has published on major GIS and tourism journals about spatial analysis of tourism phenomena.

Yingzhi Guo

Yingzhi Guo, PhD, is a professor and supervisor of graduate students at Tourism Management of Fudan University in Shanghai, China. Her research interests include tourism behavior, tourism and MICE marketing, and tourism planning. She had been a Fulbright visiting scholar in USA. She had led and involved some projects such as China National Nature Science Funds, Great Key Project of China National Social Science Funds, and many provincial/urban tourism market planning of local tourism development. She published over 100 papers in Chinese and English academic journals.

Sarah Gardiner

Dr. Sarah Gardiner is a lecturer in tourism and hotel management in the Department of Tourism, Sport and Hotel Management at Griffith University, Gold Coast, Australia. Sarah has a PhD in marketing from Griffith University. She has worked in industry and government and as a consultant in tourism and economic development roles. Her current research interests are consumer behavior, tourism experience design and destination marketing and development with a particular focus on youth tourism.

Rich Harrill

Dr. Rich Harrill is an accomplished chair and center director with a strong record of success and an international reputation as an expert in International Development, Economic Development, and Tourism with the University of South Carolina and Georgia Institute of Technology (Georgia Tech). He was named Research Professor and Acting School Director of USC's School of Hotel, Restaurant, and Tourism Management in 2012. Dr. Harrill holds a PhD in Parks, Recreation, and Tourism Management from Clemson University.

Ruijuan Hu

Dr. Ruijuan (Rachel) Hu is an associate professor in the School of International Trade and Economics at the University of International Business and Economics (UIBE) in Beijing. Her research interests include international tourism and service trade, logistics, and transportation economics.

Songshan (Sam) Huang

Dr. Songshan (Sam) Huang is an associate professor at the School of Management, University of South Australia. His research interests are tourist behavior, destination marketing, and tour guiding.

Scott C. Johnson

Scott C. Johnson, president and founder of Travel Market Insights Inc. He leads a global team that provides actionable marketing insight on international travel to the U.S. using primary and secondary world class research. Scott worked for the U.S. National Tourism Office starting in 1990, working on global promotion, tourism policy (representing USA at UN/WTO), and international visitor research. Partner centric, Scott works with leading global associations, businesses, and destinations and presents throughout the world.

Xin (Cathy) Jin

Dr. Xin (Cathy) Jin works at Griffith University in Australia. Her two main areas of research interests are event tourism/ management and destination marketing. He has authored or co-authored a number of publications in top tier tourism, hospitality and event journals such as Tourism Management and International Journal of Hospitality Management.

Brian King

Prof. Brian King is an associate dean in the School of Hotel and Tourism Management at the Hong Kong Polytechnic University. He specializes in tourism marketing research with an emphasis on cultural dimensions and emerging Asia-Pacific markets. He has examined the intersection between tourism and social phenomena including migration and international education. He has published papers about the international student phenomenon and experience, particularly relating to Chinese students and tourism and hospitality education provision.

Byron W. Keating

Prof. Byron W. Keating is the Director/Head of the Research School of Management at the Australian National University. Byron's research focuses on helping organizations to understand, design, and manage complex service systems in wide range of operating contexts. The multidisciplinary nature of this work has resulted in publications in leading international journals in the fields of information systems (e.g., European Journal of Information Systems), marketing (e.g., Journal of the Academy of Marketing

Science), operations management (e.g., Journal of Supply Chain Management), service management (e.g., Journal of Service Management), and specific service contexts (e.g., Cornell Hospitality Quarterly).

Gang Li

Dr. Gang Li is a professor in Tourism Economics in the School of Hospitality and Tourism Management, University of Surrey, UK. His research interests are in the areas of tourism demand modeling and forecasting using econometric approaches. His publications have appeared in Annals of Tourism Research, Tourism Management, Journal of Travel Research and International Journal of Forecasting. He co-authors a research monograph Advanced Econometrics of Tourism Demand.

Xiang (Robert) Li

Xiang (Robert) Li, PhD, is a professor and Washburn senior research fellow at the School of Tourism and Hospitality Management, Temple University. His research mainly focuses on destination marketing and tourist behavior, with special emphasis on international destination branding, customer loyalty, and tourism in Asia. Robert's research findings have appeared in numerous top-tier tourism, business, leisure, and hospitality journals.

Xiangping Li

Xiangping Li, PhD, works as an assistant professor at Tourism College, Institute for Tourism Studies, in Macao, China. Her research interests include tourism planning and development, destination marketing, and tourist behavior.

Yann-Jou Lin

Dr. Yann-Jou Lin is a professor at the Department of Horticulture and Landscape Architecture, National Taiwan University. His research interests focus on recreation demand analysis and forecasting, valuation of natural resources, and landscape assessment.

Hongbo Liu

Ms. Hongbo Liu is a PhD student in the School of Tourism and Hospitality Management, Temple University. She received her Bachelor's degree in Tourism Management from Fudan University and Master's Degree in International Hospitality and Tourism Management from University of South Carolina. Her research interests focus on destination marketing and consumer behavior.

Hsi-Lin Liu

Dr. Hsi-Lin Liu is the deputy director-general, Tourism Bureau, Ministry of Transportation and Communication. His research interests focus on tourism policy, international tourism marketing, tourist behavior, landscape planning, and youth travel.

Xing Liu

Xing Liu, is a graduate student of Tourism Department, Fudan University, China. She earned her bachelor degree in Management from Fudan University. Her graduate dissertation is the Information Sharing Behavior of Users on Sina Weibo. Her research interest is consumer behavior in tourism through big-data analysis.

Wei Liu

Wei Liu is a PhD candidate from the Department of Tourism, Sport and Hotel Management, Griffith University. Prior to her commencement of her PhD program, she was a research assistant in the School of Hotel and Tourism Management, Hong Kong Polytechnic University. Her work has appeared in journals such as Tourism Management and International Journal of Contemporary Hospitality Management. Her research interests are customer service experience, customer satisfaction and social media.

Iris Mao

Dr. Iris Mao is a lecturer at the School of Business of Edith Cowan University in Australia. Her research interest focuses on consumer behavior in hospitality and tourism.

Fang Meng

Fang Meng, PhD, is an assistant professor in the School of Hotel, Restaurant and Tourism Management and a Research Associate in the SmartState Center of Economic Excellence in Tourism and Economic Development at the University of South Carolina, USA. Her areas of research interest include destination marketing, tourist behavior and experience, and international tourism.

Berenice Pendzialek

Berenice Pendzialek (MBA) is a Chinese Outbound Tourism Specialist working as a freelancer in Hamburg, Germany. She recently finished her doctoral dissertation in the Catholic University of Eichstätt-Ingolstadt (KU-Department of Tourism Geography) with the dissertation topic: "Performing tourism: Chinese outbound organized mass tourists on their travels through German tourism stages." Her research interests are tourism geography, China tourism and travel market, and Chinese social studies.

Philip L. Pearce

Philip L. Pearce, PhD, is the foundation professor of Tourism, School of Business, James Cook University, Australia. He is fundamentally interested in the behavior and experience of tourists and tourist–host interaction.

Hanqin Qiu

Dr. Hanqin Qiu is a professor of international tourism. She received her BA from Nankai University in Mainland China, MA from University of Waterloo in Canada, and PhD from University of Strathclyde in the UK. Her research interests are tourism studies, consumer behavior, and China's hotel and tourism development and policy issues.

Han Shen

Dr. Han Shen has a background in marketing and was educated in China and the UK. She is currently an associate professor at Fudan University, China. Her main research area is consumer behavior, tourism marketing, and service management. Dr. Shen has conducted many grants from National Social Sciences Foundation of China, Ministry of Education of China, etc. She has extensive research and consultancy experience for destination marketing at national, provincial and city levels.

Li Shen

Dr. Li Shen is an assistant professor at the Department of Social and Regional Development, National Taipei University of Education. Her research focuses on destination image, heritage tourism, film tourism, and cultural landscape.

Kevin Kam Fung So

Dr. So is an assistant professor in the University of South Carolina. His research interests focus on service brand management, with an emphasis on customer engagement, electronic word-of-mouth, and service experience and evaluation. His work has appeared in leading journals such as Journal of Hospitality and Tourism Research, International Journal of Hospitality Management, and Journal of Travel Research. Before entering academia, he worked for several international hotels including the Sheraton Gold Coast and Sheraton Perth.

Beverley A. Sparks

Prof. Sparks works at Griffith University and focuses her research on customer experience issues including service interactions, satisfaction, and consumer-generated communication. Her research is applied to tourism/hospitality contexts and incorporates some cross-cultural investigations. Her papers are widely published in tourism/hospitality and services marketing journals. She has conducted many projects and received more than $1 million (AUD) in grant funding.

Tony S. M. Tse

Dr. Tony Tse is an assistant professor at School of Hotel and Tourism Management, The Hong Kong Polytechnic University. He is a member of the Board of Directors at the Travel Industry Council of Hong Kong (2012–2016). His research interests are Chinese outbound tourism and hospitality education. Tony has a Bachelor degree in Social Sciences from the University of Hong Kong, an MBA degree from Macquarie University, and a PhD degree from Southern Cross University.

Dan Wang

Dan Wang, PhD, is an assistant professor in the School of Hotel and Tourism Management, The Hong Kong Polytechnic University. She is interested in information communication technology and travel information service, mobile marketing in tourism and hospitality, and the impact of new media communication on tourist behavior.

IpKin Anthony Wong

Dr. IpKin Anthony Wong is an assistant professor at the Institute for Tourism Studies, Macau. His current research interests include tourism and hospitality marketing, service quality, green marketing, event management, and casino and gaming management. He has more than 100 academic publications (journal articles, books/book chapters, and conference proceedings). He is currently serving on several editorial boards including *International Journal of Hospitality Management, Cornell Hospitality Management,* and *International Journal of Contemporary Hospitality Management.*

Mao-Ying Wu

Mao-Ying Wu, PhD, works for the Tourism Program, School of Management, Zhejiang University, China. She is interested in the well-being of tourism communities, tourists in emerging markets, and tourist-host interaction in cross cultural contexts.

Ying Wang

Dr. Wang works at Griffith University in Australia and focuses her research on Chinese tourism as well as technology enabled service experience and customer responses to marketing stimulus in a tourism and hospitality context. Her work has appeared in journals such as *Annals of Tourism Research* and *Journal of Travel Research, and Tourism Analysis*. Prior to academia, she worked in hotel and business sectors in China.

Xin Yang

Xin Yang, Office of International Exchange and Cooperation, Huaqiao University. She graduated from School of Hotel and Tourism Management, The Hong Kong Polytechnic University with a Master of Science Degree. Her thesis focused on the use of social media by Chinese outbound independent tourists.

Wan Yang

Dr. Wan Yang is an assistant professor in the Collins College of Hospitality Management, California State Polytechnic University, Pomona, USA. She earned her PhD in Hospitality Management from the Pennsylvania State University. Her focal research areas include luxury consumption, services marketing, and cross-cultural research. Her work has been published in top tier journals such as *Cornell Quarterly, International Journal of Hospitality Management, International Journal of Contemporary Hospitality Management*, and *Journal of Hospitality and Tourism Research*.

Honggen Xiao

Honggen Xiao, PhD, is an associate professor in the School of Hotel and Tourism Management at The Hong Kong Polytechnic University (honggen.xiao@polyu.edu.hk). His research interests include knowledge development, leisure and society, and tourism and culture.

Pei Zhang

Pei Zhang is a PhD candidate in the School of Hotel, Restaurant and Tourism Management at the University of South Carolina, USA. Her areas of research interest include tourist behavior and experience, international tourism, and destination marketing.

Guangrui Zhang

Prof. Guangrui Zhang is an honorary director and founder of Tourism Research Centre of Chinese Academy of Social Sciences (CASS), China. He had been the editor-in-chief of the annual report China's Tourism Development: Analysis and Forecast (known as China's Green Book of Tourism) ever since its publication in 2001 up to 2013. His major fields of study include international tourism trends, tourism policy, planning and development, and world history of tourism and hotel.

Lingyun Zhang

Dr. Lingyun Zhang is a Professor and Vice Dean in the Tourism Institute, Beijing Union University, China. He serves as the Second Academic Committee Member of China Tourism Academy, Executive Editor of *Tourism Tribune*, Special Researcher of Tourism Research Center of the Chinese Academy of Social Sciences, Invited Expert of 5A Scenic Spot Inspection Group of the National Tourism Administration. His research interests include tourism economics, geography of tourism, and tourism management.

LIST OF ABBREVIATIONS

ADS approved destination status
AIC Akaike's Information Criterion
BBC British Broadcasting Corporation
CAFES Chung-Hua Association for Financial and Economic strategies
CAGR compound annual growth rate
CART Classification and Regression Tree
CS/D customer satisfaction/dissatisfaction
CEPA Closer Economic Partnership Arrangements
CHAID Chi-squared Automatic Interaction Detector
CNTA China National Tourism Administration
CNNIC China Internet Network Information Center
CRS constant returns to scale
CTA China Tourism Academy
DA destination attachment
DEA data envelopment analysis
DMO Destination Marketing Organization
DMUs decision-making units
DSEC derived from Macao Statistics and Census Service
ECFA Economic Cooperation Framework Agreement
ETC European Travel Commission
EU European Union
eWOM electronic Word of Mouth
FIT foreign independent travel
GDP gross domestic product
GPT group packaged tours
HHOG Historic Highlights of Germany
IVS Individual Visit Scheme
JNTO Japan National Tourist Organization
MGTO Macao Government Tourism Office
MI multiple imputation
MICE Meetings, Incentives, Conferences, and Exhibitions
MTR Mass Transit Railway
NTO National Tourism Organizations
OTAs online travel agencies
OTTI Office of Travel and Tourism Industries
QUEST Quick Unbiased Efficient Statistic Tree

RMB	Renminbi
SAR	special administrative region
SARs	severe acute respiratory syndrome
SARs	special administrative regions
SIAT	Survey of International Air Travelers
SJM	Sociedade de Jogos de Macau
SNS	social network sites
STDM	Sociedade de Turismo e Diversoes de Macau
UAE	United Arab Emirates
UGC	User Generated Content
UNESCO	United Nations Educational, Scientific and Cultural Organization
UNWTO	United Nations World Tourism Organization
VFR	visiting friends and relative
WHM	Working Holiday Maker
WOM	word of mouth
WTO	World Tourism Organization

INTRODUCTION[1,2]

XIANG (ROBERT) LI, PhD

To say Chinese outbound tourism grows fast would be an understatement.

Between 1994 and 2013, the number of outbound trips by Mainland Chinese citizens increased 16-fold from 6.1 million to 98.2 million, with an average annual growth rate of 15.7% (Chen et al. 2015; Zhang, 2015). China is not only the world's fastest growing source market (UNWTO, 2013), but also now the largest source market in terms of both visitation amount and travel spending. In 2014, every 1 out of 10 border crossings in international travel was made by the Mainland Chinese (Arlt, 2015), making China one of the most sought-after tourism markets.

After 2 decades' continuous growth, Chinese outbound tourism shows no sign of slowing down, yet the market is clearly experiencing some fundamental changes. For international tourism marketers who just celebrated their first "west-meets-east" moments, the good or bad news is a Second Wave of Chinese tourists is already showing up at their door. This new batch of Chinese tourists is more confident, value conscious, and ready to explore unfamiliar territory. They are more comfortable with traveling, consuming, and communicating with the world. Their presence will redefine what being "China ready" means.

This book is about these new Chinese tourists and the latest development of Chinese outbound tourism. It attempts to reflect on the trajectory of Chinese outbound travel development, reports new trends and issues, and provides practical insights and recommendations.

CHINESE OUTBOUND TOURISM 2.0: WHAT AND WHO?

Back to about 15 years ago, when Chinese outbound tourists first caught foreign marketers' attention, some industry veterans immediately noted their resemblance to the Japanese tourists; in the 1980s, the fast-growing Japanese outbound travel also drew worldwide attention. From the highly structured group tours, the low foreign language proficiency and little interactions with locals, to the tendency to snap photos, the first groups of Chinese tourists behaved just like their Japanese counterparts, except that Chinese tourists are also famous for their fondness of Chinese food and

[1]The authors gratefully acknowledge Ms. Yuna Hayakumo, Hongbo Liu, Yuan Wang, and Dr. Kevin So for their support during the preparation of this chapter.

[2]A small portion of this chapter was originally published as an invited commentary for the China Daily European Weekly (Nov. 28, 2014, p. 9) under the title of "Optimists have upper hand with tourism." Changes have been made subsequently.

shopping. On the other hand, the approved destination status (ADS) agreements generally mandated Chinese tourists to visit foreign destinations in group package tours. Therefore, Chinese tourists' destination choices were largely determined by product availability (e.g., whether a destination country has obtained ADS) and tour operators' business interests (Li et al. 2011, 2013). This era, featuring "mass-tourism style" group tours and a partially supply-driven market, could be called Chinese outbound tourism 1.0.

An important, yet not well-understood, fact about Chinese outbound tourists is that most of them are young, inexperienced customers unfamiliar with many travel/leisure products (e.g., cruise, all-inclusive resort) to which Western tourists have taken for granted. In other words, many of them are not only first-time brand consumers (e.g., visiting a particular destination for the first time), but also rookie product users (e.g., traveling abroad for the first time in their life). This means their decision-making and consumption process could be full of challenges, because not only are they inexperienced, few people around them can help either. Culture and education reasons aside, some Chinese tourists simply do not know how to behave "properly" (using their Western hosts' standard) when traveling overseas. To the extreme, some Chinese tourists' "uncivilized behavior", like spitting, speaking loudly, and lack of respect to host culture, has earned them and their country a bad reputation. Presumably, Chinese tourists must go through a "collective learning curve", during which the initial anxiety and discomforts will be steadily replaced by sophistication and more predictable consumption patterns.

Thanks to the technology development, social media, and globalization, this learning curve seems to be shorter than expected. Two important indicators of the sophistication of the Chinese market are the growing repeat travel rate (in terms of both revisiting specific destinations and traveling abroad for multiple times) and higher penetration rate of independent travel (Jiang, 2014). As for the former, the latest statistics showed that over 60% Chinese outbound tourists already had previous overseas travel experiences, and up to 70% of Hong Kong visitors and 50% of the United States visitors from the Mainland China are repeaters (Jiang, 2014). As for the latter, independent outbound tourism, although still in its infancy, has been outpacing outbound travel in general (Skift, 2013) with two-thirds of today's Chinese outbound travelers preferring to travel independently (Hotels.com, 2014). The increasing customer sophistication has led to a shift of preference for less-structured, more diversified, flexible, and personalized travel products, a trend toward what could be termed "Chinese outbound tourism 2.0."

Demographically, the new wave of Chinese tourists also shows some differences. Today, 54% of Chinese outbound tourists were born after 1980s (the Chinese Generation-Ys) (Jiang, 2014). Compared with previous generations of Chinese people, the post-80s grew up in a different environment and are better educated and more global and consumption-driven. Many of them are able to communicate in English (or other foreign languages), and many have some overseas education or working experiences. They are also quite fluent with information technology: 53% of Chinese travelers are now booking their hotel accommodation either online or via mobile apps (Hotels.com, 2014). These younger, well-traveled, globally connected, and technology-savvy travelers are less impressed by cursory sightseeing or photo-taking at iconic destinations and more interested in visiting off-the-beaten-track destinations and enjoying in-depth cultural experiences.

A number of external factors could have also affected the emergence of Chinese outbound tourism 2.0 (Arlt, 2013; Hotels.com, 2014). Same as in the 1.0 era, rising incomes, rampant consumerism, simplified procedures in obtaining passports and foreign currency, marketing efforts by international destinations, and the easing of visa regulations in many destination countries have continued spurring travel interests. Unlike the 1.0 era, some new factors helping shape new travel preferences and habits include the following: Chinese tourists' increased confidence to travel overseas and explore new destinations, easier travel information availability and online booking options, greater flexibility and more creative services assisting self-organized or half-organized tours, new routes offered by low-cost airlines, and the skyrocketing number of Chinese students and Chinese diasporas studying and living in foreign countries. Further, Chinese government's new tourism-related regulations and policies and strategic interests in building China's "soft power" through outbound tourism have also made a difference.

Chinese outbound tourism 2.0 is not merely about a new wave of Chinese tourists; strategic developments by the supply side are also an integral part of this emerging phenomenon. Western tourism suppliers have been actively involved in China's unprecedented market development, ranging from Mercedes-Benz kicking off its premium travel service brand ("Mercedes-Benz Travel") in Shanghai, to Inter Continental designing of the Hualuxe brand dedicated specifically for Chinese consumers. Meanwhile, Chinese companies have also been increasingly proactive in capitalizing on the growth of Chinese outbound tourism and their own expertise in serving Chinese consumers. China's Wanda Hotels & Resort recently launched 3 hotel brands of its own and is building a 5-star hotel in London (Montlake, 2013). In October 2014, a Chinese insurance company acquired one of America's most iconic hotels, New York's Waldorf Astoria (Cole, 2014). In a way, they are following Western hotels' footprint from a couple of decades ago, when Marriott and Hilton started building hotels in China to cater to Western guests.

Considering the size of China and substantial intracountry disparity in socioeconomic development, Western marketers have recognized that a cookie-cutter "China strategy" will not work well (Atsmon et al. 2011). Recent statistics have shown that a growing number of Chinese tourists are from outside of the most developed Bohai Sea Rim, Yangtze River Delta, and Pearl River Delta regions (Li et al. 2010). Marketers hence have to be aware of the subtle differences among Chinese consumers of different regions and classes. It is entirely plausible that while tourists from China's highly commercialized and industrialized eastern coast area are leading the transition into the "2.0" era, residents of the country's less-developed middle inland and western interior just start catching up. As such, readers are reminded that Chinese outbound tourism 1.0 and 2.0 will likely coexist for a while.

CONTINUED OPTIMISM FOR THE FUTURE: WHY?

The foregoing discussion certainly paints a rosy picture of continuous development of Chinese outbound tourism. Yet, after over 20 years' spectacular growth, how much longer the Chinese travel market will keep growing becomes a perfectly legitimate question. Indeed, the sustainability and potential of the Chinese outbound travel market have been constantly underestimated. In 1995, the United Nations World Tourism Organization made what many at the time regarded as an outlandish prediction: China would generate 100 million visits by 2020. Many pundits around the world, including some in China, were highly skeptical. That milestone has essentially been reached recently—more than 5 years ahead of schedule. When China and the United States signed the ADS agreement on promoting Chinese group leisure travel to the United States in late 2007, the official estimate was that 578,000 Chinese would arrive in the United States in 2011. That forecast turned out to be off by nearly 90%: 1.09 million Mainland Chinese visited the United States that year.

Recently, when a new China–US visa agreement was announced, amid all the hoopla, the same skepticism again surfaced. This agreement extends the 1-year limit on visas between the 2 countries to up to 10 years. A White House announcement predicts that this change will result in 7.3 million Chinese travelers traveling to the United States by 2021, a 4-fold increase on the 2013 figure.

Despite all the enthusiasm about the agreement among the public and travel trade, some experts question how much longer the strong performance of Chinese outbound tourism can continue. Some healthy skepticism never hurts, yet there are good reasons to remain optimistic.

First of all, a huge market remains untapped. In 2013, 98.2 million outbound trips were made by Mainland Chinese tourists. A common misinterpretation is that these 98.2 million outbound trips were made by 98.2 million different individuals, hence the conclusion that more than 7% of the Mainland Chinese population is traveling abroad. The percentage is likely to be much lower than 7% because many Chinese tourists make more than 1 trip a year; for example, some Guangdong residents frequently visit Hong Kong and Macao.

This observation is corroborated by the country's low passport possession rate. Although an accurate number on Chinese passport holders is hard to find, the consensus is that about 3%–5% of those in the Mainland Chinese population have passports, comparing poorly with countries like Canada (70%), New Zealand (75%), or USA (about 40%). Considering a market of at least 1.2 billion people (i.e., over 95% of the Chinese population) completely untouched, the Chinese outbound tourism growth thus far may be just an appetizer, with the main course yet to come.

Second, understanding the cultural importance of travel helps instill more confidence. Arlt (2013) suggested that travel is not merely for fun for Chinese; rather, it is considered an investment in oneself, family, and social network, practically an alternative education. Empirical research also shows that "learning/discovery" is one of the top motivations for Chinese traveling abroad (Burnett et al. 2008), dovetailing with the Chinese wisdom that knowledge comes from extensive reading and traveling. Travel is linked to broadening one's horizon, gaining insight about the world, and earning respect from peers as a result. This explains why many Chinese take their children abroad at a very young age, and why much shopping that Chinese tourists do is for their friends, colleagues, and relatives. The elevated importance of travel in the Chinese value system bodes well for continuous growth in the market.

Finally, for real optimists, the stage is already set, but the curtain has not even been rung up yet. According to the Boston Consulting Group, a country's economic growth will not be fully translated into long-haul travel demand until its gross domestic product (GDP) per head reaches about $15,000 a year (BCG, 2006). China's GDP per head is now about $6,000, so there is a long way to go. Notably, this rule of thumb seems to hold true in some of China's wealthiest regions, where outbound travel is racing ahead at full speed.

A similar S-shaped curve can be found in global aviation. China's air passenger penetration rate is still fairly low, corresponding to a fairly low-income level. The country's demand for air travel is likely to accelerate, Goldman Sachs indicates, once "a certain income threshold is crossed" (Goldman Sachs, 2010, p. 19). A report, last year, by Airbus said Chinese are only making 0.26 trips a year by air per head now, but that this will grow to 0.95 trips per head by 2032 (Airbus, 2013).

If history is any indication of the future, China's outbound travel is still at the starting point of the evolution curve. The new visa deal between China and the US government could shake things up in China's outbound tourism development, as many other Western countries could follow America's move, and easy accessibility and ready availability of travel products will become the "new normal." From this perspective, the China–US visa arrangement could be a milestone for Chinese outbound tourism 2.0, signaling a fundamental shift of Chinese outbound tourism from being partially supply-driven to entirely demand-driven. For foreign destinations and services, the preseason is over, and the real competition for Chinese outbound tourists is about to begin.

ABOUT THIS BOOK

Underpinned by the world's largest economy and population, China is set to change the world's tourism landscape. Observing the growth of Chinese outbound tourism is a rare, probably once-in-a-career opportunity, which motivated this book project. A number of world's leading tourism researchers share their most cutting-edge findings and thoughts on Chinese outbound travel market and tourists, to whom the editor is deeply indebted for their support.

The book is organized in 4 parts. Section I offers an overview of the development, policy, and research of Chinese outbound tourism. Wolfgang Arlt's Chapter 1 provides a comprehensive overview of China's outbound tourism development, which he divided into 4 stages (1983–1997, 1997–2005, 2005–2011, and 2012–today). Quantitatively, China's outbound tourism features remarkable growth in the number of travelers and their spending; qualitatively, both the structure of tourists and their demands are changing. For Arlt, China's outbound tourism has entered a new era, shifting from package tour to self-organized tour, from sightseeing to experience, and from brand to lifestyle. Looking to the future, he proposed that a third wave of China's outbound travelers with emerging niche segments and diverse demand is on the horizon.

In the next chapter, Guangrui Zhang unveils the role of policy evolution underlying China's outbound tourism development. First, key terms related to outbound tourism in China's unique policy environment were introduced, including "outbound", "self-paid" and "public paid" travels, "border tourism", and "approved destination status". Zhang summarized the characteristics of China's outbound tourism such as the relatively late occurrence, a long dormant period, the rapid growth market in recent years, the shift from public paid travel to self-paid travel, and the country's increasingly expanding tourism deficit.

The last chapter of Section I is a detailed bibliometric review on Chinese outbound tourism-related academic research by Ying Wang and Xin Jin. Based on their review of Chinese outbound tourism papers on top-tier journals in the past 15 years, the authors observe that China outbound tourism research is increasing in recent years and is dominated by ethnically Chinese researchers focusing on destinations in Asian countries, Oceania, and North America. The majority of studies employed existing theories from psychology, marketing, and business management, yet fell short of making theoretical or methodological breakthroughs. Moreover, the Second Wave of China outbound tourism has been understudied.

In Section II, researchers around the world report the latest development of Chinese outbound tourism to Hong Kong, Macao, Taiwan, other parts of Asia in general, Australia, Europe, and the United States. Despite the variances in product offerings, market readiness, and strategic focuses, the aforementioned transition from the 1.0 to 2.0 era seems to be taking place across various regions.

In Chapter 4, Tony Tse shared the Hong Kong experiences with Mainland Chinese outbound tourism development. Mainland China now represents Hong Kong's largest source market in terms of both arrivals and spending. Notably, the large volume of Chinese Mainland visitors, especially the same-day visitors, has caused severe concerns, even conflicts, in the local community. Tse suggested increasing Hong Kong's physical carrying capacity through improving infrastructure, taking proper visitor management strategies, and adjusting emotional tolerance level of the host community through education campaigns.

Chapter 5 by Xiangping Li offers the Macau perspective. Similar to Hong Kong, Mainland China is the major source market of Macau. However, the past decade has seen a decline of this market which could be attributed to the insufficient tourism resources, an overreliance on gaming, and the small size of Macau. To maintain and boost the Mainland China market, Macau should cooperate with Hong Kong and Guangdong and diversify its tourism industry structure through developing non-gaming tourism resources. Interestingly, unlike most other destinations receiving Chinese tourists, Macau is seeing an increase of package tourists from the Mainland.

Chapter 6 by Li Shen, Chia-Kuen Cheng, Yann-Jou Lin, and Hsi-Lin Liu provide a comprehensive discussion about the outbound tourism of Mainland China to Taiwan. For years, the cross-strait tourism has been closely regulated by the 2 sides and largely affected by the political relations. Despite the strict restrictions, tourist arrivals from Mainland China have increased dramatically since 2008. Although Mainland Chinese tourists have made a considerable contribution to Taiwan's economy, concerns have been expressed on the adverse social and environmental impacts due to cultural differences and the limited capacity of Taiwan; the "crowding out" effect of Chinese tourists could make Taiwan excessively reliant on the Mainland China market.

Chapter 7, prepared by Hanqin Qiu and Lei Fang, is an insightful discussion on China's outbound tourism to the rest of Asia. Overall, Asia is Chinese outbound tourists' top destination region. Market potential varies across different Asian countries, with East Asia and Southeast Asia being more popular than other Asian countries. Notably, safety and security concern is a major obstacle for Chinese tourists traveling to some Asian countries. It is suggested that more open policies should be implemented, political and social problems should be addressed to guarantee the safety of tourists, and service quality should be improved in selected Asian destinations.

Iris Mao and Songshan (Sam) Huang's Chapter 8 focuses on China's outbound tourism to Australia. China is Australia's most valuable and second largest source market. Given the importance of this particular market, Australia has made tremendous efforts to attract Chinese tourists, in terms of consumer research, government policies, and industry campaigns. It is reported that most Chinese tourists visiting Australia are for the purpose of education and visiting friends and relatives (VFR), which explains the high percentage of repeat visitors. It is suggested that improving the food quality for group tourists and providing Chinese language service and signage will enhance their travel experience in Australia.

Chapter 9 by Berenice Pendzialek mainly discusses China's outbound tourism to European countries. The majority of Chinese tourists travel to Europe in group packaged tours, but in recent years, the number of group members has been decreasing and their routes are shorter—Chinese tourists are traveling to fewer countries in a single trip, which highlights the transition from organized mass tourism to free independent travel and in-depth travel. Although the number of Chinese arrivals to Europe has been growing steadily, this market is still challenged by visa restrictions and a lack of adaptation to Chinese needs.

Chapter 10 was prepared by Hongbo Liu, Xiang (Robert) Li, and Scott Johnson, focusing on Chinese outbound travel to the United States. Compared with the past, leisure travel (as opposed to business travel) has become the major travel purpose for Chinese visiting the United States. Similar to many other destinations, recent trends in this market include growing independent travel and repeat visitation, fewer group packaged tours, and fewer destinations involved in a single trip. The new visa extension policy is expected to be a game changer. On the other hand, business travel to the United States has been contracted partly due to Chinese government's recent anti-corruption campaign.

Section III presents a collection of interesting perspectives and case studies on Chinese outbound tourism. From gaming and shopping behavior, luxury consumption, to social media usage, Chinese outbound tourists have clearly demonstrated sophistication in travel behavior and attitude. Although most studies use Chinese outbound tourism mainly as a context, some researchers have been acutely aware of the potential of China and Chinese outbound tourism as sources for new, context-specific conceptualizations. For instance, in Chapter 11, Kevin Kam Fung So, Wei Liu, Ying Wang, and Beverley Sparks highlighted the role of cultural values in understanding service expectations and proposed 4 key cultural values affecting Chinese tourists' expectations: face, harmony, group orientation, and interdependence. Their conceptual model, by integrating these 4 cultural dimensions and more general factors of service expectations, sensitize Western marketers to take these values into consideration when formulating service strategies.

Philip Pearce and Mao-Ying Wu (Chapter 12) attempt to gain new insights about the minds and psychological world of the Chinese visitors. In particular, they used an orchestra model to explore the principal sensory, emotional, cognitive, activity-based, and relationship-influenced components relating Chinese tourists' experience at Florence, Italy. Their findings suggest the new wave of Chinese outbound tourists enjoyed their visual experience most, they were emotionally relaxed and happy with their experience at Florence, and they liked to react positively to everyday features of the destinations. Also, Chinese tourists were found to be interested in observing, understanding, and making sense of tourist sites' themes and stories. Facilities for physical comfort and better translation were among Chinese tourists' key international travel concerns.

Ipkin Anthony Wong (Chapter 13) reported that the market of Chinese casino tourists had developed into different segments with diverse preferences and demands. Correspondingly, casino operators have developed several themed casinos to target different Chinese tourist segments. Luxurious physical environments, integrated entertainment facilities, and service tailored for Chinese needs seem to appeal to the new wave of Chinese casino tourists. Wong also suggested that the surging demand among Chinese for casino tourism will continue and gaming destinations adopting Chinese culture into their service experiential attributes will gain competitive advantages in the future.

Chinese tourists are recognized worldwide as big spenders; however, knowledge about Chinese tourists' shopping behavior is still limited. The next 2 chapters tackle this from different perspectives. Fang Meng and Pei Zhang (Chapter 14) conducted interviews and questionnaire surveys on Chinese citizens' overseas tourism shopping experience. They found that young and middle-aged adults from middle- and upper-class background have been the most active outbound shopping travelers. Trustworthiness, genuine brands, high quality, and good price are the most important shopping attributes valued by Chinese outbound tourists, and language and limited shopping time are reported to be the most significant external barriers.

Wan Yang (Chapter 15), from a luxury consumption perspective, introduced the "Luxury 4Ps" framework into the research on Chinese outbound tourists' luxury purchases. Following the "Luxury 4Ps" typology, luxury consumers can be categorized into 4 groups: Patricians, Parvenus, Proletarians, and Poseurs, varying in their degrees of wealthy and need for status, and thus product preferences. The author recognized that most current Chinese outbound tourists are Parvenus—affluent and high in need for status. They tend to purchase conspicuous luxury products to signal social status and spend comparatively less on luxury hotels and restaurants. Due to the national anti-corruption campaigns in China and the emergence of younger outbound tourists, a new wave of luxury consumers is coming, with lower need for status (Patricians), but higher expectations on luxury travel products such as hotels, restaurants, and authentic activities.

Another hot topic is technology, particularly social media usage among Chinese outbound tourists. The next 2 chapters, one as an overview, the other a case study, touch upon this. Han Shen and Xing Liu's chapter (Chapter 16) provides a synopsis of the unique social media environment of China and the integration of social media in Chinese tourism. Three social media platforms most relevant to tourism were particularly introduced: Sina Weibo, online travel communities, and WeChat. This chapter also discussed the social media development in tourism and the usage of social media by Chinese outbound tourists. Particularly, the authors divided travel experience of Chinese tourists into pre-travel, on-travel, and after-travel stages and discussed tourists' social media usage corresponding to each stage.

Recognizing the remarkable role of independent tourists and technology use in the Second Wave of China outbound tourism, Xin Yang, Dan Wang, and Brian King (Chapter 17) conducted semi-structured interviews with independent tourists to examine their social media usage. Three dimensions of travel information search strategy—source, degree, and spectrum of social media use—were distinguished. The results indicate that social media have significantly penetrated into Chinese outbound independent tourists' travel planning process, and social media channels were used for searching flight and accommodation deals and destination activity arrangements. Also, these tourists spent at least 40% of overall planning time on social media channels and used social media as initiated search channels for both browsing and specific search.

Byron Keating and Marg Deery's Chapter 18 provides a novel case study on how region-specific characteristics affect the region's efficiency in dispersing international visitors. In their study about Chinese outbound tourists in Queensland, the authors conducted a Data Envelopment Analysis (DEA), using appeal, access, and affordability of regions as destination-specific inputs and visitors and visitor nights as outputs. Keating and Deery's analysis identified more efficient regions based on their capability of producing higher outputs with same destination-specific inputs. This study has important implications to regions interested in improving their capability and efficiency in attracting Chinese visitors.

Focusing on the influence of political crisis on travel intention, Yingzhi Guo and Yun Chen investigated Chinese citizens' travel intention to Japan after the recent Diaoyu Island dispute. The results showed Chinese tourists varied in their intention in terms of age, number of times to Japan, monthly income, and marital status. More specifically, the "20–29 years old" and "50 and above" groups, visitors with more travel experience to Japan, high-income groups, and single tourists were more likely to visit Japan as planned than other groups.

Section III concludes with a chapter by Brian King and Sarah Gardiner, showing the significant role of education in the China outbound tourism phenomenon to Australia. Their surveys to 1,400 Chinese students and 6 focus groups showed that short-term trips to capital cities and surrounding tourist destinations during the course of their studies were popular among Chinese student travelers studying in Australia. As cautious travelers, Chinese student travelers are concerned more about safety than their Western counterparts, and they viewed the trip itself as representing adventure and risk. Also, these students attracted a large number of friends and families to Australia, and they acted as tour guides on their visitors' travels around Australia. King and Gardiner argued that understanding the travel behavior of Chinese students studying in Australia is important groundwork for satisfying future-generation Chinese tourists.

The fourth and final section (Section IV) wraps up the book with 3 chapters. Chapter 21 was prepared by Stanley Chan, a veteran marketing consultant with decades of first-hand experiences in Chinese outbound tourism research. Chan observed that 2 emerging segments—young and repeated travelers in first-tier cities and new tourists from second- and third-tier cities—characterize the new wave of Chinese outbound tourists. China outbound tourism 2.0 challenges researchers to adopt innovative methods such as using participant observation to investigate the travel decision-making process of Chinese young middle class in gateway cities and conducting focus groups with experimental design components to explore travel preference of residents in the second- and third–tier cities. Also, individual-based pre- and post-trip survey, rather than annual tracking survey, is believed to be more appropriate to monitor the change of tourists' preferences and perception.

Rich Harrill, Xiang (Robert) Li, and Honggen Xiao (Chapter 22) add a philosophical inquiry to the predominantly positivistic discourse on Chinese outbound tourism phenomenon. Drawing on Habermas and Foucault, this chapter provides an interpretation of Chinese outbound tourism through the lens of critical tourism studies. Different perceptions of gambling, sexuality, and guns culture among Chinese outbound tourists were interpreted from the critical approach with consideration of social and political conflicts in this cross-cultural instance. Arguably, Chinese tourists' experiences with gambling, sexuality, and gun can not only become self-empowering and power-challenging, but may also be critical, communicative, and pragmatic.

The last chapter by Li Jason Chen, Gang Li, Lingyun Zhang, and Ruijuan Hu summarized key market trends and forecasted the size and flow of Chinese outbound travelers by 2020. Chinese outbound tourists were becoming more experienced and independent travelers, with diversified needs and looking for destinations of high accessibility. The authors forecasted that Chinese outbound tourism will grow at an annual rate of 9.41% from 2015 to 2020; specifically, the United States and France will be the only 2 non-Asian destinations in the top 10 destinations for Chinese travelers, with Thailand, the United States, and France moving up on the list, whereas Russia and Germany gradually losing their popularity among Chinese outbound tourists.

Numerous colleagues and friends provided tremendous support to the preparation of this book. In addition to all the chapter contributors, the editor would like to extend gratitude to the book series editor, Dr. Mahmood A. Khan, and Dr. Simon Hudson (University of South Carolina), Dr. Muzzo Uysal (Virginia Tech), Ms. Helen Marano (WTTC) and Mr. Geoffrey Breeze (WTTC) for their help. I ow a great deal to the hard work of my talented graduate students and/or assistants, including Yuna Hayakumo, Hongbo Liu, Emma McAfee, Tiana Vinciguerra, and Yuan Wang. This book project would not be possible without the excellent editorial support from Mr. Ashish Kumar and his team, including Rakesh K, Sandra Sickels, and many other members whom I may or may not have directly interacted with. Finally, a heartfelt thanks to my parents and sister, my wife, and my two kids. Thank you all for making my work truly meaningful!

Chinese outbound tourism has drawn increasing attention from scholars and practitioners around the world. Going forward, one of the key challenges researchers need to face is to go beyond applying existing theories and concepts to the Chinese context and develop more "China-specific" conceptualizations. Further, most studies have focused on the demand side, yet a better understanding of the suppliers and market structure is clearly warranted. Finally, much of the current research on Chinese outbound tourism is conducted in developed economies, whereas more research is needed on Chinese outbound tourism to developing countries in South

America, Africa, and South Asia, where the impacts of Chinese outbound tourism to local economy and culture are presumably more substantial and salient. In this sense, the 23 chapters of this book have demonstrated impressive scope and depth of Chinese outbound tourism research today and herald important new directions for future research. For practitioners around the world (e.g., destination policymakers and marketers, travel and tourism service providers, owners, and managers), it is hoped this book will provide timely market intelligence and hands-on guidance on understanding tourists from Mainland China. For tourism scholars, educators, and students, this book attempts to offer basic yet essential knowledge on Chinese outbound travel market and tourist behavior, and point out important future directions for both conceptualizations and empirical research.

REFERENCES

Airbus. *Future Journeys: Global Market Forecast 2013–2032*. Airbus: France, 2013.

Arlt, W. G. *The second wave of Chinese outbound tourism*. *Tour. Plan. Dev.* 2013, 10(2), 126–133.

Arlt, W. G. *Chinese Outbound Tourism: History and Current Development* In *Chinese Outbound Tourism 2.0*; Li, X., Ed.; Apple Academic Press: Waretown, 2015.

Atsmon, Y.; Kertesz, A.; Vittal, I. Is your emerging-market strategy local enough? *McKinsey Q.* 2011.

BCG. *Is the Impact of China and India on Future Long-Haul Travel Exaggerated? Meeting the New Challenges of the Airline Industry*. The Boston Consulting Group: Boston, 2006.

Burnett, T.; Cook, S.; Li, X. *Emerging International Travel Markets: China (2007 Edition)*. Travel Industry Association: Washington, DC, 2008.

Chen, L. J.; Li, G.; Zhang, L.; Hu, R. *Market Trends and Forecast of Chinese Outbound Tourism*. In *Chinese Outbound Tourism 2.0;* Li, X., Ed.; Apple Academic Press: Waretown, 2015.

Cole, M. *Chinese Insurer Buys Waldorf Astoria for a Record $1.95B*. 2014. http://www.forbes.com/sites/michaelcole/2014/10/06/chinese-insurer-buys-waldorf-astoria-for-a-record-1-95b/.

Goldman Sachs. *The Rise of the BRICs and N-11 Consumer*. Goldman Sachs Asset Management: Singapore, 2010.

Hotels.com. *Chinese International Travel Monitor Report 2014*. 2014. http://press.hotels.com/content/themes/CITM/assets/pdf/CITM_UK_PDF_2014.pdf.

Jiang, Y. *Interpreting 'China Outbound Tourism Annual Report 2014'*. 2014. http://travel.sohu.com/20140610/n400660767.shtml.

Li, X.; Harrill, R.; Uysal, M.; Burnett, T.; Zhan, X. Estimating the size of the Chinese outbound travel market: A demand-side approach. *Tour. Manag.* 2010, 31(2), 250–259.

Li, X.; Lai, C.; Harrill, R.; Kline, S.; Wang, L. When East meets West: an exploratory study on Chinese outbound tourists' travel expectations. *Tour. Manag.* 2011, 32(4), 741–749.

Li, X.; Meng, F.; Uysal, M.; Mihalik, B. Understanding the Chinese long-haul outbound travel market: An overlapped segmentation approach. *J. Bus. Res.* 2013, 66(6), 786–793.

Montlake, S. *China's Wanda Invests $1 Billion to Launch Hotel Brand in London. Forbes*. 2013. http://www.forbes.com/sites/simonmontlake/2013/06/19/chinas-wanda-invests-1-billion-to-launch-hotel-brand-in-london/.

Skift. *Chinese Tourists are Young, Educated, and here to Stay*. 2013. http://skift.com/2013/01/30/chinese-tourists-are-young-educated-and-here-to-stay/.

UNWTO. *The Chinese Outbound Travel Market 2012 Update*. World Tourism Organization: Madrid, 2013.

Zhang, G. *Evolution of China's Policy of Outbound Tourism*. In *Chinese Outbound Tourism 2.0*; Li, X., Ed.; Apple Academic Press: Waretown, 2015.

SECTION I

OVERVIEW

CHAPTER 1

CHINA'S OUTBOUND TOURISM: HISTORY, CURRENT DEVELOPMENT, AND OUTLOOK

WOLFGANG GEORG ARLT

CONTENTS

All the ripping down of the old and throwing up of the new across China today cannot be explained as a matter of economic necessity alone. It is too frantic to be considered so simply. The visible Chinese project is, among much else, a signifier of an almost compulsive psychological urge to self-reinvent – to make "being Chinese" mean something new (Smith, 2010).

1.1 INTRODUCTION

The China's spectacular economic growth over the past 3 decades has brought back China to the position of one of the leading economies in the world, a position that China enjoyed as the source of up to a third of the global gross domestic product (GDP) (The Economist, 2010) until the rise of the "West" 5 centuries ago (Huff, 1993).

Since 2012, China can also claim to be the biggest international outbound tourism source market. In 2014, 1 out of 10 border crossings in international travel took place across the border of Mainland China, albeit in more than half of the cases already ending in the special administrative regions (SARs) of Hong Kong and Macau. Statistics on tourism almost always have to be treated with caution; this is even more so the case for China. However, even keeping in mind that all numbers connected with Chinese outbound tourism have to be taken as indicators rather than exact figures, there is no doubt that the waves of Chinese tourists, spending more per person and day than almost all other customer groups and exploring ever more exotic locations, have just started. The quest for knowledge, prestige, and self-actualization; the search for education and business opportunities; and the creation of escape routes away from Chinese institutions by affluent Chinese for both themselves and their money are among the ever more differentiated motives for the Chinese temporarily leaving their motherland.

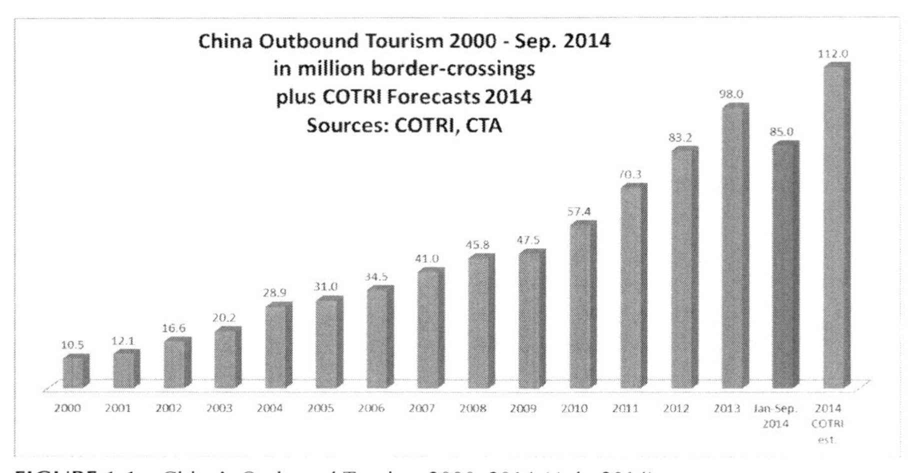

FIGURE 1.1 China's Outbound Tourism 2000–2014 (Arlt, 2014).

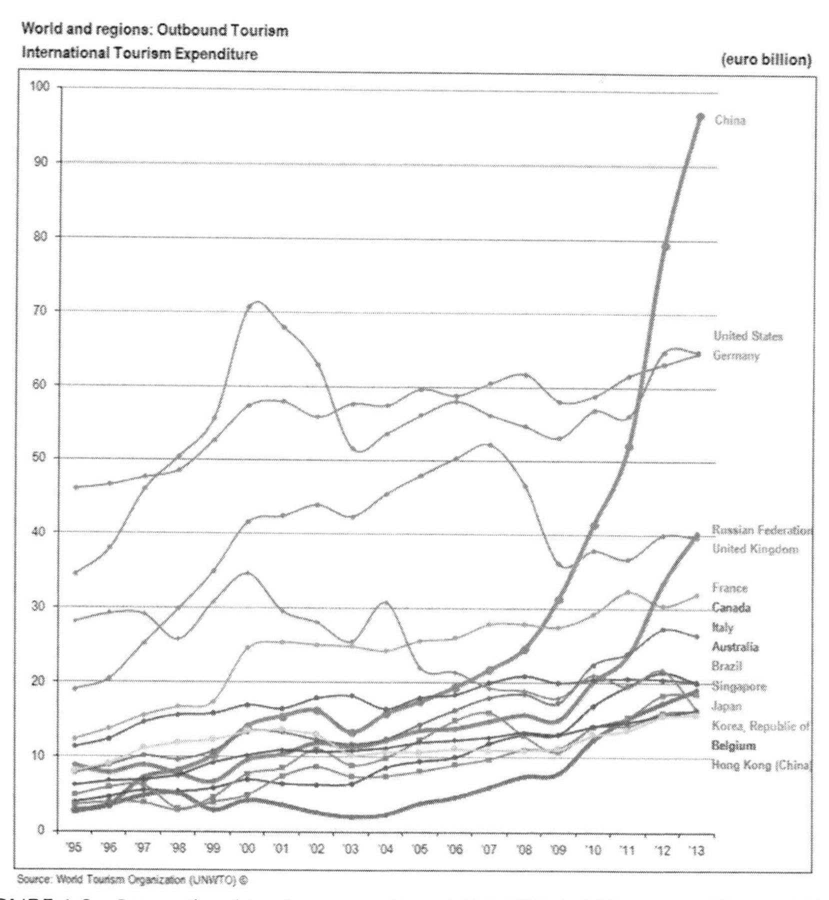

FIGURE 1.2 International tourism expenditure 1995–2013 in billion euros (Kester, 2014).

Almost every person waking up in the morning in China can expect to have enough food to eat before the day is over, a roof over his or her head, as well as, in most cases, electricity supply, a street leading to the wider world, and basic educational and health services provided. This is a big achievement and a first in the history of China. Nevertheless, for the vast majority of Chinese, a trip beyond the borders of China is way beyond their means. Even now, only about 5% of all Chinese citizens are reported to own a passport (The Economist, 2014), which is needed to travel beyond Greater China. If the 98.2 million border crossings in 2013 can be considered to have involved about 65–70 million different persons, as many Chinese travel more than once during a year, this would represent about 5% of the Chinese population. In addition, Chinese statistics do not disclose the composition of the 97 billion euros spent on international travel in 2013, but even if this number includes international airfares, it is still way ahead of the following international tourism source markets of USA and Germany, with about 65 billion euros' spending for each of these (Arlt, 2014). Divided by the total population of China, this amount of 97 billion euros represents a spending of less than 80 euros per head, compared with 870 euros per head spent on international tourism by every German and even 105 euros per head spent by every Brazilian (United Nations World Tourism Organization [UNWTO], 2014). Based on these numbers, it seems safe to assume that we are still seeing just the beginning of Chinese outbound tourism.

Domestically, Chinese travel, on an average, more than twice per year, but the 3.3 billion trips in 2013 include daytrips and represent, with a total expenditure of about 340 billion euros (Chien, 2014), an expenditure of approximately 100 euros per trip compared with the average expenditure of approximately 1,000 euros per trip per outbound travel. Hence, although in terms of trips, the ratio between domestic and outbound travel is about 33:1, in terms of expenditure, it is only 3.4:1. Combined with the diminishing number of foreigners traveling to a China increasingly perceived as polluted, overcrowded, expensive, and not very welcoming, it is no surprise that Chinese tour operators are paying increasingly more attention to the outbound sector.

They have, however, to face the fact that many, especially of the more affluent and frequent travelers, tend to move beyond the package tour toward customized tours for a given group of friends, family, or colleagues or even do not bother anymore to use outbound tour operators at all but trust "travel clubs" or private contacts as go-betweens, if not fully organizing the tour by themselves.

Chinese long-haul outbound travel is not about *holidays*; travelling for relaxation takes place inside China or to nearby destinations like South Korea or Thailand. Outbound travel beyond East and Southeast Asia is first of all an *investment*, in prestige, in gaining knowledge, in social standing, as well as in self-esteem. For this reason, even the slowdown in the growth of the Chinese economy will not result in a major decrease of the growth of outbound travel. In 2014, the GDP growth decreased to little more than 7%, but spending on outbound and international trips still increased by almost 20% (China Outbound Tourism Research Institute [COTRI], 2014).

More importantly, the segmentation of the market and the shift from sightseeing to experience and from brand to lifestyle will continue to create an ever-bigger number of niche markets. Each of them is small from a Chinese perspective but of interesting size from the point of view of the destinations and service providers involved, as they offer business opportunities for smaller destinations and specific activities.

The "Second Wave" of Chinese outbound tourism therefore offers opportunities for new places, new activities, new months of the year, and new services to sell compared with the consumption pattern of traditional "Western" visitors, bringing not only more business to destinations and companies but also a different kind of business. If offered the right products and services, Chinese visitors can help to reduce seasonality and bring tourism business to destinations off the beaten track of "Western" tourists.

1.2 HISTORICAL DEVELOPMENT

Considering China's vast territory, traveling from one place to another has always been an important part of life. Rulers traveled for military, commercial, religious, representative, educational, and administrative reasons, whereas the ruled needed to flee from war, famine, and disasters or had to act as soldiers or brigands. Furthermore, both rulers and ruled took part in short- and long-distance pilgrimages to temples, shrines, and holy mountains.

"Travelling for ten thousand li equals reading ten thousand volumes of books" has been a commonly accepted truth for many centuries in China. Already in its opening paragraph, the Analects, the most important collection of Confucius' writings compiled by his pupils after his death in 497 BC, contain the statement, "Is it not a joy to have friends come from afar?" (Lau, 1989), a sentence which even today greets the foreigners queuing up at the immigration counters of Beijing Capital Airport Terminal 3.

Outbound travel to areas beyond the realm of the Han culture, however, did not exist in China except for small colonies of Chinese migrants in neighbouring countries, singular trips of Buddhist monks like Fa Xian and Xuan Zang in the first millennium, or the voyages of the fleet of Zheng He, the Chinese Columbus in the 15th century. Admiral Zheng He, at the beginning of the Ming Dynasty, famously undertook seven journeys that brought him as far as East Africa with a fleet far superior to the Portuguese ships that arrived there a century later (Levathes, 1994; Dreyer, 2006). Xu Xiake (1587–1641), China's most famous travel writer, was only interested in China. There is no counterpart existing for Ibn Battuta, Marco Polo, or Matteo Ricci, the famous visitors to China from the Muslim and the Christian worlds, looking for their fellow believers, for either adventure and trade or trying to spread the gospel (Arlt, 2006). Nowadays, places like the Zheng He Temple in Malacca and the Zheng He Hall inside the Ibn Battuta Shopping Mall in Dubai are tourist attractions for outbound Chinese travelers.

The desire of the Chinese people to leave the Central Kingdom can be connected to the changes of the 19th century, when the notion of China as the center of the earth was shattered by the defeats in the Opium Wars. Even then, it was rather the ruled who went looking for a better life in the goldmines of Australia and the railway construction camps of North America. Unlike Meiji Japan, the rulers in China continued to see their culture as vastly superior, developing during the self-strengthening movement of the late 19th century the slogan "Chinese values for fundamental principles, Western learning for practical application" (Teng and Fairbank, 1979), which dominates Chinese thinking until today.

After the fall of the Qing Dynasty in 1911, ending 2 millenniums of Imperial rule, a brief period of development of outbound tourism occurred, with some early *nouveau riche* Chinese copying the leisure travel behavior they could study from the growing number of international tourists visiting China, facilitated by new shipping lines and the Trans-Siberian Railway. The occupancy of the bigger part of China by Japan in the 1930s and the civil war of 1945–1949, however, quickly stopped these activities.

For the Chinese living in the former British colony of Hong Kong, in the former Portuguese territory of Macau, and on the island of Taiwan, tourism has been a source of income as hosts and as a basis of personal enjoyment and business as travelers for many decades since 1949. The last restrictions on foreign travel were removed for Taiwanese passport holders in 1979.

In the People's Republic of China, however, leisure travel even within China was considered a wasteful, bourgeois practice before 1978. The desire to travel abroad was seen as committing treason or trying to defect to the enemy. Pilgrimages, which still could be practiced more or less openly in the 1950s, ceased during the Cultural Revolution, when countless religious sites were destroyed. The small amount of inbound travelers from Eastern Europe and the Soviet Union dwindled to practically zero during the 1960s and 1970s.

With the beginning of the "Reform and Opening" policy in 1978, inbound tourism started to be promoted as a fast and easy way to earn foreign currency. Domestic tourism, however, reemerged against the desire of the Chinese government during the 1980s. Only from the 1990s, domestic tourism became acknowledged as an important part of the service industry, acting as a crucial element in the ideological switch from rural socialism to urban consumerism in China (Arlt, 2006).

1.3 CHINESE OUTBOUND TOURISM: A DEMAND-DRIVEN DEVELOPMENT

The gates to outbound travel were opened only very reluctantly by the Chinese government in several stages, not the least under the impression of the Berlin Wall – and subsequently the Eastern Bloc communist governments – falling under the chants of *Reisefreiheit* (Freedom of travel). Four distinguishable phases of China's outbound tourism development can be identified.

1.3.1 FIRST PHASE 1983–1997

The first phase of China's modern outbound tourism started in 1983 with so-called "family visits," first to Hong Kong and Macao and later to several Southeast Asian countries, ostensibly paid by the receiving side. This policy provided the opportunity for the development of clandestine outbound leisure tourism by offering a way to get the passports, foreign currency, and visa necessary for a view of the world outside the People's Republic of China. At the same time, the beginning of China's integration into the world economy resulted in a growing number of delegations traveling to the leading economic countries to attend fairs, business talks, training programs, etc. Almost all of these trips comprised a touristic element, while many were in fact simply pleasure trips in disguise paid by public or government money or, in some cases, arranged by the foreign business partner.

1.3.2 SECOND PHASE 1997–2005

The second phase started in 1997 with the official recognition of the existence of outbound leisure tourism (as opposed to family reunions and business trips) in the "Provisional Regulation on the Management of Outbound Travel by Chinese Citizens at Their Own Expense" and the signing of the first approved destination status (ADS) agreements with Australia and New Zealand. The ADS system is based on bilateral tourism agreements whereby a government allows self-paying Chinese tourists to travel for pleasure to its territory within guided package groups and with a special visa. Only ADS countries can openly be promoted as a tourism destination in Chinese media. With ADS, the Chinese government tried to uphold at least the appearance of a state-controlled and orderly development.

Many more ADS agreements were signed in the years after 1997, especially after 2003. Germany was added to the ADS list in 2003 as the first important European country. With USA and Canada joining in 2009 and 2010, all major destinations arrived within the ADS system, which had become almost obsolete already. The regulations for visits to Hong Kong, Macau, and Taiwan were relaxed at the same time in several steps to help the tourism industries of these two newly formed SARs and in the light of the improved cross-Strait relations (Arlt, 2010).

A stormy development unfolded, with the increases in outbound travelers far outstripping the growth rates planned by the Chinese government. Even though the government policy still emphasized "moderate, carefully managed growth" and close links to the number of inbound tourists, in reality, a tripling of the number of outbound travelers occurred between 1999 and 2004, from 9 million to 29 million travels. A chaotic and mostly unregulated situation developed, with many travel groups organized by nonauthorized agencies in the form of "zero-dollar tours," in which tours were offered at a low price and tourists are then coerced into buying goods and additional services for inflated prices during the trip. For some destinations, including Hong Kong and Thailand, even "minus-dollar tours" became quite common, with inbound companies paying to the tour organizers to receive the groups in addition to providing food, local transport, and accommodation (Arlt, 2006).

Beside the wish to control the spatial movements of its citizens, the loss of hard currency motivated the official reluctance to fully address the issue of the outbound market during this period. As a measure to rein in its growth, an exit tax was discussed at the beginning of the new century. However, some experts defended the outbound tourism and pointed out that most of the money taken by the Chinese tourists out of Mainland China stays in the SARs Hong Kong and Macau, which are part of the bigger Chinese economy (Zhang, 2013).

1.3.3 THIRD PHASE 2005–2011

For a number of reasons, 2005 can be considered the beginning of the third phase of China's modern outbound tourism.

First, the increase of outbound travels between 2005 and 2008 remained below 15% on average, even though 2005 and the following years saw no outbreaks of SARS, similar health hazards, or other internal problems; nor did wars or other major external developments stop Chinese travelers from visiting foreign countries. In contrast, many newly opened ADS countries provided a wider range of destinations to choose from.

Second, a number of concerns and irritations ended the "honeymoon period" on the supply side. In July 2005, the Schengen countries within the European Union (EU) reintroduced the pre-ADS methods of interviewing a percentage of all visa applicants and of insisting on proof by the tour operators of the return of all members of ADS groups. In a parallel move, the Australian ADS scheme witnessed a complete overhaul in the summer of 2005 to stop zero-dollar practices and Chinese visitors disappearing into the local Chinese community.

Third, the waning enthusiasm was also recognizable on the demand side. Reacting to the increased difficulties in obtaining Schengen visa, Chinese tour operators developed new programs to bring some of their customers to African instead of European destinations.

Fourth and last, but certainly not the least, the Chinese government changed its official stance on the question of outbound tourism. During 2005, The China National Tourism Administration (CNTA) adopted the position that for a really strong tourism country, the outbound sector also has to be developed. Instead of trying to limit the total number of outbound travels, the government started to take measures to bring the chaotic situation under control through more detailed regulations. In spring 2008, CNTA officials hinted at an upcoming introduction of new comprehensive regulations for outbound tourism, even though it took until October 2013 before the Tourism Law finally came into force. In autumn 2009, the Chinese government officially declared tourism to be a "pillar industry," and in 2011, China started to celebrate an annual Tourism Day on May 19. Since then, outbound tourism is considered a "soft power" tool to demonstrate to the world the might and power of China (Tse, 2013), in line with mega events such as the Olympic Games 2008 and the Expo 2010, the establishment of Confucius Institutes around the world, a rover landing on the moon, and the establishment of English language radio and TV programmes in competition with CNN, BBC, and Al Jazeera.

During the period 2005–2011, Chinese outbound travelers also developed as the biggest growth market in luxury and brand shopping in all major destinations, because the sales of many luxury brands to Chinese customers, both home and abroad, became the biggest part of their overall sales.

The total number of border crossings increased during this phase, from 31 million in 2005 to 70 million in 2011, more than double the amount within 7 years.

1.3.3 FOURTH PHASE 2012–TODAY

In the year 2012, China overtook Germany and the USA as the biggest outbound tourism source market in the world in terms of both border crossings and spending. Ever since, the interest of global media, as well as destinations and tourism service providers around the world, for the Chinese market has considerably increased.

The new Chinese government emphasizes the support for outbound leisure tourism by Chinese citizens for the first time explicitly. In January 2013, the Chairman of the CNTA, Shao Qiwei, was quoted as saying during a visit to UNWTO in Madrid: "Outbound tourism will boost China's development in the long-term. The government, and particularly CNTA, will continue to promote the travelling of Chinese people abroad as we believe in the mutual benefits of collaboration" (COTRI, 2013). Even more importantly, in April 2013, the Annual Conference of the Boao Forum for Asia started with a keynote speech of President Xi Jinping, who said at the opening ceremony that China in the next 5 years will invest US$500 billion in foreign countries and have probably over 400 million tourists traveling abroad. This was the first time ever that a Chinese communist party leader spoke internationally – and in a positive way – about Chinese outbound tourism. Since then, Xi has, at several other international occasions, expressed his support for Chinese traveling abroad. In September 2014, while visiting Maldives, he even urged Chinese tourists, "We should also educate our citizens to be civilized when traveling abroad. Don't litter water bottles, don't destroy their coral reef. Eat less instant noodles and more local seafood" (Garst, 2014).

The Chinese Tourism Law of October 2013 marks another step in the process of reining in the excesses of "zero-dollar tours" by not only giving detailed regulations about the protection of the rights of consumers but also demonstrating the idea of Chinese outbound travelers as "ambassadors of China" who are advised to behave in a "cultured" way to avoid the creation of a bad image of China in the world (COTRI, 2013).

On the host side, the lure of the Chinese tourist dollar and considerable lobbying from the tourism and retail industry resulted in the easing of visa restrictions in many destinations. About 40 mostly smaller countries, including many island states, had by the end of 2014 already adopted a visa waiver policy. The Schengen countries and the UK are in the process of simplifying procedures and the development of multiple-entry visa for frequent travelers, especially after the USA announced on November 10, 2014, that visa for the Chinese business and leisure traveler will increase in validity from 1 to 10 years – the longest validity possible under the US law (Welitzkin, 2014).

1.4 CHINA'S OUTBOUND TOURISM – QUANTITATIVE DEVELOPMENT

The number of Chinese visitors has seen a steady increase in almost all destinations in the past 15 years (Fig. 1.1). Even in 2009, when the global economic crisis resulted in shrinking tourism markets all over the world, Chinese outbound travels increased, albeit with a temporarily weakened speed. Decreases in arrival numbers for some destinations for certain periods of time were mostly due to political problems in bilateral relations between China and the destinations, as in the cases of Japan in 2013 or Vietnam, Malaysia, and Thailand in 2014. In the latter case also, Singapore as the regional hub suffered even though the city-state itself was not involved in any political or territorial dispute with China. For several destinations, the number of Chinese arrivals doubled several times year on year, and China became the most important tourism source market. The arrival of Chinese in Thailand jumped from 1.7 million in 2011 to 4.7 million in 2013. In the period January 2013 to June 2014, South Korea developed as the rising star, overtaking Thailand with 7 million Chinese arrivals within 18 months compared with 6.5 million Chinese arrivals to Thailand.

By far, the most important destinations for Chinese outbound travelers are the two SARs Hong Kong and Macau. They are still enjoying growing arrival numbers, but their market shares are decreasing continuously. Within 1 year, the market share for Hong Kong in the second quarter of 2014 decreased to 38% from 43% for the same period the previous year; for Macau, the respective numbers were 18% in 2014 and 20% 1 year before (COTRI, 2014).

In terms of spending, China moved from the eighth position in international tourism spending in 2003, with US$15 billion, to third position, with US$55 billion in 2010, conquering the top position only 2 years later with US$102 billion in 2012. For the USA, in comparison, the corresponding amounts were US$58, US$76, and US$84 billion. China's spending grew almost 7-fold within the decade 2003–2012, whereas US spending not even doubled (UNWTO, 2014; Fig. 1.2).

1.5 CHINESE OUTBOUND TOURISM – QUALITATIVE DEVELOPMENT

The rapid market segmentation in recent years has made it impossible nowadays to talk about "the" Chinese traveler. The profile of the traditional package tour traveler is now complemented by a range of other travel modes, which reflect a growing diversification of preferences and maturing travel behavior.

Ever since the Chinese government started allowing its citizens to travel abroad for leisure in the 1990s, the format of group travel packages, organized by tour operators, dominated the scene, as under the ADS system, this was almost the only option for leisure travel. However, it was also the preferred option for inexperienced Chinese travelers, unfamiliar with the destination and its language, to travel abroad. Although in China, the ratio of group versus independent travel is clearly shifting in favor of the latter (at least for domestic and Asian destinations), the coming years will still see huge numbers of first-time travelers, especially from second- and lower-tier cities, opting to join tours for their first trips abroad, before shifting to self-organized travel.

In terms of number of customers, package tours will therefore remain for the foreseeable future the biggest segment of the market based on low-price, low-quality offers (Fugmann and Aceves, 2013).

However, with the easing of visa policies and the growing number of experienced Chinese repeat travelers, organized group travel is no longer the only option. Travel agencies have started selling "semi-self-organized" packages that generally cover flight and hotel, as well as some guided sightseeing for the first day(s) of the trip. Another important part of this segment is the form of customized tours, with varying degrees of flexibility of itinerary during the trip.

With the growing travel experience of the market and also on the level of the individual traveler, such offers have been gaining in market share.

The biggest trend, constituting the "Second Wave" of China's outbound tourism, is the growth of fully self-organized travel. Tour packages can no longer satisfy the diversity in demand, as travelers prefer to travel with people they know, instead of with strangers, and do not want to be bound to the schedule of the entire group (Xiang, 2013). Family travel, honeymooning, nature travel, outdoor activities, self-driving, gourmet shopping, health visits, luxury shopping, cruises, island trips, rail travel, and luxury destinations are all elements of trends in the ever-increasing segmentation of the overall market.

The relative importance of self-organized travel varies among different markets: from 0% for destinations that cannot be reached without an organized tour (for instance, Antarctica) to more than 70% for "easy" destinations such as Hong Kong and South Korea. The scrapping of visa requirements is supporting the trend for self-organized tours even further.

Chinese nationals residing temporarily outside China are also more likely to travel self-organized. This group of Chinese travelers is often overlooked, but students and expatriates working abroad form an important market for many destinations. Estimates of Chinese students studying abroad in 2013–2014 stand at close to 700,000 (UNESCO Institute for Statistics, 2014), with the total number of all overseas Chinese students since the start estimated at about 3 million persons. In earlier years, many of these remained in the foreign country, but in the past few years, more than 80% have returned to China to pursue their careers in their home country. Almost all affluent families send their children abroad for studies. In the USA alone, about 275,000 Chinese students are counted (Institute of International Education, 2014). The students themselves often travel during vacations, taking advantage of their language skills and their experience gained by living abroad. In addition, many students will have other family members visiting them or parents will search for a suitable university beforehand.

Expatriate persons are the other group of Chinese passport holders who appear in the nationality- and accommodation-based tourism arrival statistics of most destinations as Chinese visitors, even though they might live in the country or region. Often well paid, but living without their family, they will not only travel on business but also use their spare time for self-organized leisure trips, frequently together with their colleagues.

1.6 OUTLOOK: THE SECOND WAVE, THE THIRD WAVE, AND CHINA OUTBOUND 2.0

The high-speed development of China's outbound tourism has taken the world by surprise. When the author of this chapter published a press release in 1999 with the headline "The Chinese are coming" to alert the tourism industry during the ITB Berlin fair, everybody – at least in Europe – agreed that this was a bunch of baloney. When the author started the COTRI in 2004, this topic was still seen as a rather exotic, if not esoteric, subject for a research institution. For many destinations in Asia, Oceania, and the Indian Ocean, China has become one of the most important or even *the* most important tourism source market, while other parts of the world have just woken up to the development. The caricature of Chinese tourists who arrive by the busload, stop only for a quick photo at each sight, eat only Chinese food, are afraid of the number 4, insist on water cookers in their hotel room, and are generally loud, bad-behaved, and unwilling to spend much money on anything but shopping is still dominating the imagination of big parts of the tourism industry and is often replicated with relish by the media. Chinese do not like to spend much time on the beach? – so for 3S locations, there is no need to bother to learn about the Chinese market. Chinese travel only to the most important sights in the most important cities? – we have no Eiffel Tower, so let us concentrate on our traditional markets. Such ideas are still to be found, even though especially since China was named by the UNWTO as the biggest international tourism source market, destinations and companies from the Azores to Western Australia and from Finland to Kenya are paying more attention to the opportunities arising from the new generation of Chinese travelers.

China's outbound travel market has indeed already developed much further, and the differentiation and segmentation of the market will continue. The mass-market package tour segment is set to continue to dominate in terms of numbers of travelers, especially in the first-time traveler market based on travelers from second-, third-, and even lower-tier cities in China. Currently, more than half of the total market for trips going beyond Greater China is still organized in package tours, offering very limited profit margins. Customized organized tours, as well as semi- and self-organized tours including Meetings, Incentives, Conferences, and Exhibitions (MICE) travels, however, will further increase and probably represent already more than 50% of the market in terms of turnover and even more in terms of profits.

The "Second Wave" of Chinese outbound travelers is not performing according to the caricature mentioned above. Most of these travelers are middle-aged, with high education and income levels and increasingly possessing the experience of traveling internationally since a number of years. They are "money rich" – and increasingly also "experience rich" – but "time poor"; they have to maximize their investment of time, effort, and money while abroad. They expect their hosts to show respect toward the Chinese culture, so even if they do understand English, they will demand to find information in Chinese. They might eat toast, bacon, and eggs for breakfast while traveling in Western countries, but they will think it only proper for a hotel to provide Chinese breakfast items on the buffet.

Research conducted by COTRI and many other surveys show that experienced Chinese visitors are expecting to be treated not only on an equal level with other travelers, but also better than anybody else. Most Chinese are aware that their country is not only the most populous but also, by now, the biggest source market for international travel in the world; so they are likely to insist that service providers clearly indicate that they go the extra mile for Chinese customers both in words and in deeds.

Almost nobody in China could inherit wealth in the People's Republic of China in its first 65 years of existence. Most outbound travelers are working hard during most days of the year. Accordingly, they are looking for intense and fast experiences without waiting time between activities, which is one of the many aspects of necessary adaptation of existing tourism offers.

For the same reason, accessibility is an important demand of the "Second Wave" guests from China. The example of Mauritius, an island that offers visa-free access and direct nonstop flights from China, demonstrates that product adaptation and accessibility are more important than low prices for this growing section of the Chinese outbound market. From less than 10,000 arrivals in 2011, Chinese guest numbers increased to 21,000 in 2012, doubling to 42,000 in 2013, with already 50,000 in the first 9 months of 2014 alone (Statistics Mauritius, 2014). With this development, Chinese arrivals were responsible not only for almost the total increase of arrivals in Mauritius, but also for the considerable increase in spending per day and per person compared with European tourists, as they are not content to spend most of the visit on the beach but will go – and pay – for sport, culture, and shopping activities.

Therefore, while the global tourism industry is starting to comprehend the "Second Wave" of Chinese outbound travelers, the "Third Wave" is already on the horizon. One important trend will be the widening of the age band of Chinese outbound travelers. International travel being less and less perceived by frequent travelers as special or dangerous, more children can be expected to join their parents or travel in youth travel groups with their classmates or friends. As China's outbound tourism started about 10 years ago for real, there are also some teenagers joining the game who have been traveling internationally with their parents since childhood.

At the other end, destinations can start to welcome the Chinese affluent pensioners. Most of them will still be in the age group of 55–65 years rather than above and will probably still be involved in a bit of trading real estate and playing the stock markets, but unlike the majority of middle-aged affluent travelers, they are not "time poor" and can afford to go on trips that are less frantic, last several weeks, and are less driven by the question of what others might think about their trip. The huge increase in the number of Chinese going on cruises can be partly explained by the rising demand for convenience, safety, and less stressful travel.

New niches and products will develop, following existing ones like traveling to Tasmania to buy a teddy bear, based on a single blog entry by a celebrity (Arlt, 2013), or acquiring a driving license during a holiday in South Korea (COTRI, 2014), or traveling to Portugal, Cyprus, or Malta to buy real estate with an EU passport as a bonus.

Destinations and tourism service providers who invest time and effort into understanding and preparing for the needs and expectations of the different segments of the Chinese outbound market will be able to profit from the opportunities provided by the emergence of new customer groups that can be offered activities, destinations, and seasons of the year that are different from those of traditional, mostly Western, markets.

They will have to make sure, however, to offer the right type of reasons to visit, which can range from novelty, exclusivity, intensity, a connection to China, endorsements by celebrity or travel bloggers, and specialized quality labels to hype on Chinese Social Media. They will further need to tell the right story and to give the opportunity to retell it right away to the peers of the travelers back home (Kristensen, 2013), who – with the help of smartphones and WiFi – are copresent during most times of the trip and sometimes will be even consulted to decide what to do, which dishes to order in a restaurant, etc. Certification and documentation of the activities of the Chinese guests will likewise support the appeal of travel products.

From a quantitative point of view, barring *Black Swan* events, the affluent top decile of Chinese society, which is able to spend money on international tourism, will not be influenced by the slowing Chinese economic growth in a way that would substantially decrease their spending power as far as relatively small purchases like an international trip are concerned. Even the anticorruption and antihedonism campaign of the new Chinese government under President Xi Jinping has not targeted international travel and consequently has not negatively influenced the outbound sector.

International travel will remain an important part of the consumption pattern of affluent Chinese. At the same time, at the lower end of the market, package tours might become more expensive as a result of the Tourism Law and similar future regulations; however, this will rather result in shorter trips to not-too-distant destinations than in travels not taken at all.

That does not mean, however, for the individual destination that growth rates are guaranteed, as, for instance, Thailand experienced a growth rate of 102% for the first half of 2013 but saw a decline of 23% in the first half of 2014 (COTRI, 2014).

The COTRI forecast for the remainder of the decade considers a Compound Annual Growth Rate (CAGR) of 16% for the number of border crossings and a slightly stronger CAGR of 18% for spending on international tourism. Based on the official Chinese statistics for 2013 of 98 million border crossings and international travel expenditure of 97 billion euros, this would result in the number of departures being slightly above 200 million in 2020 and a spending of more than 230 billion euros for international travel in the same year. The bigger part of this growth will focus on new destinations both in terms of countries and in terms of destinations and activities within the countries.

The relative importance of Hong Kong and Macao as destinations will continue to decrease and fall below 50% of the total border crossings before 2020. Inner-Asian travel will continue to be by far the most important element, but destinations outside of Asia will gain more relative importance, almost doubling their share from 10% in 2013 to close to 20% by 2020.

One out of seven international border crossings starting from Mainland China, four times more Chinese visitors outside Asia in 2020 compared to 2013 – indeed, when it comes to Chinese outbound tourism: *You Ain't Seen Nothin' Yet*!

REFERENCES

Arlt, W. G. *China's Outbound Tourism*; Routledge: London, 2006.

Arlt, W. G. (ed.) *China Tourism Academy: Annual Report China Outbound Tourism Development 2009/2010 (English Edition)*; Profil Verlag: München, 2010.

Arlt, W. G. How Did Tasmanian Lavender Bears Turn into a Social Media Sensation in China. Blog Entry Dec 16, 2013. http://www.forbes.com/sites/profdrwolfganggarlt/2013/12/16/social-media-bring-tasmanian-lavender-bear-into-chinese-bedrooms-china-outbound-tourism-best-practice-example/ (accessed Dec 22, 2014).

Arlt, W. G. Attracting and Satisfying the Second Wave of Chinese Outbound Tourists. ITB Berlin Webinar Dec 9, 2014. https://www.youtube.com/watch?v=tdBbmXz5VnI (accessed Dec 22, 2014).

Chien, L. H. China's Domestic and Outbound Tourism to Keep Growing in 2014. WantChinaTimes Jan 6, 2014. http://www.wantchinatimes.com/news-subclass-cnt.aspx?id=20140106000006&cid=1202 (accessed Dec 22, 2014).

China Outbound Tourism Research Institute [COTRI] (ed.). *China Outbound Market Intelligence*, Edition September 2013; COTRI: Heide, 2013.

COTRI (ed.). *China Outbound Market Intelligence*, Edition December 2014; COTRI: Heide, 2014.

Dreyer, E. L. *Zheng He: China and the Oceans in the Early Ming Dynasty*; Pearson: London, 2006; pp 1405–1433.

Fugmann, R.; Aceves, B. Under control: performing Chinese outbound tourism to Germany. In tourism planning & development. *Special Issue: China Outbound Tourism*. 2013, 10 (2), 159–168.

Garst, W. D. Chinese tourists are golden harbingers. *China Daily USA*. [Online] September 26, 2014. http://usa.chinadaily.com.cn/opinion/2014-09/26/content_18664813.htm (accessed Dec 22, 2014).

Huff, T. E. *The Rise of Early Modern Science. Islam, China and the West*; Cambridge University Press: New York, 1993.

Institute of International Education (ed.). Open Doors 2014: International Students in the United States and Study Abroad by American Students are at All-Time High. http://www.iie.org/Who-We-Are/News-and-Events/Press-Center/Press-Releases/2014/2014-11-17-Open-Doors-Data (accessed Dec 22, 2014).

Kester, J. G. C. World Tourism Trends, Presented at IPK 22st World Travel Monitor Forum, Pisa, Italy, October 28–29, 2014.

Kristensen, A. E. Travel and social media in China: from transit hubs to stardom, in: tourism planning & development. *Special Issue: China Outbound Tourism*. 2013, 10 (2), 169–177.

Lau, D. C. (ed.). *Confucius: The Analects*; Penguin Classics: London, 1989.

Levathes, L. *When China Ruled the Seas: The Treasure Fleet of the Dragon Throne*; Simon & Schuster: New York, 1994.

Smith, P. *Somebody Else's Century: East and West in a Post-Western World*; Knopf Doubleday: New York, 2010.

Statistics Mauritius (ed.). International Travel and Tourism Jan–Sep 2014. http://statsmauritius.govmu.org/English/Publications/Pages/Tourism-Jan-Sep-2014.aspx (accessed Dec 22, 2014).

Teng, S. Y.; Fairbank, J. K. *China's Response to the West: A Documentary Survey, 1839–1923*; Harvard University Press: Cambridge, MA, 1979.

The Economist. Hello America. China's Economy Overtakes Japan's in Real Terms, August 16, 2010. http://www.economist.com/node/16834943 (accessed Dec 22, 2014).

The Economist. Coming to a Beach Near You. How the Growing Chinese Middle Class is Changing the Global Tourism Industry, April 19, 2014. http://www.economist.com/news/international/21601028-how-growing-chinese-middle-class-changing-global-tourism-industry-coming (accessed Dec 22, 2014).

Tse, T. S. M. *Chinese Outbound Tourism as a Form of Diplomacy*. In Tourism Planning & Development, *Special Issue: China Outbound Tourism*. 2013, 10 (2), 149–158.

UNESCO Institute for Statistics (ed.). Global Flow of Tertiary-Level Students. http://www.uis.unesco.org/Education/Pages/international-student-flow-viz.aspx (accessed Dec 22, 2014).

United Nations World Tourism Organization (UNWTO) (ed.). *UNWTO Tourism Highlights, 2014 Edition*; UNWTO: Madrid, 2014.

Welitzkin, P. New Visa Extension will Boost US Tourism. http://usa.chinadaily.com.cn/epaper/2014-11/11/content_18897729.htm (accessed Dec 22, 2014).

Xiang, Y. X. The Characteristics of Independent Chinese Outbound Tourists. In Tourism Planning & Development, *Special Issue: China Outbound Tourism*. 2013, 10 (2), 134–148.

Zhang, G. R. *China's Tourism Development: Situation, Analysis and Future Perspectives* [Online]. COTRI: Heide, 2013.

CHAPTER 2

EVOLUTION OF CHINA'S POLICY OF OUTBOUND TOURISM

GUANGRUI ZHANG

CONTENTS

2.1 INTRODUCTION: FROM "CHINA VISIT" TO "CHINESE VISITORS"

During the past 6 decades or so since the founding of the People's Republic in 1949, China has become one of the top world international tourism destinations and tourism-generating countries, in addition to the unsurpassable size of its domestic tourism market. Indeed, great changes have taken place in tourism in the country.

2.1.1 FUNCTION CHANGE

The initial function of developing tourism defined by the government in the early 1950s was to win the understanding and friendship of the international community; that is to say, until the late 1970s, tourism was taken as a means of politics without any intention for economic gains from it. The epoch-making campaign of economic reform and opening up to the outside world, which was launched in 1978, marked another revolutionary change in the country, bringing forth great changes in the economic and political systems of China. Hence, tourism was taken as a means of economy, earning badly needed foreign exchange to support the four aspects of modernization of the country, namely, modernization of agriculture, industry, national defense, and science and technology, the long-term goal set by the government for the campaign. With the deepening of the economic reform and improvement of national economy, the function of the developing tourism sector has been readjusted since 2009 from a mere earner of foreign dollars to harbinger of both economic growth and welfare of the nationals. In the State Council's document,[1] it has been stated that the guiding strategy is "to turn tourism into a pillar industry with strategic importance to the national economy and a modern service industry that better meets the aspirations of the general public"(2009).

2.1.2 MARKET CHANGE

Accordingly, the order of priority in China's tourism market has changed in the corresponding periods. During the first 3 decades since 1949, China developed only the international brand of tourism, more precisely, only inbound tourism, with no domestic tourism, not to mention outbound tourism by the nationals. Domestic tourism started from the mid-1980s. Since then, until the end of the past century, priority of market was given to inbound tourism, followed by the domestic one, for promotion of the country's economic growth. Outbound tourism, in a real sense, hardly existed before the late 1990s, and the outbound tourism market had been controlled cau-

[1]The State Council of People's Republic of China, The Opinions of the State Council on Speeding up the Development of the Tourism Industry, December 1, 2009.

tiously until the turn of the new millennium. Unexpectedly, the outbound market has expanded rapidly in the new century, faster than the other two markets ever since, in stark contrast with most other countries of the world at the time of economic depression from 2008 onward.

Generally speaking, the world's focus on China's tourism has been shifting, gradually in the past 30 years or so, from "China visit" (China as an attractive destination) to "Chinese visitors" (China as an emerging source market).

2.2 CHINA'S OUTBOUND TOURISM DEVELOPMENT: AN OVERVIEW

Nowadays, "tourism" has become a catchword everywhere in the world, although it is one of the phenomena that are very difficult to define. It is true to what Yeoman (2008) writes: tourism "is a phenomenon that has been created and is difficult to define because of the complexity." As known to all, China's tourism development follows an unconventional road owing to its own historical, political, economic, and cultural reasons, and the issue of outbound tourism in the country is even more complicated than that in many other countries in the world. Therefore, to better understand this issue, explanations for some specific terms used in the country might be necessary beforehand.

2.2.1 CHINA'S OUTBOUND TOURISM BY DEFINITION

OUTBOUND TRAVEL

Outbound travel indicates (a) travels to foreign countries and regions, (b) travels to the two Special Administrative Regions (SARs): Hong Kong and Macau within China, and (c) travels to Taiwan, which is part of China's territory. The travel documents used for these three destinations are different. Passports are used for all foreign countries and regions, and specific passes are used for Hong Kong, Macau, and Taiwan. For the purpose of statistics, all travels to foreign countries and to Hong Kong, Macau, and Taiwan are considered outbound travels, and only travels to foreign countries are regarded as travels abroad.

SELF-PAID AND PUBLIC-PAID TRAVELS

In English, there is a special term for business travel including both travels for public official and commercial affairs. However, in China, for a long time before the economic reform, there were few cases considered private overseas travels for leisure, and only business travels were made by the government officials and state-owned enterprise leaders. Customarily, two pairs of terms are used for outbound travels: outbound travel for private affairs and for public affairs; and self-paid outbound travels and public-paid travels. For the former pair of terms, it may indicate what kind of travel documents are granted, business passport (issued by the Ministry of Foreign Affairs) or private passport (issued by the Ministry of Public Security), whereas the latter pair of terms indicate who may pay the related expenses for the trip. In fact, the public-paid tours may include real business travel and junkets. For a long period, some public-paid tour abroad, or junket, has been a common form of corruption by government officials and state-owned enterprise leaders, which has been seriously examined and dealt with in recent years.

OUTBOUND TOURISM AND BORDER TOURISM

As a large country, China features multiple neighboring countries, sharing borders with 16 countries on land. Due to many reasons, such as easy procedures, low cost, short distance, and easy access, border tourism is one of the popular forms for outbound travels, playing an important role in local tourism development, local economy, and bilateral relations. However, this type of tourism is very different from outbound tourism in regular forms.

APPROVED DESTINATION STATUS

Approved Destination Status (ADS) is a very special bilateral arrangement for leisure travels abroad that is only practiced in China. An ADS agreement signed between China and a destination country aims to confirm that the signed country is an approved destination for Chinese tour groups. This is an innovation of China's outbound tourism policy in the specific period, with no precedent known in the world. According to the government regulations, Chinese tour operators are not allowed to organize tours to non-ADS destinations; however, an ADS destination is not obligated to grant tourist visas to the individuals. From a long-term point of view, this is only a transitional measure. As the number of countries signing ADS agreement increases, this type of arrangement may lose its original significance.

2.2.2 CHINA'S OUTBOUND TOURISM: FACTS AND FIGURES

Outbound tourism may be traced back to the early 1980s; however, the official tourism statistics about this market were first released in the *Statistics Report on China's Tourism Industry in 1994,*[2] with only two sets of simple statistics: departures of Chinese passport holders and outbound tourists organized by travel agents. With the

[2]The National Tourism Administration of the People's Republic of China started to publish its annual *Yearbook of China Tourism Statistics* in 1986, and later in 1991, it began to release *Statistics Report on China's Tourism Industry* with the date of the previous year.

rapid growth of the outbound tourism sector in the country in the following years, the related statistics officially released has improved accordingly; however, the information relating to outbound tourism is still inadequate, inconsistent, and of dubious validity. However, it is believed that a little is better than none. Tables 2.1–2.3 show the changes in the Chinese market in terms of both number of departures and the corresponding spending in the past few decades, as well as part of change in destination for Chinese outbound visitors.

TABLE 2.1 Departures of Chinese Residents (1992–2013)

Year	Departures in Total	For Private Purposes	
		Departures	% in Total
2013	98,185,200	91,969, 000	93.7
2012	83,182,700	77,056,100	92.6
2011	70,250,000	64,117,900	91.3
2010	57,384,500	51,507,900	89.8
2009	47,656,000	42,209,700	88.6
2008	45,844,400	40,131,200	87.5
2007	40,954,000	34,924,000	85.3
2006	34,523,600	28,799,100	83.4
2005	31,026,300	25,140,000	81.0
2004	28,850,000	22,980,000	79.7
2003	20,221,900	14,810,900	73.2
2002	16,602,300	10,061,400	60.6
2001	12,134,400	6,946,700	57.2
2000	10,472,600	5,630,900	53.8
1999	9,232,400	4,266,100	46.2
1998	8,425,600	3,190,200	37.9
1997	8,175,400	2,439,600	29.8
1996	7,588,200	2,423,900	31.9
1995	7,139,000	2,053,900	28.8
1994	6,106,000	1,642,300	26.9
1993	3,740,000	1,466,200	39.2
1992	2,928,700	1,119,300	38.2

Source: Data from China National Tourism Administration's (CNTA) Yearbook of China Tourism Statistics in the related years.

TABLE 2.2 Balance of China's International Tourism Service (1982–2013)

Year	Balance	Revenue	Spending
2013	−72,000,000	48,000,000	120,000,000
2012	−51,900,000	50,000,000	102,000,000
2011	−24,121,050	48,464,000	72,585,050
2010	−9,066,030	45,814,000	54,880,300
2009	−4,026,670	39,675,000	43,701,670
2008	4,686,000	40,843,000	36,157,000
2007	7,446,950	37,233,000	29,786,005
2006	9,627,300	33,949,000	24,321,700
2005	7,536,930	29,296,000	21,759,070
2004	6,589,700	25,739,000	19,149,300
2003	2,218,730	17,406,000	15,187,270
2002	4,986,584	20,385,000	15,398,416
2001	3,883,174	17,792,000	13,908,826
2000	3,117,313	16,231,000	13,113,687
1999	3,233,970	14,098,450	10,864,480
1998	3,396,300	12,601,740	9,205,440
1997	1,907,500	12,074,414	10,166,640
1996	5,726,000	10,200,000	4,474,000
1995	5,042,000	8,730,000	3,688,000
1994	4,287,000	7,323,000	3,036,000
1993	1,886,000	4,683,000	2,797,000
1992	1,435,000	3,947,000	2,512,000
1991	2,329,000	2,840,000	511,000
1990	1,784,000	2,218,000	470,000
1989	1,431,000	1,860,000	429,000
1988	1,614,000	2,247,000	633,000
1987	1,458,000	1,845,000	387,000
1986	1,223,000	1,531,000	308,000
1985	936,000	1,250,000	314,000
1984			150,000
1983			53,000
1982			66,000

Unit: US$ '000.
Source: State Information Center: China statistics of balance of international payments.

TABLE 2.3 The Ten First-Stop Destinations for Chinese Outbound Departures (2000–2013)

Year	The Ranking of First-Stop Destinations for the Chinese Mainlander Departures									
	1	2	3	4	5	6	7	8	9	10
2000	HK	Macau	Thailand	Japan	Russia	S. Korea	USA	S'pore	N. Korea	Australia
2001	HK	Macau	Thailand	Japan	Russia	S. Korea	USA	S'pore	N. Korea	Australia
2002	HK	Macau	Japan	Russia	Thailand	S. Korea	USA	S'pore	N. Korea	Malaysia
2003	HK	Macau	Japan	Russia	Vietnam	S. Korea	Thailand	USA	S'pore	Malaysia
2004	HK	Macau	Japan	Russia	Vietnam	S. Korea	Thailand	USA	S'pore	Malaysia
2005	HK	Macau	Japan	Vietnam	S. Korea	Russia	Thailand	USA	S'pore	Malaysia
2006	HK	Macau	Japan	S. Korea	Thailand	Russia	USA	S'pore	Vietnam	Malaysia
2007	HK	Macau	Japan	S. Korea	Vietnam	Russia	Thailand	USA	S'pore	Malaysia
2008	HK	Macau	Japan	Vietnam	S. Korea	Russia	USA	S'pore	Thailand	Malaysia
2009	HK	Macau	Japan	S. Korea	Vietnam	Taiwan	USA	Russia	S'pore	Thailand
2010	HK	Macau	Japan	S. Korea	Taiwan	Vietnam	USA	Malaysia	Thailand	S'pore
2011	HK	Macau	S. Korea	Taiwan	Malaysia	Japan	Thailand	USA	Cambodia	Vietnam
2012	HK	Macau	S. Korea	Taiwan	Thailand	Japan	Cambodia	USA	Malaysia	Vietnam
2013	HK	Macau	S. Korea	Thailand	Taiwan	USA	Japan	Vietnam	Cambodia	Malaysia

Source: China National Tourism Administration, Statistics Report on China's Tourism Industry (of the related years).

2.2.3 FEATURES OF CHINA'S OUTBOUND TOURISM DEVELOPMENT

From the view of development process, China's outbound tourism may have some obvious features as follows:

LATECOMER

China's tourism development started in the early 1950s, and tourism as an industry was recognized in the late 1970s as a result of the economic reform; however, outbound tourism developed in the real sense only in the late 1990s, later than most of the countries in Asia, much later than its neighboring countries like Japan and Korea, not to mention Europe and North American countries.

LONG-TERM PREPARATION

Owing to the specific political and economic system and the social and economic conditions, Chinese outbound tourism had been tightly controlled. The epoch-making economic reform and opening-up campaign have made China more open to the outside world. Tourism has been made one of the important sectors to develop; however, outbound tourism had not been well considered. It took nearly 20 years after the reform to have the first regulation, *Interim Measures for the Management of Self-paid Tourism Abroad,* in 1997. In response, the world opening to China also has experienced the same process.

SUDDEN RISE AND SUSTAINED RAPID GROWTH

China's outbound tourism has maintained a sustained rapid growth ever since the restrictions were relaxed. It took approximately 17 years for the departures of Chinese citizens to reach the first 10 million, from 1983 to 2000, and then the second 10 million mark was reached in only 3 years' time, and a more rapid rate has been maintained ever since. During the 5 years from 2007 to 2012, the departure increased from about 41 million to 83 million, with an annual growth of more than 15%, whereas the average annual growth of the world international tourism remained only about 2% during the same period. The number of Chinese departures crossed the 100 million mark in 2014.

FREQUENT CHANGES IN VARIOUS ASPECTS

During the past 2 decades, great developmental changes have taken place in China's outbound tourism sector, in addition to the rapid growth. Some of the outstanding changes include the following:

MARKET STRUCTURE

The year 2000 is the dividing point when the proportion of overseas trips for private affairs (leisure included) surpassed that for public affairs, and since 2011, overseas trips for private affairs have made up more than 90% of the grand total each year.

EXPANDING DESTINATIONS

Before 2000, only nine destinations (including Hong Kong and Macau) were designated officially as destinations for Chinese outbound tourism, and the tour combining Singapore, Malaysia, and Thailand, the earliest three officially designated destinations, has remained the most prime product for ages. As of 2012, the number of designated destinations has increased to 115, covering all continents in the world, and more than 140 countries have signed the ADS agreement with China for this purpose.

SURPRISING OVERSEAS SPENDING

As the largest developing country with a huge population and rigid economic system, China was known in the world as a poor and backward state for long in the past. In fact, before the reform, few of the ordinary people could afford a trip abroad by themselves even if they were allowed to. However, the spending power and the behavior of Chinese visitors in overseas destinations in the new century constitute a big surprise to the world. In 2012, the Chinese overseas tourism spending made a new record of more than US$102 billion. According to the Chinese Luxury Traveler White Paper (2013), which was released by the firm Hurun Report Inc. during the International Luxury Travel Market Asia Conference (Shanghai, June 3–5, 2013), it was the third consecutive year that Chinese tourists ranked first in the global shopping market, with an average consumption of US$1139 per trip in 2012.

RAPID EXPANSION OF INTERNATIONAL TOURISM DEFICIT

As a result of sluggish inbound tourism and booming outbound tourism in China in the years since 2009, the balance of international tourism has changed from surplus to deficit, and the gap continues to grow extensively. The trade deficit that occurred for the first time ever in 2009 in the country was slightly more than US$4 billion and then jumped up to US$76.9 billion in 2013; it might increase to US$100 billion, more than double the foreign exchange earned from the trade for the same time.

POLITICAL CONCERN

Finally, China's outbound tourism development is highly related to the government's political concern, which remains unchanged. In general, China's policy toward outbound tourism is rather conservative, cautious, and passive compared with that toward other sectors of tourism. In addition, the outbound tourism policies are more sensitive and highly related to bilateral political relationship.

There might be many ways to interpret the features of the development of outbound tourism in China. The obvious factors may include, generally speaking, the economic growth of the country and increase in overall income of the citizens, long-term accumulation of overseas tourism demand, bigger income gap among residents, external pull effect of destination marketing, and internal push effect of the crowd consumption psychology.

2.3 EVOLUTION OF CHINA'S OUTBOUND TOURISM POLICIES

In this chapter, discussions of the outbound tourism policy mainly focus on the tours and travels of the Chinese citizens for leisure, rather than the official business travels in general. Therefore, the past 40 years or so may be roughly divided into three periods in terms of policy.

2.3.1 POLICY EXPERIMENTAL PERIOD (1983–1996)

The visit to Hong Kong and Macau by the residents of Mainland China was one of the first breakthroughs of the China's outbound tourism market after the economic reform of the late 1970s. In late 1983, before Hong Kong and Macau returned to their motherland, the central government, on a trial basis, approved Guangdong Province, which is close to the two regions, to organize visiting friends and relative (VFR) tours to Hong Kong and Macau on the condition that all the travel arrangements should be made by the officially designated travel agents and all expenses for the trip be paid by the overseas hosts to meet the demands of the mainlanders, in addition to a daily quota of participants. Up to 1996, the averaged daily quota was 900. Later, this travel arrangement was expanded to the Fujian province at home, as well as to Singapore, Malaysia, and Thailand externally – where a large number of overseas Chinese and Chinese descendants live.

Another breakthrough of the outbound tourism about the same time was border tourism, that is, border-crossing excursions by the local residents in the border regions to the border regions of the neighboring countries. This arrangement was first made between China and the Democratic People's Republic of Korea (North Korea) in 1984 and then extended to the border regions of the neighboring countries in the north (such as Russia and Mongolia) and south (such as Vietnam, Laos, and Myanmar).

However, these two special forms of outbound tourism were launched in accordance with the specific purpose to save foreign exchange and meet the increasing demand. Again, these arrangements had contributed to the gaining of experience and laid the foundations for further opening of outbound tourism in the country.

On the whole, during this period, the general policy of outbound tourism mainly was to control its growth, although the government statements were often not very clear-cut, such as to develop the outbound tourism in "an appropriate" or "a planned" way. The government measures were reflected in many ways such as the tight control of travel documents for outbound travels, limited foreign exchange allowance, annual quota for overseas outbound travelers, and operational qualifications for the business operators, in essence, a state monopoly.

2.3.2 POLICY BREAKTHROUGH PERIOD (1997–2008)

The first ever government document governing outbound tourism – Interim Measures for Managing the Self-paid Outbound Tourism by the Chinese Citizens – was issued by the State Council in 1997, marking the beginning of the new era of outbound tourism in China. Since then, a series of regulations relating to this sector were made, putting outbound tourism under the control of the central government and helping the outbound tourism market in China to develop faster.

REGULATIONS: FROM TRANSIENT TO FORMAL

Some 3 years later in 2000, the *Interim Measures for Managing the Self-paid Outbound Tourism by the Chinese Citizens* was replaced by a new document entitled *Measures for Managing the Outbound Tourism by the Chinese Citizens*. The change in title means that "measures" are no more "Interim" but are formal, weakening the limits between "self-paid" or "non-self-paid." The new document did not reaffirm the prior stance to develop the self-paid outbound tourism sector in "an organized, planned, and controlled way" but proposed more measures to simplify the related procedures.

TRAVEL DOCUMENTS: FROM APPROVAL TO APPLICATION

China had exercised a very strict approval system over the travel documents for overseas visits for a long time since 1949. In 2003, starting from Beijing, then extending to other administrations, a new application system replaced the former approval system for private passports of Chinese citizens, removing many complex bureaucratic procedures such as approval by local government or employer and other proof materials for the given trips.

TRAVEL PATTERNS: FROM GROUP TO INDIVIDUAL

According to the regulations set by the State Council, all the Chinese outbound tourists had to join in tour groups organized by the designated travel agencies, departing and entering the country through the specified port in the form of the entire group. Even the ADS agreements with foreign destination countries stressed this principle for the package tours. However, the September 11 incident in 2001 and the Asia severe acute respiratory syndrome (SARS) outbreak in 2003 heavily hit the tourism business in Hong Kong and Macau. To support the economic development of the two special administrations in such a situation, the central government signed Closer Economic Partnership Arrangements (CEPA) with Hong Kong and Macau in 2003. According to the arrangement, the individual visit scheme was launched: the mainlanders may visit these two regions individually, not necessarily in package tour groups. At the same time, more and more overseas destinations try to simplify visa procedures to woo the Chinese visitors; "visa on arrival" and "visa exemption" treatments were granted by some countries, leading to a consequent growth in numbers of individual travelers.

FOREIGN EXCHANGE: FROM TIGHT CONTROL TO FREE PURCHASE

One of the early constraints for outbound tourism was shortage of foreign currency in the country; therefore, tight control of foreign currency was a national policy for years. However, thanks to the successive sound economic growth since the reform, the lack of foreign reserve is no longer a problem for the country; instead, by 1996, China's foreign exchange reserves increased to more than US$100 billion, and by 2008, up to US$1900 billion. Starting from 2005, Chinese residents can purchase foreign exchange in a number of commercial banks at home, and credit cards and debit cards issued to Chinese residents are widely accepted in foreign lands. In addition, the appreciation of the Chinese currency against most foreign currencies has stimulated the outbound travels of the Chinese for sure.

HOLIDAY SYSTEM REFORM: MORE POSSIBILITIES FOR OVERSEAS TRAVELS

Holiday system reform since the mid-1990s has not only offered more free time for the Chinese residents, but, what is more important is it has helped them to reconsider seriously the meaning of holiday, leisure, and tourism, more profoundly, the real meaning of life and the goal to pursue. The direct effect of the holiday rearrangements, as expected by the government, is to promote tourism in all fields, including outbound travels. Particularly, the 7-day holidays in China known as the "Golden Week," matchless in the world, has served as a primary impetus behind the rapid growth of national tourism, and outbound tourism may benefit more from it.

2.3.3 RAPID GROWTH PERIOD (2009 AND BEYOND)

In terms of China's tourism policy, 2009 serves as a milestone. In preparing the 12th Five-Year Program for China's Economic and Social Development (2011–2015), the Central Government worked out a series of significant strategic adjustments and changes in the development modes, some of which are directly related to tourism development in general. Since 2009, many specific policies and legislations have been released toward tourism development in the future.

The document *The Opinions of the State Council on Speeding up the Development of the Tourism Industry* issued on December 1, 2009, was an important document by the State Council, which presented the government policy toward tourism in a most clear, comprehensive, and detailed way. This is the first time ever that the State Council pledged to "turn tourism into a pillar industry with strategic importance to the national economy and a modern service industry that better meets the aspirations of the general public." It could be understood that tourism is very important to the entire nation's economic development, and more importantly, in the long run; at the same time, the social function of the sector is equally important to better meet the increasing demand of the general public for tourism consumption as part of quality of life. For the first time ever, national tourism, including both domestic and outbound tourism sectors, has been placed in an important position. This document also put forward that outbound tourism should be developed in an orderly way, promoting further opening to the outbound tourism business operation, advocating civilized behaviors, and ensuring the safety of the outbound travelers. In general, tourism is not only an economic matter, but its development should also be conducive to the satisfaction of people's needs and for improvement of international relations.

CONSTANT SUPPORT TO THE TWO-WAY TOURIST FLOW CROSSING THE TAIWAN STRAITS

As an important part of outbound tourism, the central and local governments have been very supportive in related policy making. Since Taiwan is open to the Mainland tourists, up to July 2010, all inland provinces and autonomous regions have organized package tours to Taiwan and more direct flights linking the two sides with each passing year. The individual travel arrangement to Taiwan from 2011 has upgraded the visits to Taiwan by the residents of the mainland to a new level. Within 5 years, from 2008 to 2013, the mainland visitors to Taiwan reached more than 6.2 million, becoming the third largest market of the destination.

ACTIVE AND EFFICIENT ASSISTANCE MEASURES TO THE TWO SARS

As mentioned above, the Chinese outbound travels started with visits to Hong Kong and Macau, and after returning to the motherland, these two SARs have been the destinations of most of the Chinese visitors. Numerous policies for increasing tourist arrivals from the mainland to the two SARs have been introduced. Since 2008, Hong Kong and Macau have been entitled to set up their own wholly owned travel agencies in some mainland cities to organize tours to the two regions. By 2012, the scope of individual travel mode to Hong Kong and Macau has been extended to some 49 cities in 22 provinces and autonomous regions in the mainland. As a result, Hong Kong and Macau have remained the top two destinations of the mainlanders for years, and the mainland provides more than two-thirds of the overnight visitors to these two regions annually.

TOURISM COOPERATION AS A NEW DIPLOMATIC HIGHLIGHT

Since the economic reform and opening up, the tourism industry has been one of the growth sectors in the national economy; however, tourism as an important part of the country's diplomatic strategy has been maintained ever since. With the increasing numbers of Chinese departures and expanding of their spending overseas, China's outbound tourism sector plays an active part in international relations and cooperation with other countries. Of recent years, in almost all talks between top Chinese and foreign leaders, either in China or elsewhere, tourism cooperation has been one of the major topics. In 2010 alone, according to the published data, top Chinese leaders visited more than 50 countries in the world, including close Asian neighbors and faraway African countries, well-developed European and American economies, and emerging economies; as a result, tourism collaboration and cooperation were widely stressed, particularly in the joint communiqués with visited countries, such as India, Germany, Russia, Greece, Italy, Turkey, France, Kyrgyzstan, Tajikistan, and Uzbekistan, among others. From 2009 to 2013, China had launched a good many bilateral events, such as the Year of Travel or the Year of Cultural Exchange with Russia, Korea Republic, Italy, Austria, and others, promoting bidirectional flow of international tourists as a common goal.

THE OUTLINE FOR NATIONAL TOURISM AND LEISURE (2013–2020) AND THE LAW OF TOURISM

In early 2013, two other significant documents on tourism were issued. In the *Outline for National Tourism and Leisure (2013–2020)* approved by the State Council, a goal for tourism development set by 2020 is to basically put in place a paid annual leave system for employees, to keep a substantial growth of urban and rural residents' consumption in tourism and leisure, to let healthy, civilized, and environment-friendly ways of tourism and leisure be widely accepted by the public, to help the quality of national tourism and leisure improve remarkably, and to set up a modern national tourism and leisure system required by an initially prosperous society. In short, citizens may have more time and facilities for travels both at home and abroad. The Law of Tourism, the first ever in China's history, came into effect from October 1, 2013. In its "General Provisions," the Law states that "this law is made to protect the legitimate rights and interests of the tourists and tourism operators, regulate the order of the tourism market, protect and reasonably utilize tourism resources, and promote the sustained and healthy development of tourism," and "the law applies to the tours, vacations, leisure activities and other forms of tourism activities organized within the territory of or to go beyond the territory of the People's Republic of China, and the business operations providing relevant tourism services." The implementation of the law may ensure that China's outbound tourism, like other forms of tourism, develops in accordance with legal procedures.

SOME OPINIONS OF THE STATE COUNCIL ON PROMOTING THE REFORM AND DEVELOPMENT OF THE TOURISM INDUSTRY

In August 2014, the Chinese State Council issued another important document for tourism development in the country. The document reconfirms the implementation of paid vacation system for employees and the improvement of regional and sub-regional international tourism collaboration systems by such major initiatives such as the Silk Road economic belt, the 21st Century Maritime Silk Road, and others. These projects will further promote China's outbound tourism in the times to come.

2.4 CONCLUSION AND DISCUSSION

During the past 3 decades since the reform and opening up, China's outbound tourism has developed by leaps and bounds, leading to the change of the world focus on China's tourism from "China visit" to "Chinese visitors." However, at home, debates and controversies around the outbound tourism policies have never stopped, providing the basis for the government to adjust its related policies. Some of the arguments are as follows.

2.4.1 INBOUND TOURISM VERSUS OUTBOUND TOURISM: WHICH IS MORE IMPORTANT?

In the Western countries, this seems not a problem, and both inbound and outbound tourism modes are developed in line with the actual market demands. However, China's tourism has developed in an unconventional way. For a long period, owing to the lower rate of economic development and shortage of foreign currency, the government policies often tended to be in favor of inbound tourism for earning foreign dollars rather than favoring outbound tourism, which spends them. When the international tourism balance had become negative and the deficit had increased, proposals were made to take some restrictive measures such as departure tax or tariff increase on imported goods. So far, the government has hardly done anything to check the growth of outbound travels, thanks to the understanding that it is necessary to meet the demand of the residents for outbound tourism, and that it is better to maintain a good balance of international tourism by increasing the revenue from inbound tourism rather than by curbing residents' consumption of outbound tourism. To achieve the goal of building a powerful tourism country, both inbound and outbound tourism branches of the country should be well developed.

2.4.2 ECONOMICS VERSUS POLITICS: WHICH IS MORE IMPORTANT?

Generally speaking, tourism has been discussed in the world within the realm of economy, and international tourism, both inbound and outbound, is often considered to improve the international balance of payment, earning foreign currency as service export or spending the foreign exchange as service import. Globally speaking, countries like Japan that would encourage their residents to travel overseas to spend money are very rare; in contrast, most other countries, including developed economies, strive to woo foreign visitors for foreign exchange earnings. However, one issue that should not be ignored is that international tourism, particularly the outbound tourism, is a type of effective lubricant and activator for international or bilateral diplomatic relationship. In recent years, Chinese leaders on many different occasions have expressed that international tourism is the best way to strengthen friendliness among different peoples and have stressed the significance of tourism in promoting mutual understanding and advancing strategic relationship with other countries, especially the neighboring countries. China pays great attention to the significance – both economic and political – of outbound tourism development. For this, China so says and so does, while other countries may do so, but not like to say so. However, the recent frequent readjustments of visa policies toward Chinese tourists by many governments may interpret the truth from a perspective.

2.4.3 ON THE GLOBAL STAGE, CHINESE VISITORS MAY FACE THE "LOVE AND HATE" TEST

Known to all, on the global tourism stage, the Chinese visitors are latecomers. The growth of the Chinese outbound tourists and the ability of their spending power are exciting and unexpected news to the tourism destinations. The market potential, great and attractive, especially at the time of world economic depression, is an opportunity no one would like to miss. At the same time, many of the destinations, particularly the destinations in the developed world, are not really ready to offer heartfelt welcome or willingness to the newly emerging unfamiliar market. From the very beginning up to now, Chinese tourists have been encountering the "love and hate" test. This may be reflected in their dubious visa policy toward Chinese tourists and their sardonic or even unfriendly attitude to Chinese customers in both words and acts. To maintain a better image of the country, the Chinese government has done a great deal in educating the inexperienced Chinese travelers to behave well while abroad, and the China National Tourism Administration (CNTA) published the *Tourism Etiquette Rules for Chinese Citizens Travelling Abroad* on October 2, 2006, and republished it on July 31, 2013. The Law of Tourism clearly stipulates that the state build a risk warning system for the safety of tourism destinations, and Chinese outbound tourists have the right to ask the local Chinese official institution for assistance and protection within its scope of responsibilities when stuck in a plight overseas. Consequently, more measures and efforts have been extended in protecting the lawful rights, interests, and security of the Chinese outbound tourists while traveling overseas.

There is good reason to believe that Chinese outbound tourism is set to maintain a high growth rate for the years to come, though the travel and consumption patterns may change, and the market may become more diversified accordingly and government policies may remain further supportive, relaxed, and liberalized to the sector.

Working closely with the overseas destination administrations and the industry, the Chinese government may spare no efforts to make the Chinese residents travel more conveniently, safely, and in dignity. China's tourism, both international and domestic, is about to embark on a new stage of conventional development.

REFERENCE

Yeoman, I. *Tomorrow's Tourist: Scenarios & Trends*; Elsevier: London, 2008; p 11.

CHAPTER 3

REVIEW OF CHINESE OUTBOUND TOURISM RESEARCH: STATUS QUO AND FUTURE DIRECTIONS

YING WANG and XIN JIN

CONTENTS

3.1 INTRODUCTION

The Chinese outbound travel market has been one of the fastest-growing international markets for destinations. Over the past 2 decades, China witnessed significant and fast transition from a planned economy to a market-driven economy, political liberalisation, and the resulting changes and diversification in both the economy and social and cultural values (Cai et al., 2008). Associated with these transformations is the emergence of the First and Second Wave of Chinese outbound tourists. While the First Wave is dominated by package tourists from first-tier cities in China, the Second Wave is characterized by the increasing shares of (a) young, "latte-drinking," "avant-garde," and experience-savvy travelers; (b) independent, self- or half-organized tours; (c) the Internet, BBS, forum, travel blogs, as well as friends and relatives as main information sources (CTA, 2014; Arlt, 2013). This extends opportunities for service providers outside conventional streams (Arlt, 2013). Yet, knowledge about this market is primarily gained from the first generation of Chinese tourists, with its applicability to the second generation unclear.

Two studies (Cai et al., 2008; Keating and Kriz, 2008) have reviewed the research on Chinese outbound tourism before the acknowledgment of the Second Wave of China outbound tourism. Keating and Kriz (2008) focused exclusively on the destination choice of Chinese tourists. Cai et al. (2008) reviewed 30 studies published in 20 top-ranked tourism and hospitality journals up to 2006. The authors posited that the academic community outside China appeared to be unexcited about conducting research into this fast-growing market with its unique consumer characteristics (Cai et al., 2008).

As China's economy continues to rise, an increasing number of the Chinese are able to afford overseas travel. The economic, sociocultural, environmental, and ethical impacts of this market on destinations will be increasingly significant. A deeper understanding of outbound tourism will assist destination planners and policy makers in market development and destination management in mitigating any detrimental impacts.

The purpose of this study is to determine the status of research on Chinese outbound tourism. This study adopted a hybrid design that incorporated the characteristics of both the narrative and systematic quantitative review methods. Using the systematic method, we document the geographical spread of the studies by author, year, destination of interest, research methods, data analysis techniques, and primary topical areas. The categorization of topics for the systematic review was based on Weaver and Lawton's (2009) model of multidisciplinary linkages within tourism studies, which includes the categories of geography, history, law, ecology, sociology, psychology, business management, anthropology, marketing, agriculture, political science, and economics. The discussion within each of the topical areas followed a narrative review style to indicate research production in the field, investigate emerging themes and methods, and identify knowledge gaps for future research directions.

We reviewed 61 journal articles published from 2000 to May 2014 in six top-tier tourism and hospitality journals (Table 3.1). Key words used for the data search were China, Chinese, outbound, travel, tourism, tourist, and combinations of these words. Year 2000 was selected because it was a milestone in the development of Chinese outbound tourism, and for the first time, the volume of Chinese outbound travelers exceeded 10 million (UNWTO, 2003). A number of studies, primarily econometric analyses of demand, where China was of minor interest, were excluded. We acknowledge that in this study, articles published outside the six journals were reviewed. There are also articles published in languages other than English. However, the review of those articles is beyond the scope of this study.

TABLE 3.1 Number of Articles by Journals and Time Periods

	ATR	CHRAQ	IJHM	JHTR	JTR	TM	Total
2000–2003	0	0	0	0	0	0	0
2004–2006	0	0	0	1	2	4	7
2007–2009	0	0	1	1	2	5	9
2010–2014	6	1	4	6	9	19	45
Total	6	1	5	8	13	28	61

Abbreviations: ATR, Annals of Tourism Research; CHRAQ, Cornell Hotel and Restaurant Administration Quarterly; IJHM, International Journal of Hospitality Management; JHTR, Journal of Hospitality and Tourism Research; JTR, Journal of Travel Research; TM, Tourism Management.

3.2 RESULTS

3.2.1 NUMBER OF PUBLICATIONS BY JOURNAL

Table 3.1 presents the number of publications in the selected journals cross-referenced to the four time periods: 2010–May 2014, 2007–2009, 2004–2006, and 2000–2003. Research has increased significantly since 2010, with a total of 45 articles published thereafter. The journals *Tourism Management* and *Journal of Travel Research* each published a significant number of articles.

3.2.2 AUTHORS, INSTITUTIONS, AND COUNTRIES OR REGIONS HAVING MOST PUBLICATIONS

The most productive researchers in this field based on total number of publications in the six journals are given in Table 3.2. Three of the seven researchers are currently affiliated with the Hong Kong Polytechnic University. Three researchers are ethnically non-Chinese. Many studies have at least one Chinese-speaking author, possibly attributable to the familiarity of these authors with the study context. In future, studies completed independently by academics of non-Chinese origin may provide interesting and potentially different insights from a host perspective.

TABLE 3.2 Most Productive Researchers with Number of Publications (2000–2014)

Author	No. of Publications	Affiliation
Xiang (Robert) Li	7	University of South Carolina*
Samuel Seongseop Kim	6	The Hong Kong Polytechnic University
Cathy H.C. Hsu	5	The Hong Kong Polytechnic University
Hanqin Qiu Zhang	4	The Hong Kong Polytechnic University
Rich Harill	3	University of South Carolina
Songshan (Sam) Huang	3	University of South Australia
Beverley Sparks	3	Griffith University

* Now Temple University

3.2.3 DESTINATIONS EXPLORED IN THE STUDIES REVIEWED

As shown in Table 3.3, research seems to respond to market development, as Asia (primarily Hong Kong) is the most investigated destination, corresponding to its status of having the largest share of the market (CTA, 2012). Notably, research interest has extended from Asia to Oceania and then on to North America. This trend aligns with the geographical expansion of the Approved Destination Status (ADS) program. Europe, despite being the principal Western destination for Chinese outbound tourism, remains an unexplored continent in research, possibly because of its non-English academic environment and relatively smaller number of academics with Chinese background. The most studied regions and/or countries are Hong Kong, Australia, and the United States, which, as identified by Cai et al. (2008), are the same regions or countries studied prior to 2006. Research effort should be extended to other popular destinations, such as Southeast Asia and Europe. Although most studies consider special administrative regions of Hong Kong and Macau as outbound destinations in accordance with official Chinese statistics, there are exceptions such as study by Li et al. (2010) in which Hong Kong and Macau do not qualify as outbound destinations.

TABLE 3.3 Region of Interest Explored by the Studies Reviewed

	Asia	Multiple/Non-Specific Destinations	North America	Oceania	Total
2004–2006	6	1	0	0	7
2007–2009	4	2	0	3	9
2010 –2014	25	6	6	8	45
Total	35	9	6	11	61

3.2.4 METHODS UTILIZED IN THE ARTICLES REVIEWED

In Table 3.4, various qualitative and quantitative approaches adopted by the 61 articles, together with their sample size, are outlined. Focus group and interview are prominent in qualitative studies, whereas questionnaire is the most commonly used data collection technique in quantitative studies. Popular data analysis techniques include thematic content analysis for qualitative research and factor analysis, regression analysis, and structural equation modeling for quantitative studies. Types of respondents include actual tourists, potential tourists, service providers and government representatives, students, and residents. The views of residents and tourism employees on issues related to Chinese outbound tourism are underrepresented. Less used research designs or sample collection approaches are observation, using online and print materials, and experimental design. We call for future research to use more creative and innovative techniques, online and print media materials, and longitudinal data.

TABLE 3.4 Research Methods and Sample Size Utilized in the Publications Reviewed[a]

Methods	Frequency	Sample Size of Interview or Focus Group	Frequency	Sample Size of Field Survey	Frequency
Conceptual/ review	2	Smaller than 10	4	Smaller than 200	8
Mixed methods	7	11–30	7	201–400	10
Qualitative	15	31–40	2	401–600	9
Quantitative	37	Larger than 41	3	Larger than 601	16
Total	61	Total	16	Total	43

[a]For focus groups, sample size is the number of participants, not the number of sessions.

3.2.5 PRIMARY TOPICAL AREAS DISCUSSED

The model of multidisciplinary linkages within tourism studies by Weaver and Lawton (2009) was used as the framework for analysis to categorize the 61 articles reviewed. These articles were classified into following categories: psychology (23 articles), marketing (13), business management (19), economics (1), sociology (1), and political science (2). The remaining two articles were of a cross-disciplinary nature. For each discipline, the key topical areas covered in the articles were identified. Three most studied disciplines were analyzed according to their current research status, knowledge gap, and future research directions.

PSYCHOLOGY – TOURIST MOTIVATION, EXPECTATIONS, AND TRAVEL BARRIER

In most of the 23 articles, travel motivations, expectations, and barriers, primarily adopting an etic approach that is based on researcher-imposed frame of references, were investigated. Common motivations were established across multiple destinations, such as knowledge, relaxation, novelty, prestige, escape, and self-development (Hsu et al., 2010; Li and Cai, 2012). A common practice for developing instruments is to employ a combination of measurements from previous studies (e.g., Hsu et al., 2010). The studies gave insufficient consideration to the uniqueness of the Chinese market.

The findings of Li et al. (2011) could represent the expectations of Chinese outbound tourists. Chinese tourists highly valued attributes of cleanliness, safety, and value for money, and they were concerned about being taken advantage of and desired genuine respect and hospitality. Other major expectations were related to food and accommodations. Tour guides played a critical role and were expected to be bilingual, friendly, professional, and culturally and historically knowledgeable. However, the study fell short in explaining what might underpin tourist expectations. One possible moderating factor is culture, as Hoare et al. (2011) posited that core cultural values influenced how Chinese tourists interacted with service intermediaries and impacted on their dining experiences. This study by Hoare et al. has focused primarily on travel expectations rather than service expectations, particularly in relation to interpersonal interactions at various service venues, demanding more research efforts in this regard.

Another group of studies focused on travel barriers that inhibit the Chinese from traveling abroad (e.g., Sparks and Pan, 2009). Sparks and Pan (2009) identified a range of external factors with a moderate influence on travel to Australia: currency, flight time, media warnings, language barriers, safety and risk, negative media coverage of the bilateral relationship between China and the destination country, and visa regulations. A similar range of structural constraints were identified by Lai et al. (2013) for Chinese travelers to the USA. Reflecting an emerging trend among independent travelers to Australia, the study by Wu (2014) on drive tourism indicated on-the-road concerns with unfamiliarity in vehicles, road conditions, driving rules, and accommodation. These concerns were compounded by personal factors such as language skills, driving experience, confidence and stress, and physical conditions. Tourists coped with these concerns by greater preparation to increase familiarity and take various precautions.

Much of the literature on motivations, expectations, and barriers focused on the development potential of the market through a better understanding of its facilitators and constraints. A trend in tourism and hospitality literature is to apply social psychology theories to understand cognition, affection, experience, and behaviors of travelers (Tang, 2014), which has been observed in the literature of Chinese outbound tourism. Another underrepresented area in this body of literature is the "self", that is, how the development of the self by Chinese tourists impacts on their motivation, expectations, and concerns.

Psychology is acknowledged as being culturally bound or relative; thus investigations into behavior and experience shall be conducted through different cultural lenses (Pearce and Packer, 2013). To enrich such studies, a potential direction is to investigate "the self", by extending discussions in Western literature on the desire of tourists to seek authenticity and performativity to the Chinese context. Future research can also address several questions that remain unanswered. First, what social, cultural, and psychological factors underlie motivations, perceived barriers, and expectations of Chinese travelers? Second, do demographic, geographic, and psychographic differences exist in the formation of motivations and perceived barriers and expectations? Third, how and to what degree do these motivations, barriers, and expectations affect behaviors of tourists? Fourth, how does the degree of impact differ across different segments within the Chinese market? And finally, what role can cross-cultural analysis play to contribute to the understanding of Chinese tourists in these regards? To obtain the complete frame of psychological reference of Chinese outbound tourists, an emic approach is recommended for future studies.

MARKETING – DESTINATION IMAGE, POSITIONING, AND MARKET SEGMENTATION OR SEGMENTS

Destination image and positioning are of great interest to researchers. Selected topics and researched destinations include effects of product perceptions on destination image (Lee and Lockshin, 2012, Australia); destination image richness, evenness, and dominance (Stepchenkova and Li, 2012, the United States); destination image held by long-haul travelers (Li and Stepchenkova, 2012, the United States); and preference and positioning analysis (Kim et al., 2005, multiple destinations; Li et al., 2015, the United States against five other Western destinations).

Most destination image-related studies on Chinese outbound tourists focused only on images attached to particular destinations and did not investigate issues such as the formation of destination image or the relationship between destination image and other constructs, such as loyalty and destination selection. Destination-specific studies focused almost exclusively on the United States (e.g., Li and Stepchenkova, 2012; Stepchenkova and Li, 2012, 2014). Li and Stepchenkova (2012) indicated that Chinese travelers perceived the United States as a highly urban, economically and technologically advanced destination with an open and democratic system. Their images were largely constructed on the basis of attractions on the east and west coasts. Stepchenkova and Li (2012) compared the image of the United States among four groups of travelers based on their travel experiences. The groups who had traveled outside Asia perceived the United States as a friendly, open, democratic, and free society that offers a relaxing experience, whereas the less-experienced groups based their image of the United States on tangible attributes and perceived the country as economically developed, scenic, and beautiful. In another study by Stepchenkova and Li (2014), the structure of brand association was analyzed based on Chinese travelers' top-of-mind images of the United States. Notably, some of these studies adopted frameworks or methods from fields outside social science. These studies offered new solutions to analyzing big data that are increasingly available to researchers.

Targeting and positioning were relatively neglected areas of Chinese outbound tourism research. Only two studies – Li et al. (2015) and Kim et al. (2005) – investigated the relative preference and positioning of selective destinations in the Chinese outbound market. Li et al. (2015) positioned the United States against 5 other Western destinations and concluded that the United States holds a unique position that differentiated itself from its competitors. Kim et al. (2005) analyzed the relative standings of a range of ADS and non-ADS destinations. Since the study by Kim et al. (2005), the range of destinations has expanded to include most major tourist destinations. Destinations, depending on their history of being part of the ADS scheme, may be at different stages of their destination lifecycle in the Chinese outbound market. Therefore, a renewed effort to identify the relative standing of destinations is of interest to both tourism practitioners and those researching the subject. Targeting and positioning are strategically important topics and are closely related to destination image and market segmentation, deserving more attention in the future.

Another group of studies, although not necessarily classified as such in the marketing discipline, focused on specific market segments and types of products, such as backpackers to Macao (Ong and du Cros, 2012), film tourism (Kim, 2012) and medical tourism to Korea (Yu and Ko, 2012; Han and Huang, 2013), gaming tourism to Macau (Liu and Wan, 2011; Wong and Rosenbaum, 2010; Zeng and Prentice, 2014; Shi et al., 2014; Wan et al., 2013), and drive tourism in Australia (Wu, 2014; Wu and Pearce, 2014). Generally, this type of research into specialist tourist groups and product types is gaining greater interest as the Chinese market is becoming increasingly sophisticated in its transition toward the Second Wave of Chinese outbound tourists. Industry has developed a diverse range of tourist products catering to specific markets (e.g., www.tourismaustralia.com.au). Academic research needs to respond to this increased sophistication, by focusing on independent, self- or half-organized tours and specialist travels, to provide a knowledge foundation for effective marketing and product development.

Future studies can give consideration to the identification of green segments within the Chinese markets and how destinations can develop and promote socially and environmentally responsible products to the Chinese travelers. Second, research effects can be directed to delineate marketing process (e.g., image building and branding) and evaluate marketing effectiveness, in particular, how the deployment of electronic marketing can facilitate or undermine marketing effectiveness on the young Chinese market. Issues such as website development, online transaction and distribution, and social media marketing have emerged as viable topics in general tourism and hospitality literature (Oh et al., 2004), but e-marketing has not been pursued earnestly in Chinese outbound tourism research. For instance, the fast transition to mobile device-based information search and consumption behavior require research attention, as one recent report – Insights into China Mobile Internet Users – suggested that 77.3% of social communication and 51.9% of shopping and booking were via mobile devices (ADER, 2013). Third, attention should be given to the role of Chinese popular culture, media, and social media in marketing practices. The relationship between tourism, popular culture, and the media is lacking in general tourism literature (Long and Robinson, 2009) as well as in the Chinese outbound tourism studies.

BUSINESS MANAGEMENT – SHOPPING, UNETHICAL PRACTICES, AND MANAGEMENT OF TOURIST EXPERIENCES

Another prominent area to emerge from our review was business management, including 19 articles on shopping, tourist expenditure, and unethical practices, as well as management of experiences of tourists. Five studies investigated the shopping component of Chinese outbound trips using constructs and scales derived from Western literature. Xu and McGehee (2012) investigated for what and why Chinese travelers shop in the United States. Lloyd et al. (2011) concluded that customer-perceived value has a substantial and direct impact on customer satisfaction and behavioral intention. They found that Chinese travelers visiting Hong Kong exhibit shopping behaviors distinctively different from local shoppers. For example, service quality and product quality were the most influential dimensions for local Hong Kong shoppers, whereas perceived risk, price, and product quality were the most influential dimensions for Chinese tourists. Lew and Ng (2012) revealed that the spending behaviors of Chinese travelers varied on the basis of visitor demographics. These studies provided insights allowing destinations to maximize revenues, emphasizing the enormous economic impact of this market (UNWTO, 2013), as well as the importance of offshore shopping to Chinese outbound tourists. However, studies failed to consider incorporating culturally related constructs (e.g., cultural identity and values) to explain the distinctiveness of shopping behavior of Chinese tourists. More research effort in this regard is required.

One study (Zhang et al., 2009) discussed unethical business practices (zero-fare tours) related to Chinese outbound tourism, reflecting the operation in the industry. Zero-fare tourism, which originated in Southeast Asian destinations, is particularly relevant to package tours, where operators at destinations underprice tour packages and then during the tour recoup the lost revenue from miscellaneous shopping and entertainment options, which are often offered at inflated prices (Zhang et al., 2009). These authors attempted to demystify the mechanism behind the operation of such tours and identified nine factors, such as misleading information, deceptive and bullying language, exorbitant prices, and ineffective post-tour complaint handling mechanisms.

It is worth noting that Australian researchers acknowledged the prevalence of unethical practices in the inbound Chinese market (e.g., King et al., 2006; March, 2008) outside the six journals. While these studies delineated the business mechanism behind this operation, the cultural roots and underlying political drivers have not been investigated. Regulatory efforts have been implemented to curb unethical business practices, such as the *Tourism Law* of the People's Republic of China, effective on October 1, 2013. The effectiveness of the regulation seems to be short-lived in the case of Gold Coast of Australia, as operators quickly found loopholes in the law and reinvented packages following the same old unethical practice. In addition, academic attention to the issue is largely restricted to Hong Kong and Australia, although the media have repeatedly exposed unethical practices in other destinations. Comparative studies of different destinations (such as Southeast Asia, Europe, and North America) will enable investigation of the historical, cultural, and political derivation of the phenomenon and inform effective regulatory efforts.

Limited research investigated on-site experiences and post-visit reflection with the exception of a number of studies on food services and food preferences of tourists at destinations (e.g., Chang et al., 2010; Guillet and Tasci, 2010; Hoare et al., 2011; Kim et al., 2010; Law et al., 2008). Mainstream tourism and hospitality research emphasizes tourism complaint behavior, service failure, and recovery (Oh et al., 2004), but this pattern is not observed in the research of Chinese outbound tourism.

Although the review of Cai et al. (2008) acknowledged some concerns of destinations regarding the revival of mass tourism owing to the influx of Chinese tour groups, research is still almost nonexistent with regard to the sociocultural impact of Chinese tour groups on the host destination. Recent media coverage (Liu, 2012) on the sociocultural interaction and conflicts between Chinese tourists and local residents in Hong Kong highlight the need for academic research to address these issues to assist in tourism planning, policy making, and management. Research into the environmental impacts of Chinese outbound tourism is also lacking. Further, how visitation to destinations by both package and non-package tourists impact on their own social, cultural, and psychological composition are yet to be investigated.

3.3 DISCUSSION AND DIRECTION OF FUTURE RESEARCH

In this research, 61 studies were reviewed in an effort to identify significant trends in the research of Chinese outbound tourism and to set an agenda for future research. Credit shall be given to research efforts manifested in several areas such as travel motivation, expectations, travel barriers, destination image and preference, market segmentation, and business management issues. In general, these studies applied existing theories to the Chinese market but fall short of making a significant breakthrough beyond the widely adopted Western models. As such, knowledge generated is to some extent fragmented and context-confined. Long-term strategic development requires recognition of the complexity and reality of the Chinese outbound tourism phenomenon.

Future research needs to address the challenges brought about by the scale, rapid growth, and changing characteristics of the Second Wave of Chinese outbound tourism to destinations. None of the 61 articles specifically referred to the "Second Wave" or "second generation" of Chinese outbound tourists, but some studies started to question the current relevance of knowledge generated from the early years of market development (e.g., Hsu et al., 2010). There has been increasing effort to address the diversification of Chinese outbound tourism, manifest in investigations into various specialist markets; however, none has consciously differentiated the two waves of tourists in terms of their distinctive needs, characteristics, and preferences. The Second Wave is characterized by increased independent travel, yet the sheer size of the Chinese market and its reliance on mass tourism infrastructure make Weaver's (2012) call for converging toward sustainable mass tourism development significant, which provides directions for the development of tourism.

The heterogeneity of the Chinese market has become increasingly evident, requiring corresponding diversification in product development and marketing strategy. Future studies should explore a wider spectrum of specialized tourism forms, such as backpacking, adventure tourism, film-induced tourism, wine tourism, and heritage tourism. Such studies would provide invaluable insights for product and destination development, beneficial to both tourists and service providers. Another aspect related to product development and marketing is how the Chinese make use of technologies. China differs substantially from Western markets in terms of economic and technological development, cultural values (e.g., avoidance of risk, group orientation), and political governance, which may influence the interaction of Chinese travelers with technologies such as the websites, social media platforms, and mobile phones. Topics of e-marketing, e-distribution, e-satisfaction, e-WOM, as well as their determinants and moderators are opportunities for theory building.

Western literature highlights motivations such as the desire for authenticity and experiencing "otherness" and the tourist gaze, which are constructed in the discourse of colonization. The Chinese, although with the oldest continuous civilization, recently experienced century-long history of being semi-colonized. What constitutes the other and how the Chinese experience the other should be investigated to contribute to the development of non-Western paradigms for tourism research. Future research should also extend to authenticity, performativity, embodiment and affect, relationship of host and guest, and mediatization. Such research needs to incorporate the aspect of "the self" from both guest and host perspectives.

REFERENCES

ADER. Insights into China Mobile Internet Users. 2013. http://adermob.renren.com/s/images/index/periodical/ader_report_en.pdf (accessed June 2014).

Arlt, W. G. The second wave of Chinese outbound tourism. *Tourism Plann. Dev.* 2013, 10 (2), 126–133.

Cai, L. A.; Li, M.; Knutson, B. J. Research on China outbound market: a meta-review. *J. Hosp. Leisure Mark.* 2008, 16 (1–2), 5–20.

Chang, R. C.; Kivela, J.; Mak, A. H. Food preferences of Chinese tourists. *Ann. Tourism Res.*, 2010, 37 (4), 989–1011.

CTA. *Annual Report of China Outbound Tourism Development.* China Tourism Academy: Beijing, 2012.

CTA. *Annual Report of China Outbound Tourism Development.* China Tourism Academy: Beijing, 2014.

Guillet, B. D.; Tasci, A. D. Travelers' takes on hotel – restaurant co-branding: insights for China. *J. Hosp. Tourism Res.* 2010, 34 (2), 143–163.

Han, H.; Huang, J. Multi-dimensions of the perceived benefits in a medical hotel and their role in international travelers' decision-marking process. *Int. J. Hosp. Manage.* 2013, 35, 100–108.

Hoare, R. J.; Butcher, K.; O'Brien, D. Understanding Chinese diners in an overseas context: a cultural perspective. *J. Hosp. Tourism Res.* 2011, 35 (3), 358–380.

Hsu, C. H.; Cai, L. A.; Li, M. Expectation, motivation, and attitude: a tourist behavioral model. *J. Travel Res.* 2010, 49 (3), 282–296.

Keating, B.; Kriz, A. Outbound tourism from China: literature review and research agenda. *J. Hosp. Tourism Manage.* 2008, 15 (2), 32–41.

Kim, S. Audience involvement and film tourism experiences: emotional places, emotional experiences. *Tourism Manage.* 2012, 33 (2), 387–396.

Kim, D.-Y.; Wen, L.; Doh, K. Does cultural difference affect customer's response in a crowded restaurant environment? A comparison of American versus Chinese customers. *J. Hosp. Tourism Res.* 2010, 34 (1), 103–123.

Kim, S. S.; Guo, Y.; Agrusa, J. Preference and positioning analyses of overseas destinations by mainland Chinese outbound pleasure tourists. *J. Travel Res.* 2005, 44 (2), 212–220.

King, B.; Dwyer, L.; Prideaux, B. An evaluation of unethical business practices in Australia's China inbound tourism market. *Int. J. Tourism Res.* 2006, 8 (2), 127–142.

Lai, C.; Li, X. (Robert); Harrill, R. Chinese outbound tourists' percevied constraints to visiting the United States. *Tourism Manage.* 2013, 37, 136–146.

Law, R.; To, T.; Goh, C. How do mainland Chinese travelers choose restaurants in Hong Kong?: an exploratory study of individual visit scheme travelers and packaged travelers. *Int. J. Hosp. Manage.* 2008, 27 (3), 346–354.

Lee, R.; Lockshin, L. Reverse country-of-origin effects of product perceptions on destination image. *J. Travel Res.* 2012, 51 (4), 502–511.

Lew, A. A.; Ng, P. T. Using quantile regression to understand visitor spending. *J. Travel Res.* 2012, 51 (3), 278–288.

Li, M.; Cai, L. A. The effects of personal values on travel motivation and behavioral intention. *J. Travel Res.* 2012, 51 (4), 473–487.

Li, X. (Robert); Stepchenkova, S. Chinese outbound tourists' destination image of America part I. *J. Travel Res.* 2012, 51 (3), 250–266.

Li, X. (Robert); Cheng, C-K.; Kim, H.; Li, X. Positioning USA in the Chinese outbound travel market. J. Hosp. Tourism Res. 2015, 39, 75–104.

Li, X. (Robert); Harrill, R.; Uysal, M.; Burnett, T.; Zhan, X. Estimating the size of the Chinese outbound travel market: a demand-side approach. *Tourism Manage.* 2010, 31 (2), 250–259.

Li, X. (Robert); Lai, C.; Harrill, R.; Kline, S.; Wang, L. When east meets west: an exploratory study on Chinese outbound tourists' travel expectations. *Tourism Manage.* 2011, 32 (4), 741–749.

Liu, J. Surge in anti-China sentiment in Hong Kong. *BBC News.* http://www.bbc.co.uk/news/world-asia-china-16941652 (accessed February 2012).

Liu, X. R.; Wan, Y. K. P. An examination of factors that discourage slot play in Macau casinos. *Int. J. Hosp. Manage.* 2011, 30 (1), 167–177.

Lloyd, A. E.; Yip, L. S. C.; Luk, S. T. K. An examination of the differences in retail service evaluation between domestic and tourist shoppers in Hong Kong. *Tourism Manage.* 2011, 32 (3), 520–533.

Long, P.; Robinson, M. Tourism, popular culture and the media. In *The SAGE Handbook of Tourism Studies;* Jamal, T., Robinson, M. Eds.; SAGE: Los Angeles, 2009; pp 98–114.

March, R. Towards a conceptualization of unethical marketing practices in tourism: a case-study of Australia's inbound Chinese travel market. *J. Travel Tourism Mark.* 2008, 24 (4), 285–296.

Oh, H.; Kim, B.-Y.; Shin, J.-H. Hospitality and tourism marketing: recent developments in research and future directions. *Int. J. Hosp. Manage.* 2004, 23 (5), 425–447.

Ong, C.-E.; du Cros, H. The post-Mao gazes: Chinese backpackers in Macau. *Ann. Tourism Res.* 2012, 39 (2), 735–754.

Pearce, P. L.; Packer, J. Minds on the move: new links from psychology to tourism. *Ann. Tourism Res.* 2013, 40, 386–411.

Shi, Y.; Prentice, C.; He, W. Linking service quality, customer satisfaction and loyalty in casinos, does membership matter?. *Int. J. Hosp. Manage.* 2014, 40, 81–91.

Sparks, B.; Pan, G. W. Chinese outbound tourists: understanding their attitudes, constraints and use of information sources. *Tourism Manage.* 2009, 30 (4), 483–494.

Stepchenkova, S.; Li, X. Chinese outbound tourists' destination image of America part II. *J.Travel Res.* 2012, 51 (6), 687–703.

Stepchenkova, S.; Li, X. Destination image: do top-of-mind associatons say it all? *Ann. Toruism Res.* 2014, 45, 46–62.

Tang, L. (Rebecca). The application of social psychology theories and concepts in hospitality and tourism studies: a review and research agenda. *Int. J. Hosp. Manage.* 2014, 36, 188–96.

UNWTO. *Chinese Outbound Tourism*. UNWTO: Madrid, 2003.

UNWTO. China – The New Number One Tourism Source Market in the World. 2013.http://media.unwto.org/en/press-release/2013-04-04/china-new-number-one-tourism-source-market-world (accessed June 2013).

Wan, P. Y. K.; Kim, S. S.; Elliot, S. Behavioral differences in gaming patterns among Chinese subcultures as perceived by Macao Casino staff. *Cornell Hosp. Q.* 2013, 54 (4), 358–369.

Weaver, D. B. Organic, incremental and induced paths to sustainable mass tourism convergence. *Tourism Manage*. 2012, 33 (5), 1030–1037.

Weaver, D.; Lawton, L. *Tourism Management,* 4th ed.; John Wiley & Sons: Brisbane, Australia, 2009.

Wong, I. A.; Rosenbaum, M. S. Beyond hardcore gambling: understanding why mainland Chinese visit casinos in Macau. *J. Hosp. Tourism Res.* [Online early access]. DOI:10.1177/1096348010380600. Published Online: October 21, 2010. http://jht.sagepub.com/content/early/2010/10/21/1096348010380600.full.pdf+html (assessed April 2013).

Wu, M.-Y. Driving an unfamilair vehicle in an unfamiliar country: exploring Chinese recreational vehicle tourists' safety concerns and coping techniques in Australia. *J. Travel Res.* [Online early access]. DOI:10.1177/0047287514532364. Published Online: May 5, 2014. http://jtr.sagepub.com/content/early/2014/05/04/0047287514532364.full.pdf+html (assessed May 2014).

Wu, M. Y.; Pearce, P. L. Chinese recreational vehicle users in Australia: a netnographic study of tourist motivation. *Tourism Manage*. 2014, 43, 22–35.

Xu, Y.; McGehee, N. G. Shopping behavior of Chinese tourists visiting the United States: letting the shoppers do the talking. *Tourism Manage*. 2012, 33 (2), 427–430.

Yu, J. Y.; Ko, T. G. A cross-cultural study of perceptions of medical tourism among Chinese, Japanese and Korean tourists in Korea. *Tourism Manage*. 2012, 33 (1), 80–88.

Zeng, Z. L.; Prentice, C. A patron, a referral and why in Macau Casinos – the case of mainland Chinese gamblers. *Int. J. Hosp. Manage*. 2014, 36, 167–175.

Zhang, H. Q.; Heung, V.; Yan, Y. Q. Play or not to play – an analysis of the mechanism of the zero-commission Chinese outbound tours through a game theory approach. *Tourism Manage*. 2009, 30 (3), 366–371.

SECTION II

REGIONAL OBSERVATIONS

CHAPTER 4

MAINLAND CHINESE OUTBOUND TOURISM TO HONG KONG: RECENT PROGRESS

TONY S. M. TSE

CONTENTS

4.1 INTRODUCTION

Chinese outbound tourism grew rapidly in the 10 years between 2003 and 2013. Total number of departures increased from 20.2 million in 2003 to 97.3 million in 2013, averaging a 17% annual growth. Hong Kong, as one of China's special administrative regions, has been the most popular destination among Chinese visitors, with more than 40% of the departures having Hong Kong as their first stop, including those who are visiting other destinations after Hong Kong. Figure 4.1 shows the total number of departures in China and those to Hong Kong from 2003 to 2013. The statistics show that Hong Kong has been the most-visited destination by the Mainland Chinese every year over the decade.

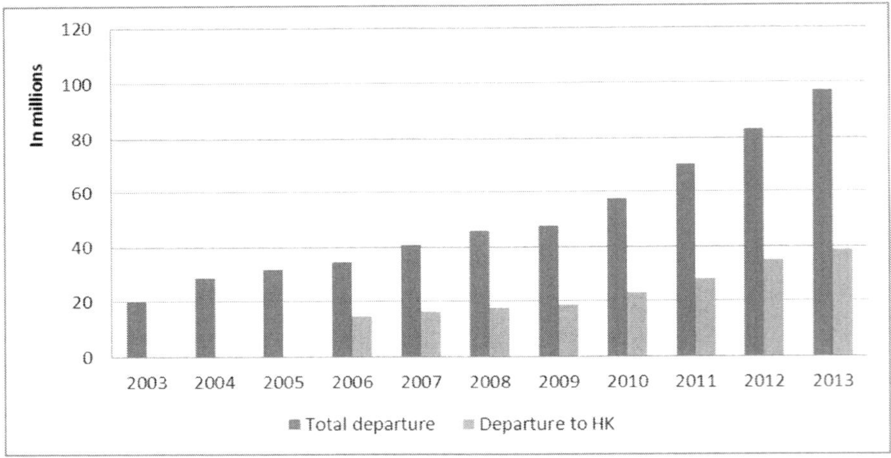

FIGURE 4.1 Chinese Outbound Tourism 2003–2013.
(From China National Tourism Administration, 2013.)

The growth of outbound tourism in China can be attributed but not limited to the more liberal travel policy adopted by the government, economic growth generating much higher disposable income, increased vacation duration as a result of reshuffling of the timing of public holidays, and relaxation in foreign exchange controls. Hong Kong as a destination has been receiving the most number of visitors from the Mainland because of geographical proximity, historical connections, family networks, convenience of traveling, and the aspiration of Hong Kong as a shopping paradise. From the destination perspective, the number of Chinese visitors to Hong Kong has been increasing steadily and reached 40.7 million in 2013, accounting for 75% of the total arrivals. Figure 4.2 shows the total arrivals and the number of Chinese visitors to Hong Kong from 2003 to 2013. It shows that Chinese visitors have become increasingly significant to Hong Kong over the decade.

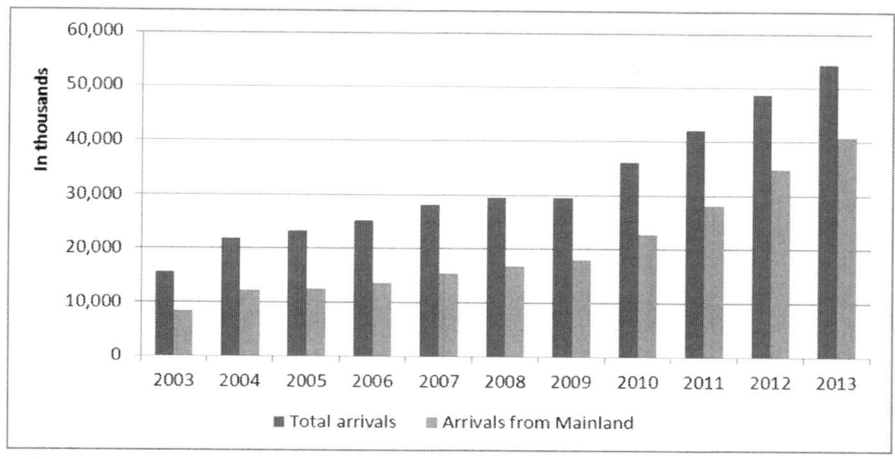

FIGURE 4.2 Visitor Arrivals in Hong Kong: Total Arrivals and Arrivals from the Mainland 2003–2013.
(From Hong Kong Tourism Board.)

The majority of Chinese visitors to Hong Kong have been from Guangdong, the province which is closest to Hong Kong. Nonetheless, the proportion of visitors from other places such as Shanghai, Beijing, Zhejiang, Jiangsu, Hubei, Sichuan, and Shandong are increasing, partly due to the introduction of Individual Visit Scheme (IVS), which is explained in detail in section 4.2, Individual Visit Scheme.

In this chapter, the phenomena of Chinese outbound tourism in Hong Kong with particular emphasis on the latest development and in relation to the unique situation in Hong Kong are analyzed.

4.2 INDIVIDUAL VISIT SCHEME

Apart from the proximity and the convenient transportation into Hong Kong, the introduction of Individual Visit Scheme (IVS) in 2003 played a very important part in increasing the number and proportion of Mainland Chinese visitors. Prior to the introduction of IVS, Chinese residents could visit Hong Kong only in group tours or with special purpose such as business, education, and cultural exchange. In contrast, IVS, for the first time in Mainland China's history, allows residents of designated mainland cities to visit Hong Kong as individuals without joining group tours. Hong Kong was the first destination in which China implemented IVS. The scheme was introduced right after the Severe Acute Respiratory Syndrome (SARS) incidence, which caused serious negative economic impact on Hong Kong. It was believed that the scheme was introduced by the Chinese government to help boost Hong Kong's economy (Tse, 2013).

Initially, IVS granted such access to Chinese residents in 10 cities, and it was gradually extended to 49 mainland cities in 2007, including all 21 cities in Guangdong Province, Shanghai, Beijing, Chongqing, Tianjin, Chengdu, Dalian, Shenyang, Jinan, Nanchang, Changsha, Nanning, Haikou, Guiyang, Kunming, Shijiazhuang, Zhengzhou, Changchun, Hefei, Wuhan and 9 cities of Fujian (Fuzhou, Xiamen, Quanzhou), Jiangsu (Nanjing, Suzhou, Wuxi), and Zhejiang (Hangzhou, Ningbo, Taizhou) (Tourism Commission, 2013). In 2010, IVS has further evolved to allow year-round multiple entries for residents of Shenzhen (China Hospitality, 2011). The number of Mainland Chinese visitors visiting Hong Kong by means of IVS as opposed to those visiting in group tours has been increasing since the introduction of the scheme. The proportion of Mainland Chinese visitors with IVS increased from 34.8% in 2004 to 67.4% in 2013 (see Table 4.1).

TABLE 4.1 Number and Proportion of Mainland Chinese Visiting Hong Kong with IVS

Year	Mainland Visitors with IVS	Percentage of Mainlanders Visiting with IVS
2004	4,259,601	34.8
2005	5,550,255	44.3
2006	6,673,283	49.1
2007	8,593,141	55.5
2008	9,619,280	57.0
2009	10,591,418	59.0
2010	14,244,136	62.8
2011	18,343,786	65.3
2012	23,141,247	66.3
2013	27,464,867	67.4

Source: Hong Kong Tourism Board.

It is believed that visitors with IVS are more sophisticated and experienced than those joining group tours. They are generally more discerning and look for personalized travel experiences. As Hong Kong becomes a more familiar and convenient destination, the number of visitors with IVS has increased steadily over the years. Due to the negative impact of SARS on visitor arrivals from source markets other than Mainland China and the introduction of IVS in 2003, the number of Mainland visitors actually overtook the number of non-Mainland visitors in that year. In fact, Mainland visitor arrivals had been the main source of growth of tourism in Hong

Kong since 2003, and the visitor arrivals continued to grow strongly in the following 10 years. All the non-Mainland markets including the United States, Europe, Australia, New Zealand, Taiwan, Japan, Korea, and Southeast Asia combined have registered very modest growth at about 3% each year, whereas the Mainland market grew by more than 17% on average annually. Figure 4.3 shows the visitor arrivals from Mainland and non-Mainland markets in 2002–2013, in relation to the increasing number of Mainland cities included in the IVS. The strong growth of Chinese visitor arrivals in Hong Kong means changing tourism landscape over the years. Chinese visitors have dominated the tourism market with the ratio of Mainland to non-Mainland visitor arrivals changed from 40:60 in 2002 to 75:25 in 2013.

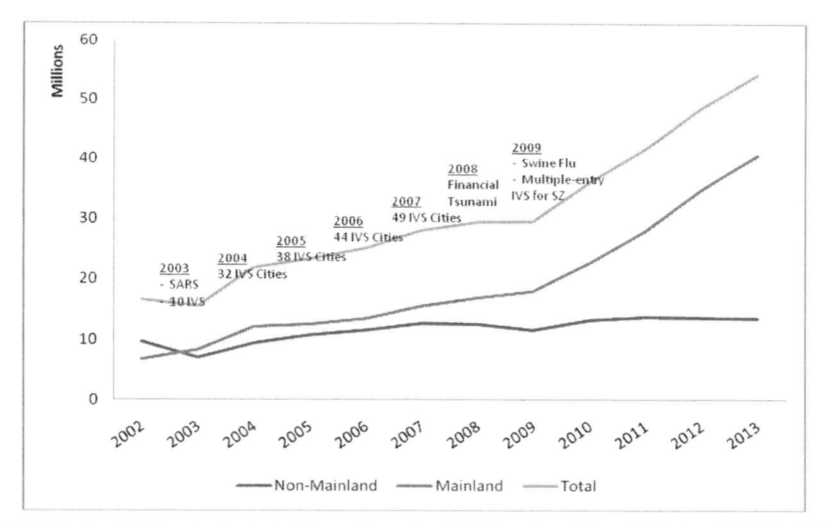

FIGURE 4.3 Visitor Arrivals in Hong Kong by Mainland and Non-Mainland (2002–2013). (From Hong Kong Tourism Board.)

Although the introduction of IVS significantly increased the number of Chinese visitors to Hong Kong, it also has some loopholes. For example, residents of non-IVS cities who wish to visit Hong Kong will need to apply for an "L-visa" and travel in groups. Some travel agencies took advantage of the procedure and made profit by sending these Mainland visitors to Hong Kong in groups, but once they have gone through customs, they travel effectively as individuals rather than in groups. These travel agencies have indicated that several hundred customers go through customs by "L-visa" each day (China Review News, 2014; Sing Tao Daily, 2014a). Some of these "visitors" are believed to travel to Hong Kong to work illegally (China Review News, 2014). Some of these Chinese travelers pretend to use Hong Kong as the "transit point" to gain entrance, and once reaching Hong Kong, they cancel the onward journey, allowing them to stay in Hong Kong for up to 7 days (Sing Tao Daily, 2014b). Some travel agencies help in arranging for such "third country visa", itineraries, and flight tickets with transits in Hong Kong. This kind of travel arrangement allows those who were previously denied entry permits to enter Hong Kong. Some Chinese visitors also used this as a way to gain access to and conduct illegal activities such as prostitution in Hong Kong (Beijing Sina, 2014).

4.3 OVERNIGHT AND SAME-DAY MAINLAND VISITORS

Apart from contributing to the significant increase in number of Mainland visitor arrivals to Hong Kong, IVS also created a unique segment of same-day visitors. Although traditionally visitors stay in a destination overnight, IVS has brought to Hong Kong a large number of visitors who return to Mainland or leave for another destination on the same day of crossing the Mainland–Hong Kong border. According to the UN World Tourism Organization's definition, these groups of same-day visitors are not classified as tourists (World Tourism Organization, 2007). Although the numbers of both overnight and same-day Mainland visitors have been increasing since 2002, the rate of growth of the same-day Mainland visitors is much higher than that of the overnight Mainland visitors. By 2011, the number of same-day Mainland visitors reached 14.5 million and actually surpassed that of overnight Mainland visitors. In 2013, there were 23.7 million same-day visitors and 17.1 million overnight Mainland visitors. Figure 4.4 shows the number of overnight and same-day Mainland visitor arrivals from 2002 to 2013 and the same-day visitors becoming a key segment. The purposes of visit by this segment of same-day visitors are mostly shopping, and some studying, working, and visiting friends and relatives. This large group of visitors has to some extent strained the carrying capacity of the destination and caused tensions in the community, which is explained in detail in section 4.9, Response of Local Community to Chinese Visitors.

The profiles of overnight Mainland visitors and same-day Mainland visitors are somewhat different, and their purposes of visit also differ. The composition of gender for overnight Mainland visitors is relatively stable, with female comprising the majority. As for same-day Mainland visitors, the composition shifts from male dominant to female dominant in 2009 onward. Although vacation is the main reason to visit Hong Kong for both overnight and same-day Mainland visitors, overnight Mainland visitors tend to travel to Hong Kong for visiting friends and relatives more than same-day Mainland visitors. As expected, there are a higher percentage of same-day Mainland visitors passing through Hong Kong en route compared with that of overnight Mainland visitors.

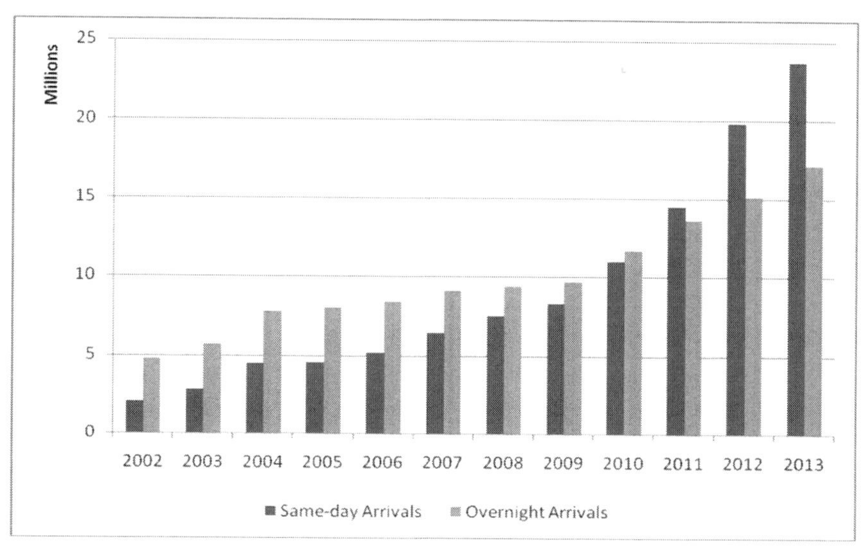

FIGURE 4.4 Overnight and Same-day Mainland Visitors to Hong Kong (2002–2013). (From Hong Kong Tourism Board)

4.4 SPENDING PATTERN

In addition to the increasing number of visitors to Hong Kong, the expenditure related to inbound tourism has been increasing. This increased from HK$77 billion (US$10 billion) in 2002 to HK$332 billion (US$43 billion) in 2013. Figure 4.5 shows the total tourism expenditure related to inbound tourism and those by mainlanders from 2002 to 2013. While Mainland Chinese visitors accounted for 75% of the visitor arrivals in 2013, they accounted for 65% of the total expenditure, that is, HK$217 billion (US$28 billion). The market share of Mainland Chinese visitors in terms of expenditure is lower than that in terms of visitor arrivals because there is a high proportion of same-day visitors, who spend much less than the overnight visitors. This phenomenon will be elaborated in section 4.5, Shopping. According to World Travel & Tourism Council (2014), tourism industry contributed 8.9% to the gross domestic product (GDP) in Hong Kong in 2013. Based on the ratio of Mainland Chinese visitors accounting for 65% of the inbound tourism expenditure, they would have contributed 5.8% of the GDP in Hong Kong in 2013.

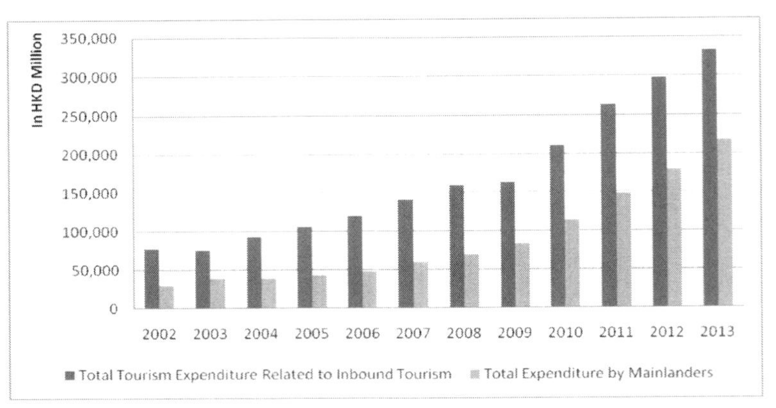

FIGURE 4.5 Tourism Expenditure Related to Inbound Tourism.
(From Hong Kong Tourism Board)

As expected, the expenditure by overnight visitors from Mainland China was much higher than that of same-day visitors, and it increased steadily over the years. Figure 4.6 shows the total expenditure of both overnight and same-day Mainland visitors from 2002 to 2013. In 2013, an overnight expenditure of HK$153 billion (US$20 billion), accounting for 70% of the total expenditure by all Chinese visitors had been recorded, whereas the expenditure for same-day Mainland visitors amounted to HK$64 billion (US$8 billion). Figure 4.7 shows differences between the expenditure per capita of total visitors and Mainland visitors for overnight and same-day visitors from 2002 to 2013. In general, the expenditure per capita of Mainland visitors is higher than that of the total visitors.

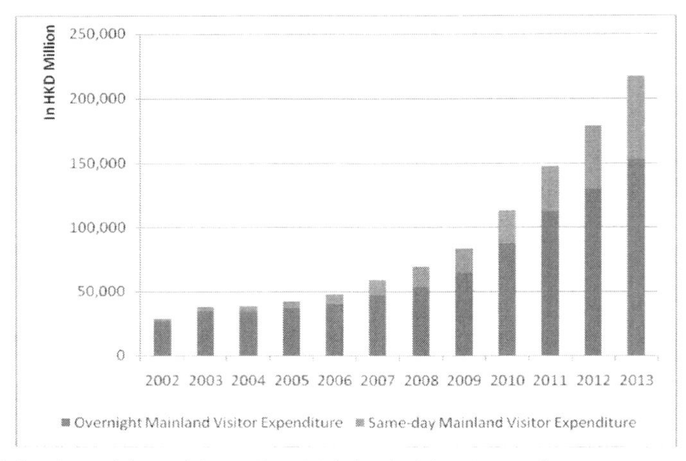

FIGURE 4.6 Overnight and Same-Day Mainland Visitor Expenditure.
(From Hong Kong Tourism Board.)

It was mentioned earlier that one of the important reasons for Mainland Chinese to visit Hong Kong is shopping, as reflected by the share of total expenditure on shopping. Figure 4.8 shows the main categories of spending by Mainland Chinese in Hong Kong in 2013, and it can be seen that shopping accounted for the bulk of their spending. Shopping accounted for 77.8%, whereas hotel bills, meals outside hotels, entertainment, tours, and other expenditures together accounted for 22.2%. The importance of shopping is even more marked among same-day visitors, with shopping accounting for 92.4% of their spending.

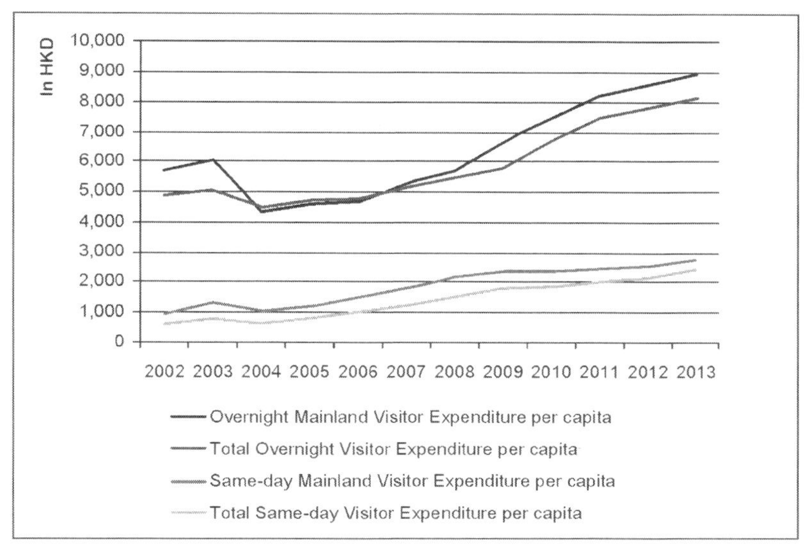

FIGURE 4.7 Visitor Expenditure Per Capita.
(From Hong Kong Tourism Board.)

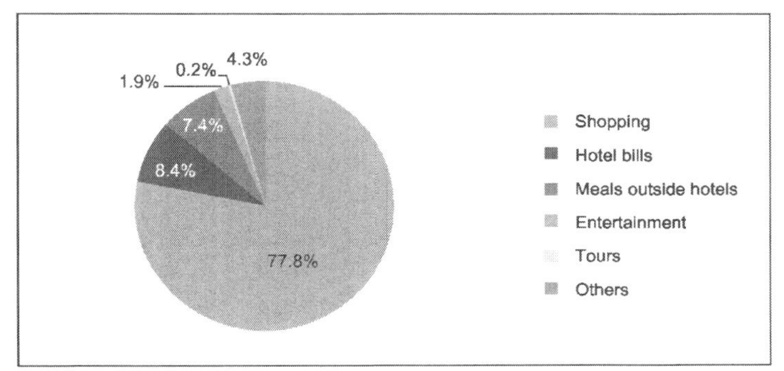

FIGURE 4.8 Spending Category of Mainland China Visitors in 2013.
(From Hong Kong Tourism Board.)

4.5 SHOPPING

Given the importance of shopping among Chinese visitors, let us look more closely at what they buy in Hong Kong. The main shopping items bought by Mainland visitors can be categorized as follow: (a) Cosmetics, skin care, and perfume; (b) food, alcohol, and tobacco; (c) jewelry and watch; (d) personal care such as shampoo and diapers; (e) electrical or photographic goods; (f) garments or fabrics; (g) leather and synthetic goods; and (h) other items.

Figure 4.9 shows the different shopping patterns of overnight Mainland Chinese visitors and same-day Mainland Chinese visitors. Overnight visitors tend to spend more in jewelry, garments, and leather goods, whereas same-day visitors are found to spend quite evenly across different items. In fact, same-day visitors spend quite a large proportion of their expenditure on food, alcohol and tobacco, personal care items, and cosmetic goods. As many same-day visitors shop for daily necessity items in Hong Kong, they are found to be competing to some extent with the local community for retail supply. Such competition has begun to create concerns in the community as explained in section 4.9, Response of Local Community to Chinese Visitors.

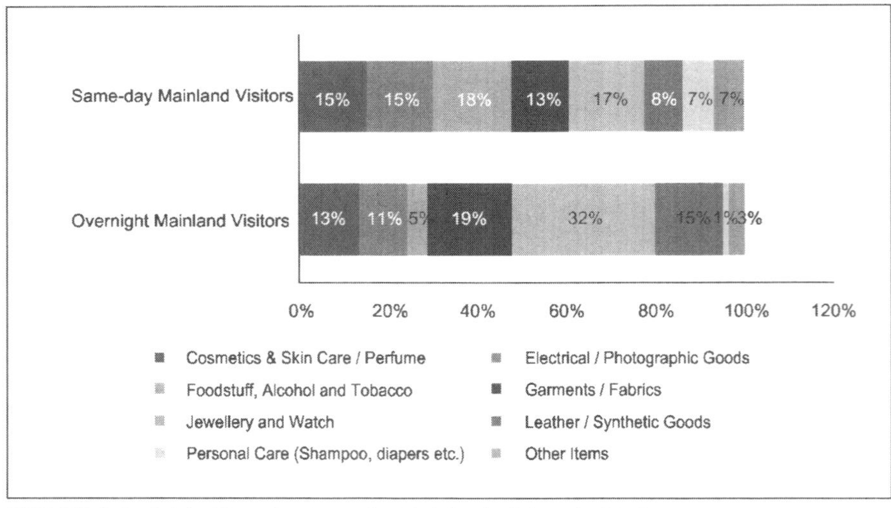

FIGURE 4.9 Main Shopping Items by Mainland Visitors in 2012.
(From Hong Kong Tourism Board.)

According to the Chinese Millionaire Wealth Report of 2013 (Hurun, 2013), there are 3 million individuals in China with assets worth more than RMB 6 million (approximately US$1 million). They spend on average RMB 1.77million (approximately US$300,000) a year, and 2% is allocated toward traveling. France, the

United States, and Singapore are the top 3 international luxury destinations. Hong Kong and Sanya are the more popular domestic destinations. It has been reported that the Chinese tour groups are decreasing in size and more tourists are looking for unique travel experiences with an increasing demand for quality accommodation and are willing to spend more on hotels. There are luxury travel agencies such as HH Travel and 8 Continents which provide luxury tour packages to Hong Kong. Their packages on average last for 3–4 days with business class flights and accommodation at hotels such as The Peninsula and Four Seasons, costing RMB13,000 (US$2,000) per person per trip. For most of the packages, other than a couple of designated activities, the itineraries are entirely left for customers to decide (8 Continents, 2014; HH Travel, 2014).

4.6 POPULAR PLACES TO VISIT

Apart from shopping, Mainland Chinese tourists who travel to Hong Kong like to visit a number of attractions. The most popular attraction is Avenue of Stars, which is a waterfront, where well-known movie star celebrities in Hong Kong cast their handprints, and Jackie Chan is one of the personalities portrayed. In addition, tourists can have a panoramic view of the Victoria Harbor while recollecting their memories of the movie stars and movies concerned. The next most popular places to visit are the 2 theme parks: Ocean Park and Hong Kong Disneyland. They are followed by the Victoria Peak and Hong Kong Convention & Exhibition Centre where the Golden Bauhinia Square is located right outside. The Square has special meaning to Mainland Chinese visitors because there is a giant 6-meter high statue of golden Bauhinia to commemorate the handover of Hong Kong to China in 1997 and the establishment of the Hong Kong Special Administrative Region. Other attractions include Ladies Market (an open-air market), Wong Tai Sin Temple (a Taoist temple), Repulsive Bay (a beach located on the southern side of Hong Kong Island), and clock tower (the remnant of a railway station in the British colonial era). As for same-day Mainland China visitors, more than 80% of the surveyed visitors stated that shopping is one of the things they have done over the trip. Only 2% have visited Avenue of Stars and Ladies Market, and 1% have visited the clock tower, Victoria Peak, Wong Tai Sin Temple, and Hong Kong Convention & Exhibition Centre. This is consistent with the earlier discussion that same-day visitors are mostly concerned with shopping, and attractions in the destination are of less interest to them.

4.7 CARRYING CAPACITY

There is no doubt that inbound tourism has made significant contributions to the economy in Hong Kong, creating jobs and stimulating tourism infrastructures. However, there were also mounting concerns expressed whether Hong Kong, as a city with population of 7 million, could handle more than 50 million visitors a year.

The Hong Kong government conducted a study on the carrying capacity of destination, and "Assessment Report on Hong Kong's Capacity to Receive Tourists" was released at the end of 2013 (Commerce and Economic Development Bureau, 2013). The report concludes that, as a free port, Hong Kong treasures the freedom to enter or leave the region, and as a small and externally oriented economy, Hong Kong cannot and should not set a limit on the overall visitor arrivals. On the basis of assessment that visitor arrivals would be greater than 70 million in 2017, the report states that Hong Kong would generally be able to receive the visitor arrivals in 2017, but hotel rooms would continue to be in tight supply.

The assessment report covers IVS, border control points, tourism attractions, hotels, public transport, and impact on the livelihood of the community. The report states that with the increasing number of people (including Hong Kong residents and tourists) crossing the borders, it is important to make certain the effectiveness of the immigration control. Currently, there are 14 immigration control points in Hong Kong, and by 2018, 4 more control points are expected to be fully operational. The government will be continually seeking to enhance the handling capacity of the control points by improving the facilities and introducing information technology. As long as adequate amount of workforce is being allocated, it is believed that the border control points can accommodate the number of visitor arrivals.

As for tourism attractions, the report covers the major tourist facilities in Hong Kong such as Hong Kong Disneyland, Ocean Park, Ngong Ping 360, Peak Tram, and the Sky Terrace. As Hong Kong Disneyland and Ocean Park will undergo various expansions, it is believed that they will be able to handle the projected number of visitors. Ngong Ping 360 and Sky Terrace are believed to have spare capacity to receive tourists, while the Peak tram will have to increase the frequency of tram service and also provide special crowd control measures during peak hours. Overall, the tourism attractions will have the ability to receive more visitors. It is noted that hotels in Hong Kong reached an average occupancy rate of 80%–90% in the past decade. At the end of June 2013, a total of 99 hotel projects involving 16,000 rooms were approved, bringing the number of hotel room supply to approximately 84,000 in 2017. However, it is believed that the supply of hotel rooms is considered to be tight, and with the continuous high occupancy rate, it will drive room rates upward and subsequently reduce the desire of visitors to stay or shorten their duration in Hong Kong.

Public transport network in Hong Kong is one of the best in world, well known for its efficiency and convenience. There are various ways for tourists to commute within Hong Kong, via Mass Transit Railway (MTR), buses, taxis, and coaches. In preparation for the projected increased number of visitors, there are projects to expand MTR services and also investments in new routes are made, in an attempt to redirect the flow of traffic to relieve the crowdedness, especially during peak hours. As for the impact on the livelihood of the community, the assessment report focused mainly on law and order. Along with the increase of visitor arrivals, the need of strengthening the frontline police officers for prevention of crime and anti-crime operations has been noted, and the police force will need to increase the workforce to cope with the demand.

4.8 COERCED AND EXCESSIVE SHOPPING

In 2013, more than 13 million Mainland Chinese visited Hong Kong as group tour travelers. Among these travelers, some of them joined "zero-fare" tours with shopping itineraries. Typically, these visitors are attracted by very cheap or sometimes free tours to visit Hong Kong. Once they have arrived Hong Kong, they are arranged to shop at designated outlets where the travel agency takes a cut of the sales revenue (Hornby, 2010) as commission or rebate. Such tours are often associated with excessive shopping and sometimes selling dubious products, inferior travel experience, and tourist complaints. The tours sometimes end up with conflicts and disputes. For example, in 2010, a female tour guide cursed and fought with Chinese Mainland tourists in Hong Kong because some of them refused to shop (Guan, 2010; Fauna, 2011), and in 2013, a travel agency in Hong Kong did not offer proper accommodation to tour groups of Mainland tourists because of disputes in shopping (Nip, 2013). This kind of "zero-fare" tour is not uncommon in emergent source markets which have a large segment of inexperienced or first-time travelers who are very budget conscious. China is indeed such a market.

With the view of protecting tourists and enhancing the tourism industry, the Chinese government issued the *Tourism Law of People's Republic of China*, which came into effect on October 1, 2013 (China National Tourism Administration, 2013). It covers various aspects such as tourist safety and behavior, unfair competition, and shopping tours. Among the 112 articles in the *Tourism Law*, Article 35 states that "Travel agencies are prohibited from organizing tourism activities and luring tourists with unreasonably low prices, or getting illegitimate gains such as rebates by arranging shopping or providing tourism services that requires additional payment." The wordings are clearly directed at the malpractice of coerced shopping and hidden commissions. It is obvious that the *Tourism Law* is set up to target the notorious trade between shops and travel agencies. Article 35, however, continues to state that the regulation does not include circumstances (which can be identified as shopping) where both sides have agreed or the tourists have requested for such arrangements and no influence is caused on the itinerary of other tourists. This means that as long as the travel agency can prove that the tourist agreed to the shopping tour, it may lawfully run such tours. Not only does this appear to contradict the crackdown on "zero-fare" tours, the provision is conducive to arguments as to whether the customer had agreed to the shopping.

If the *Tourism Law* is to have its effect in curtailing "zero-fare" tours, there should be a reduction in the number of group tours with shopping itineraries. Figure 4.10 shows that the law had immediate impact on the number of tour groups to

Hong Kong in the 4 months following the introduction of the law in October 2013. The total number of tour groups in the first 4 months was 33,882 versus 49,998 in the same period a year ago (-32%), due to a large decrease in the number of tour groups with shopping itineraries and a small increase in the number of tour groups without shopping itineraries. This indicates that some travel agencies were shying away from operating "zero-fare" shopping tours in the first 4 months. There were fluctuations in the following months, where the total number of tour groups in February, March, April, and July surpassed the total number of tour groups in the corresponding months of previous year. Overall, during the first year of the enactment of the *Tourism Law of China*, the total number of group tours decreased by 11% when compared with the previous year. The number of group tours with shopping itineraries decreased by 18% from previous year, while the number of group tours without shopping itineraries increased by 32%. This indicates that the *Tourism Law of China* has a small impact on tour groups. In addition, at about the same time, anti-Mainland Chinese visitors' sentiment was rising in the community, which may have deterred the Mainland Chinese from visiting Hong Kong. The response of local community to Chinese visitors will be explained in the following section.

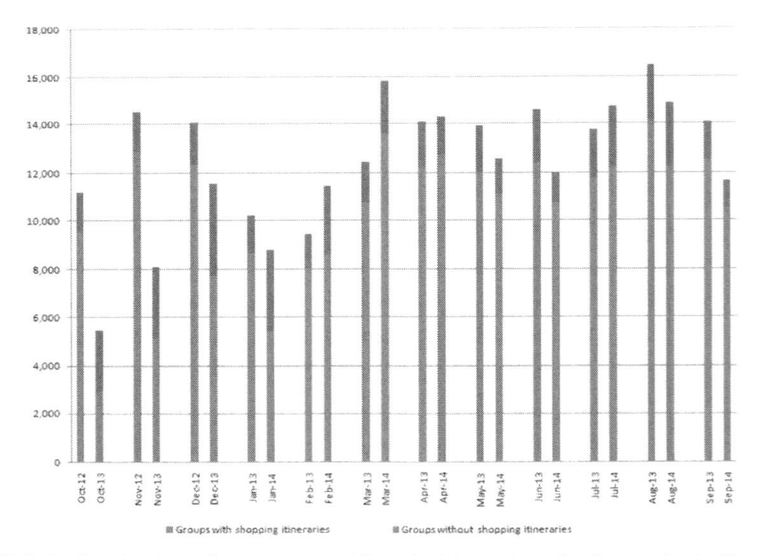

FIGURE 4.10 Number of tour groups with and without shopping itineraries in Hong Kong. (From The Travel Industry Council of Hong Kong.)

It has been reported in the press that in order to get around the ban on free or cheap tours imposed by the *Tourism Law*, organizers accepted a "travel voucher" costing RMB200 (US$32) as payment for a tour valued at RMB4,000 (US$640). Some travel agencies sought to bypass the law by offering such scam sponsorships in the form of travel voucher redemption promotions (Lee and Mok, 2014). Thus, it was observed that low-budget shopping tours within Mainland China have reappeared quickly after the enactment of the *Tourism Law* (Commercial Radio Hong Kong, 2014).

4.9 RESPONSE OF LOCAL COMMUNITY TO CHINESE VISITORS

Although the large number of visitors and rapid growth in visitors from Mainland China to Hong Kong certainly benefit the destination economy, the sheer volume, particularly of the same-day visitors, has also given rise to various concerns and conflicts in the community. It has been reported that tour buses are often seen parked along the streets with their engine running to wait for tourists as they hop on and off at the scenic spots. One of the residents voiced out that the fumes generated have made her reluctant to bring her children out (Kao, 2014). Nuisance to locals was reported about large groups of Mainland China visitors congregated on the streets, waiting for coaches after shopping. With the tourist crowd occupying the pedestrian walkway, locals resort to walking on the road, endangering their lives. A great deal of noise was also created by the tour groups, disturbing the residents in the neighborhood (Chan, 2014). Concerns have been raised with regard to vehicular congestion created by tourist coaches, overcrowding on public transport and in major shopping areas, and perceived price hikes as a result of shopping by Mainland Chinese visitors (Lu, 2013; Luo, 2013).

The overcrowding and conflicts had become more obvious after the enactment of multientry IVS for Shenzhen residents. The number of same-day visits grew rapidly, and it is suspected that many of the visitors are associated with parallel trading. Some same-day visitors cross the border several times a day to buy daily necessity items and resell them on the Mainland at a profit. The parallel trading inevitably caused shortages and price rise of everyday products in the neighborhood near the border (Lau, 2013a). Different measures have been taken such as limiting the amount of milk powder to be carried across border (SCMP, 2013) and arresting those who are suspected of involving in parallel trading (Mok, 2013). As a response to the inconvenience created, there are voices advocating the revision of the multi-entry scheme as a further measure to tackle parallel trading (Lau, 2013b). Due to different cultural backgrounds and congestion in public areas, sometimes there occur frictions between Chinese visitors and Hong Kong citizens. Recently, the negative emotions born by some extreme Hong Kong citizens have escalated and were then taken to the streets, attracting public attention toward the matter. In early February 2014, some extremists demonstrated against Mainland Chinese visitors labeling them as "locusts" and urging the government to put a cap on Mainland visitors (Siu, 2014a). In the following month, anti-mainlanders launched another protest, sarcastically urging Mainland visitors to be patriotic and shop in Mainland China instead of Hong Kong (Kao and Lo, 2014; Chiu and Tam, 2014). Such demonstrations disrupted the peace and order in Hong Kong and forced some shops to close for a short period of time. They may also have a negative influence on the views of Chinese visitors on Hong Kong as a destination. Lately, a video on Chinese parents allowing their child to relieve himself amongst the busy streets of Hong Kong was filmed and posted online (Moore, 2014). Disputes on this matter further created tension between the host community and Chinese visitors. In response to the increasing conflict, the Hong Kong government urged the radical thinkers to be more tolerant to the Mainland China visitors as there is an inevitable culture gap and stressed the need to educate them with the correct social etiquette instead of condemning them (Siu, 2014b).

Apart from the news reported, a recent survey also reflected the discontents toward Mainland visitors by some of the general public. The survey revealed that more than 60% of the citizens admitted that the current amount of Mainland China visitors have caused inconvenience toward their daily lives, whereas 57.2% believed that the number of inbound Mainland China visitors should be controlled, and 61.8% agreed to placing a cap on the number of Mainland China visitors (Hong Kong Institute of Asia-Pacific Studies, 2014).

4.10 CONCLUSION

Chinese outbound tourism has been very important to Hong Kong, particularly during 2003–2013. Chinese visitor arrivals reached 40.7 million, accounting for 75% of total visitor arrivals in 2013, and expenditure amounted to HK$217 billion (US$28 billion), accounting for 65% of the total expenditure. Chinese Mainland visitors are by far the largest source market in terms of arrivals and spending, and the trend is still growing. The Chinese outbound market has played a major role in supporting the hotel and retail sectors, creating employment, and providing impetus to tourism infrastructure development.

Chinese Mainland visitors are by no means homogeneous. While about half of them are from Guangdong Province, which is the province closest to Hong Kong, increasing number of visitors come from other provinces and cities such as Zhejiang, Shanghai, and Beijing. In addition, with the implementation of IVS since 2003, increasing number of visitors come as individuals rather than in groups. By 2013, as much as 67% of the visitors from Mainland China visiting Hong Kong were traveling as individual visitors. These individual visitors are more sophisticated and experienced than those joining group tours. They are generally more discerning, and they look for personalized travel experiences. The IVS also created a unique segment of same-day visitors, and the rate of growth for same-day Mainland visitors is much higher than that of overnight Mainland visitors. In 2013, there were 23.7 million same-day visitors and 17.1 million overnight Mainland visitors to Hong Kong. The profile and the spending pattern of same-day visitors are quite different to those of overnight visitors. Nonetheless, shopping is the most important activity among all Chinese visitors, accounting for nearly 78% of their spending.

"Zero-fare" tour remains an issue among group tourists in Hong Kong. It is interesting to note that both the Mainland Chinese government and the Hong Kong government resort to legal measures to tackle zero-fare tour issues. The Chinese government introduced the *Tourism Law* in October 2013, which prohibits travel agencies from luring tourists with unreasonably low prices and getting rebates by arranging shopping or providing tourism services that requires additional payment. It is obvious that the wording in *Tourism Law* is directed at the malpractice of coerced shopping and hidden commission. There is indication that some travel agencies were shying away from operating "zero-fare" shopping tours in the first 4 months following the introduction of the *Tourism Law*. The impact, however, dissipated in the following months. The total number of tour groups in the following months returned to about the same level compared with the same period a year ago, and there was no significant change in the total number of tour groups without shopping itineraries. In mid-2010, a spate of incidents concerning the receiving arrangements for Mainland inbound tours and suspected coerced shopping not only tarnished the reputation of Hong Kong's tourism sector, but also aroused public concern over the effectiveness of the industry self-regulatory regime in ensuring proper conduct within travel trade. The Hong Kong government chose to revamp the travel industry regulatory regime, by planning to establish an independent statutory body, namely Travel Industry Authority (Commerce and Economic Development Bureau, 2011). The new regulatory regime aims at raising the threshold for setting up travel business and instigating a more stringent licensing system for tourist guides and tour escorts rather than dealing with forced shopping directly. The Authority will adopt a two pronged approach to raise the threshold. First, all travel agents will need to fulfill a paid-up capital requirement and, in addition, deposit with the Authority a sum of guarantee money for covering the payment of any outstanding penalties imposed upon a travel agent for breaching relevant rules and regulations. Secondly, each travel agent will be required to appoint a person to be an authorized person who will be under a statutory duty to make certain that the operation of the travel business will comply with the license conditions and rules and regulations. Tourist guides and tour escorts will be required to complete designated Continuing Professional Development courses for renewal of licenses. How effective the Travel Industry Authority in Hong Kong is in dealing with "zero-fare" tours are yet to be seen.

The social impact of Chinese outbound tourism on the local community in Hong Kong has surfaced as an issue in recent years. Concerns have been raised with regard to vehicular congestion created by tourist coaches, overcrowding on public transport and in major shopping areas, perceived price hikes, and changing retail landscape catering more to Chinese tourists than local residents in the neighborhood. Recent issues have included crowding at shopping areas and in major attractions and shortages of some consumer goods such as baby milk formula. Protests against Chinese tourists have increased during 2013 and 2014 and are becoming a sociopolitical issue that has created divisions in the society of Hong Kong (Prideaux and Tse, 2014). The great volume of same-day tourists, particularly in shopping precincts, has created concerns and at times conflicts with the host community. Of particular concern is the potential for disruption of harmonious ties between Hong Kong and China and disrupting normal trade (China Daily, 2013). The different cultural and behavioral norms in Mainland China and in Hong Kong tend to aggravate the issue as well. It is observed that the issue of carrying capacity, particularly with respect to community response and tolerance, has to be addressed to make certain the sustainability of Chinese outbound tourism in Hong Kong. Issues that need to be considered include physical carrying capacity of key attractions, visitor management strategies, the need to expand existing infrastructure or build new infrastructure, and the emotional tolerance level of the host community. There may be a need to mount education campaigns to bridge any cultural gaps that may arise (Prideaux and Tse, 2014).

ACKNOWLEDGMENT

This chapter is developed with the research support provided by Ms. Veronica Tam.

REFERENCES

8 Continents. List of Packages. 2014. http://www.badazhou.com/dest/routes/dest_id/43 (accessed September 17, 2014)

Beijing Sina. Woman, claiming a tourist, arrested for prostitution (in Chinese). *Sina News Taiwan*, [Online] May 13, 2014. http://news.sina.com.tw/article/20140513/12413212.html (accessed May 30, 2014)

Chan, B. Anger towards mainland visitors likely to worsen unless problems addressed. *South China Morning Post*, [Online] April 25, 2014. http://www.scmp.com/lifestyle/travel/article/1495953/anger-towards-mainland-visitors-likely-worsen-unless-problems (accessed May 30, 2014)

China Daily. Regulate parallel trading by interception and diversion. *China Daily*, [Online] January 26, 2013. http://usa.chinadaily.com.cn/business/2013-01/26/content_16177480.htm (accessed April 30 2014)

China Hospitality. Shenzhen Extends Individual Visitor Scheme. China.org.cn, [Online] January 5, 2011. http://www.china.org.cn/travel/2011-01/05/content_21674188.htm (accessed May 30, 2014)

China National Tourism Administration. Tourism Law of People's Republic of China. 2013. http://en.cnta.gov.cn/html/2013-6/2013-6-4-10-1-12844.html (accessed May 30, 2014)

China Review News. L-visa with guaranteed cross-border services causes Hong Kong to be stormed with mutated IVS. *China Review News*, [Online] April 28, 2014. http://hk.crntt.com/doc/1031/5/5/1/103155146.html?coluid=0&kindid=0&docid=103155146 (accessed May 30, 2014)

Chiu, J. and Tam, J. 'Parody' protesters march to urge mainlanders to 'reignite their patriotism… and shop at home'. *South China Morning Post*, [Online] March 10, 2014. http://www.scmp.com/news/hong-kong/article/1444313/parody-protesters-march-urge-mainlanders-reignite-their-patriotism (accessed May 30, 2014)

Commerce and Economic Development Bureau. Assessment Report on Hong Kong's Capacity to Receive Tourists. 2013. http://www.tourism.gov.hk/resources/english/paperreport_doc/misc/2014-01-17/Assessment_Report_eng.pdf (accessed May 30, 2014)

Commerce and Economic Development Bureau. Review of the Operation and Regulatory Framework of the Tourism Sector in Hong Kong – Consultation Paper. 2011. http://www.tourism.gov.hk/english/papers/files/consultation_paper_en.pdf (accessed May 30, 2014).

Commercial Radio Hong Kong. Zero-fare shopping tours resurfaced in Mainland China during the 1st May public holiday period. *Commercial Radio Hong Kong*, [Online] May 2, 2014. http://www.881903.com/Page/ZH-TW/newsdetail.aspx?ItemId=715312&csid=261_341 (accessed May 30, 2014)

Fauna. Hong Kong tour guide curses and fights mainland tourists. *ChinaSMACK*, [Online] February 8, 2011. http://www.chinasmack.com/2011/stories/hong-kong-tour-guide-curses-fights-with-mainland-tourists.html (accessed May 4, 2014)

Guan, X. Permanent ban on 'cruel' HK tour guide. *China Daily*, [Online] September 5, 2010. http://www.chinadaily.com.cn/china/2010-09/05/content_11258272.htm (accessed February 27, 2014)

HH Travel. List of Packages. 2014. http://www.hhtravel.com/product/search?destinationGroupId=10&destinationRegionId=337&propertyType=2 (accessed on September 17, 2014)

Hong Kong Tourism Board. A Statistical Review of Hong Kong Tourism, 2002–2013. Hong Kong Tourism Board, Hong Kong.

Hong Kong Institute of Asia-Pacific Studies. Survey Findings on Views about the Individual Visit Scheme Released by Hong Kong Institute of Asia-Pacific Studies at CUHK. The Hong Kong Chinese University. 2014. http://www.cuhk.edu.hk/hkiaps/tellab/pdf/telepress/14/Press_Release_20140402.pdf (accessed May 30, 2014)

Hornby, L. China warns tourists on "forced shopping" in Hong Kong. *Reuters*, [Online] July 19, 2010. http://www.reuters.com/article/2010/07/19/us-china-hongkong-tourism-idUSTRE66I27V20100719 (accessed May 30, 2014)

Hurun. The Chinese Millionaire Wealth Report. *Hurun Report*, [Online] 2013. http://up.hurun.net/Humaz/201312/20131218145315550.pdf (accessed May 30, 2014)

Kao, E. and Lo, W. 'Patriotic' protest leaves mainland Chinese visitors bemused. *South China Morning Post*, [Online] March 17, 2014. http://www.scmp.com/news/hong-kong/article/1450190/patriotic-protest-leaves-mainland-chinese-visitors-bemused (accessed May 30, 2014)

Kao, E. Locals fume as polluting tour buses choke scenic Repulse Bay. *South China Morning Post*, [Online] January 22, 2014. http://www.scmp.com/news/hong-kong/article/1410157/locals-fume-polluting-tour-buses-choke-scenic-repulse-bay (accessed May 30, 2014)

Lau, S. Border-town residents say traders cost them dear. *South China Morning Post*, [Online] April 17, 2013a. http://www.scmp.com/news/hong-kong/article/1216247/border-town-residents-say-traders-cost-them-dear (accessed May 30, 2014)

Lau, D. Will the "two-can limit" regulation work? *Hong Kong Law Blog*, [Online Blog], February 07, 2013b. http://hklawblog.com/2013/02/07/will-the-two-can-limit-regulation-work/ (accessed May 30, 2014)

Lee, E. and Mok, D. 'Forced shopping' tours of Hong Kong back as Mainland organizers find loophole. *South China Morning Post*, [Online] April 02, 2014. http://www.scmp.com/news/hong-kong/article/1462831/forced-shopping-back-tour-firms-find-loophole-mainland-rules (accessed May 30, 2014)

Lu, C. J. Tuen Mun and Sheung Shui prices the most expensive in the New Territories because of individual visitor parallel trade (in Chinese). *Apple Daily*, 2013, p A11.

Luo, J. W. Mainland tour group occupying Tokwawan (in Chinese). *Oriental Daily*, July 10, 2013, p A24.

Mok, D. Over 90 arrested in parallel trading blitz near border. *South China Morning Post*, [Online] January 22, 2013. http://www.scmp.com/news/hong-kong/article/1133397/over-90-arrested-parallel-trading-blitz-near-border (accessed May 30, 2014)

Moore, H. How peeing in public divided China and Hong Kong. *BBC News*, [Online] April 26, 2014. http://www.bbc.com/news/blogs-trending-27163525 (accessed May 30, 2014)

Nip, A. Tour agency 3A Holidays slapped with suspension from regulator. *South China Morning Post*, [Online] February 23, 2013. http://www.scmp.com/news/hong-kong/article/1155680/tour-agency-3a-holidays-slapped-suspension-regulator (accessed February 27, 2014)

Prideaux, B. and Tse, T. UNWTO Knowledge Network Issue Paper Series I, 2014: issues arising from the rapid growth of tourists from China to Hong Kong, China between 2002 and 2013.

SCMP. Opportunities in parallel trade. *South China Morning Post Editorial*, [Online] March 12, 2013. http://www.scmp.com/comment/insight-opinion/article/1188719/opportunities-parallel-trade (accessed May 30, 2014)

Sing Tao Daily. Luring customers online, where L-visa forms for Mainland Chinese to Hong Kong can be filled at the checkpoint (in Chinese). *Sing Tao Daily,* [Online] April 28, 2014a. http://news.singtao.ca/toronto/2014-04-28/hongkong1398662674d5025600.html (accessed May 30, 2014)

Sing Tao Daily. CCTV uncovers Travel Agencies in Shenzhen and Zhuhai provide "Fake Visa" (in Chinese). *Sing Tao Daily,* [Online] May 05, 2014b. http://std.stheadline.com/yesterday/loc/0505ao09.html (accessed May 30, 2014)

Siu, P. Anti-mainlander protest urging curbs on visitor numbers tarnished city, say top officials. *South China Morning Post*, [Online] February 18, 2014a. http://www.scmp.com/news/hong-kong/article/1429558/locust-protest-urging-curbs-mainland-visitors-tarnished-city-say-top (accessed May 30, 2014)

Siu, B. So pooh-poohs idea that only hostility can bring relief. *The Standard*, [Online] April 30, 2014b. http://www.thestandard.com.hk/news_print.asp?art_id=144968&sid=42159314 (accessed May 30, 2014)

Tourism Commission. *Hong Kong: The Facts. 2013*. http://www.gov.hk/en/about/abouthk/factsheets/docs/tourism.pdf (accessed May 30, 2014)

Tse, T. Chinese Outbound Tourism as a Form of Diplomacy. *Tourism Plann. Dev.* 2013, 10(2), 149–158.

World Tourism Organization. Understanding Tourism: Basic Glossary. 2007. http://media.unwto.org/en/content/understanding-tourism-basic-glossary (accessed September 8, 2014)

World Travel & Tourism Council. Economic Impact 2014 Hong Kong. 2014. http://www.wttc.org/site_media/uploads/downloads/hong_kong2014.pdf (accessed May 30, 2014)

CHAPTER 5

MAINLAND CHINESE OUTBOUND TOURISM TO MACAO: RECENT PROGRESS

XIANGPING LI

CONTENTS

5.1 INTRODUCTION

The First Wave of Chinese private outbound travel started with visits to Chinese friends and relatives in Hong Kong and Macao in 1983. Over the past 3 decades, Macao has remained the primary destination for Chinese outbound travel along with Hong Kong. Chinese economic surge, coupled with Macao's gaming liberalization in 2002 and the Central Government's the "Individual Visit Scheme" (IVS) in 2003, has completely restructured tourism in Macao. In 2003, Mainland Chinese tourists overtook Hong Kong as Macao's largest source of visitors. A total of 5.7 million visitors from the Mainland were recorded in 2003 compared with less than 1 million in 1998. Since then, Mainland has been the top source market for Macao, constituting 63.5% of Macao's tourist arrivals in 2013. This chapter will focus on Chinese outbound to Macao with an emphasis on recent decade (2002–2013). Data used for analysis are derived from Macao Statistics and Census Service (DSEC).

5.2 MACAO AS A DESTINATION

Macao is a special administrative region (SAR) in the southeastern coast of China, bordering Zhuhai, Guangdong Province. The most recent census claims that Macao has a total population of approximately 552,503 with a total area of 29.5 square kilometers (DSEC, 2012). Macao SAR consists of Macao Peninsula, islands of Taipa and Coloane, and the Cotai Strip. Macao's economy is largely based on tourism, and Macao's tourism development has been closely linked to its gaming.

Macao's gaming industry can be traced back to the 16th century; however, it is not until 1847 when Macao's Portuguese government announced the legality of gaming for the time. Since then, small and unregistered casinos had begun to bloom, which earned Macao the fame of "Monte Carlo of the Orient". Macao's gaming industry welcomed a new era in 2002 when the casino monopoly concession granted to *Sociedade de Turismo e Diversoes de Macau* (STDM) in 1962 came to an end (Wan, 2012). The liberalization of casino licensing in 2002 increased the number of casino operators from only 1 to 3, later to 6 companies in 2007. The number of casinos jumped from 11 to 28 over the same period. By the end of 2013, the number of casinos reached 35 (Table 5.1) (Gaming Inspection and Coordination Bureau, 2014).

TABLE 5.1 Number of Casinos in Macao

Concessionaires	2002	2003	2004	2005	2006	2007	2008	2009	2010	2011	2012	2013
Sociedade de Jogos de Macau (SJM)[a]	11	11	13	15	17	18	19	20	20	20	20	20

TABLE 5.1 *(Continued)*

Concessionaires	2002	2003	2004	2005	2006	2007	2008	2009	2010	2011	2012	2013
Galaxy Casino, S.A. (Galaxy)	–	–	1	1	5	5	5	5	5	6	6	6
Venetians Macao, S.A. (Venetian)	–	–	1	1	1	2	3	3	3	3	4	4
Wynn Resorts (Macao)	–	–	–	–	1	1	1	1	1	1	1	1
Melco PBL Jogos (Macao), S.A. (Melco Crown)	–	–	–	–	–	1	2	3	3	3	3	3
MGM Grand Paradise, S.A. (MGM)	–	–	–	–	–	1	1	1	1	1	1	1
Total	11	11	15	17	24	28	31	33	33	34	35	35

[a]SJM is a subsidiary of STDM.
Source: Gaming Inspection and Coordination Bureau (2014).

Undoubtedly, gaming industry contributes tremendously to Macao's economy. Since 2006, the gaming revenue of the city accounted for almost 50% to its gross domestic product (GDP), with an ever-increasing proportion over the past decade (Table 5.2). Furthermore, in 2006, Macao surpassed Las Vegas for the first time and became the leading gaming capital with gaming revenue of MOP 55 billion (US$6.87 billion). As of 2013, the city reported a gambling revenue of more than MOP 360 billion (US$45 billion), almost 7 times that of Las Vegas, further cementing its status as "the undisputed heavyweight champion of the gambling industry" (Riley, 2014).

TABLE 5.2 Gaming Revenue and GDP[a]

	Gross Gaming Revenue	Revenue Growth	Tax Revenue from Gaming	GDP	Contribution of Gaming to GDP
2002	23,496.0	–	7,765.8	56,298.5	41.7%
2003	30,315.1	29.0%	10,579.0	63,579.4	47.7%
2004	43,510.9	43.5%	15,236.6	82,294.0	52.9%
2005	47,133.7	8.3%	17,318.6	94,471.0	49.9%
2006	57,521.3	22.0%	20,747.6	116,570.5	49.3%
2007	83,846.8	45.8%	31,919.6	145,084.8	57.8%
2008	109,826.3	31.0%	43,207.5	166,265.1	66.1%
2009	120,383.0	9.6%	45,697.5	170,171.3	70.7%
2010	189,587.8	57.5%	68,776.1	226,941.3	83.5%
2011	269,058.3	41.9%	99,656.4	293,745.0	91.6%
2012	305,234.9	13.4%	113,377.7	343,416.3	88.9%
2013	361,866.3	18.6%	134,382.5	413,471.0	87.5%

[a]**Unit:** Million MOP.
Source: DSEC (2002–2013).

In addition to its gaming resources, Macao is also a city of culture, boasting China's largest group of historical properties in urban area (Ung and Vong, 2010). After more than 450 years of peaceful coexistence and joint development before the handover in 1999, the people of China, Portugal, and many other countries have created a unique cultural landscape in Macao. The Historic Centre of Macao, the highlight of Macao's cultural assets, was inscribed on the list of the UNESCO World Heritage Sites in July 2005. The Historic Center includes 22 historic monuments and 8 public squares, providing "a unique testimony to the meeting of aesthetic, cultural, architectural and technological influences from East and West" (UNESCO, 2014). Its unique offerings from both the Eastern and Western cultures are appreciated by tourists as an "exoticised leisure space" (Ong and du Cros, 2012).

5.3 CHINESE OUTBOUND TOURISTS TO MACAO

Over the past 20 years, the number of visitors to Macao has almost quadrupled, increasing from 7,929,311 in 1993 to 29,324,822 in 2013 (Table 5.3). In 2000, the increase in visitor arrivals was 23.08% compared with that in 1999. The reasons for this increase being the handover of Macao and implementation of Golden Weeks (three 7-day holidays including Chinese New Year holiday in January or February, Labor Day holiday in May, and National Day holiday in October). However, the boom in tourism and gaming industry did not emerge until the liberalization of casino licensing in 2002 and the implementation of the Chinese central government's new visa regulations in 2003, which permitted mainland Chinese to travel to Hong Kong and Macao under the IVS. Since then, the visitor profile has also undergone significant changes in Macao.

TABLE 5.3 Visitor Arrivals to Macao from Greater China Area[a]

Year	Total		Mainland China			Hong Kong		Taiwan	
	Visitor Arrivals	Arrival Growth	Visitor Arrivals	Arrival Growth	Market Share	Visitor Arrivals	Market Share	Visitor Arrivals	Market Share
1993	7,929.3	–	272.1	–	3.5%	6,067.8	77.5%	272.3	3.5%
1994	7,833.8	−1.2%	245.2	−9.9%	3.1%	6,088.4	77.7%	244.4	3.1%
1995	7,752.5	−1.0%	543.5	121.6%	7.0%	5,617.5	72.5%	279.9	3.6%
1996	8,151.1	5.1%	604.0	11.1%	7.4%	5,205.3	63.9%	758.9	9.3%
1997	7,000.4	−14.1%	529.8	−12.3%	7.6%	4,702.5	67.2%	908.9	13.0%
1998	6,948.5	−0.7%	816.8	54.2%	11.8%	4,721.8	68.0%	816.6	11.8%
1999	7,443.9	7.1%	1,645.2	101.4%	22.1%	4,229.8	56.8%	984.8	13.2%
2000	9,162.2	23.1%	2,274.7	38.3%	24.8%	4,954.6	54.1%	1,311.0	14.3%
2001	10,279.0	12.2%	3,005.7	32.1%	29.2%	5,196.1	50.6%	1,451.8	14.1%
2002	11,530.8	12.2%	4,240.4	41.1%	36.8%	5,101.4	44.2%	1,532.9	13.3%
2003	11,887.9	3.1%	5,742.0	35.4%	48.3%	4,623.2	38.9%	1,022.8	8.6%
2004	16,672.6	40.2%	9,529.7	66.0%	57.2%	5,051.1	30.3%	1,286.9	7.7%
2005	18,711.2	12.2%	10,462.8	9.8%	55.9%	5,611.1	30.0%	1,482.3	7.9%

2006	21,998.1	17.6%	11,985.7	14.6%	54.5%	6,935.6	31.5%	1,437.8	6.5%
2007	26,993.0	22.7%	14,866.4	24.0%	55.1%	8,174.1	30.3%	1,444.1	5.3%
2008	22,933.2	−15.0%	11,613.2	−21.9%	50.6%	7,016.5	30.6%	1,315.9	5.7%
2009	21,752.8	−5.1%	10,989.5	−5.4%	50.5%	6,727.8	30.9%	1,292.6	5.9%
2010	24,965.4	14.8%	13,229.1	20.4%	53.0%	7,466.1	29.9%	1,292.7	5.2%
2011	28,002.3	12.2%	16,162.7	22.2%	57.7%	7,582.9	27.1%	1,215.2	4.3%
2012	28,082.3	0.3%	16,902.5	4.6%	60.2%	7,081.2	25.2%	1,072.1	3.8%
2013	29,324.8	4.4%	18,632.2	10.2%	63.5%	6,766.0	23.1%	1,001.2	3.4%

[a]Visitor arrivals: numbers in thousands.
Source: DSEC, 1993–2013.

Macao's tourists are mainly from the greater China area with the top 3 source markets being Mainland China, Hong Kong, and Taiwan, accounting for about 90% of the visitor arrivals to the city. Traditionally, Hong Kong was the major source market for Macao's tourism. However, since the establishment of Macao SAR in 1999, the surge in Mainland Chinese visitors has expanded the tourism industry of the city to a new level. Gaming liberalization and implementation of IVS have made Mainland China to overtake Hong Kong as the largest source market to Macao in 2003. Even with the SARS outbreak, the total visitor arrivals increased by 3.1% with the influx of tourists from Mainland China that year. With the number of Mainland Chinese visitors increasing notably, Mainland China has also secured its position as the most important source market for Macao, comprising over 50% of visitor arrivals over the years. In 2013, 63.5% of Macao tourists were from Mainland China. In the following paragraphs, the details regarding Mainland Chinese visitors to Macao from 2002 to 2013 based on data compiled by DSEC (2002–2013) have been described.

Demographic characteristics (Table 5.4) of Mainland visitors were not recorded throughout the years. Regarding gender, data from 2002 to 2007 demonstrated that in 2002, the number of males who visited Macao was twice than that of the females. However, in 2007, gender distribution was somewhat equal with 52% male tourists and 48% female tourists documented. Marital status was only compiled for the years 2002 and 2003. In both these years, 80% of Mainland visitors were married. With regards to their occupation, an average of almost one-third (32%) of Mainland visitors are in high positions in the government or companies, followed by an average of 15% of them being clerks.

Table 5.5 summarizes the characteristics of the trip by Mainland Visitors, including length of stay, travel arrangement, type of visitors, visitors coming under

IVS, and their primary purpose of visit. Every month, millions of visitors come from Mainland to Macao; however, Macao is not a destination that can keep visitors long. In comparison with the length of stay of visitors from Hong Kong and Taiwan, Mainland Chinese tend to stay a bit longer. Nonetheless, their average length of stay is still less than 1.5 days over the past years. In addition, during the past 4 years, their length of stay became shorter, and the average stay was only 1.1 days. This is also reflected in the decreasing percentage of Mainland visitors staying in Macao overnight. In 2002, 83.7% of Mainland visitors stayed in Macao overnight; yet, in 2013, the percentage declined to less than 50%. This is mainly due to the rising price of accommodations in the city, which makes Mainland tourists shorten their length of stay in Macao and instead spend the night in nearby city Zhuhai. Although the IVS started in 2003, not all provinces in Mainland were granted such privilege. In addition, from 2008 to 2013, an average of 44% of Mainland Chinese visited Macao under IVS.

TABLE 5.4 Demographic Characteristics of Mainland Visitors to Macao

	Sex (%)		Occupation (%)						
	Male	Female	Housewives, Students, Unemployed, and Retired	Legislators, Senior Officials, and Managers	Professionals	Technicians and associate professionals	Clerks	Service and Sales Workers	Others
2002	68	32	20	37	6	6	17	6	8
2003	64	36	20	39	6	7	18	5	6
2004	65	35	19	45	7	7	16	4	3
2005	58	42	24	37	6	6	18	4	5
2006	55	45	24	37	6	5	18	6	4
2007	52	48	26	34	7	5	17	6	5
2008	–	–	26	32	7	6	17	7	5
2009	–	–	24	26	11	11	14	11	5
2010	–	–	31	25	8	9	12	10	6
2011	–	–	31	27	7	9	10	9	6
2012	–	–	31	25	9	10	11	9	4
2013	–	–	28	24	9	12	12	10	5
Average			25	32	7	8	15	7	5

Note: – indicates no data available.
Source: DSEC (2002–2013).

TABLE 5.5 Trip Characteristics of Mainland Visitors to Macao

	Length of Stay	Package Tours	Type of Visitors (%)		IVS	Primary Purpose of Visit (%)				
	(Days)	(%)	Over-night	Same-Day	(%)	Vacation	VRF	Business	Gambling	Others
2002	1.6	33.6	83.7	16.3	–	82.2	4.1	8.0	3.4	2.4
2003	1.6	21.9	62.1	37.9	–	80.9	3.7	10.3	3.1	2.1
2004	1.3	21.7	54.7	45.3	–	77.4	4.4	11.0	4.3	3.0
2005	1.4	20.1	53.7	46.3	–	73.0	7.0	11.0	6.0	3.0
2006	1.2	19.1	53.2	46.8	–	76.0	6.0	9.0	6.0	3.0
2007	1.3	19.6	50.0	50.0	–	79.0	6.0	8.0	4.0	3.0
2008	1.4	28.9	45.0	55.0	56.7	79.0	6.0	7.0	3.0	5.0
2009	1.3	30.3	47.0	53.0	43.8	81.4	6.0	3.6	5.3	3.7
2010	1.0	30.7	47.7	52.3	41.5	72.0	7.0	2.0	5.0	13.0
2011	1.1	33.9	45.4	54.6	40.8	66.0	7.0	3.0	5.0	19.0
2012	1.1	38.6	47.8	52.2	42.2	64.0	6.0	2.0	4.0	24.0
2013	1.1	40.0	47.9	52.1	43.3	66.0	5.0	2.0	3.0	24.0
Average	1.3	28.2	53.2	46.8	44.7	74.7	5.7	6.4	4.3	8.8

Note: Indicates no data available.
Source: DSEC (2002–2013).

With regard to primary purpose of visit, an average of 75% of the Mainland visitors stated that their visit to Macao was for vacation, although the number decreased over the past 3 years. The percentage of business trips decreased over the past years from approximately 10% in the beginning of 2000s to only 2–3% in the beginning of 2009. A small percentage of visitors travel to Macao to visit friends and relatives. However, the most unclear part in purpose of visits is although Macao is renowned for its gaming facilities, only an average of 4.3% of visitors claimed that their primary purpose is gambling. Presumably, due to the social desirability issue, Mainland visitors are not willing to admit that they visit Macao to gamble. Interestingly, the percentage of Mainland Chinese visit Macao for "other" purpose has been increasing dramatically so much so that for the past 2 years, this accounted for nearly one-fourth of the visitors. However, "other" purposes are not investigated for its specificity.

Although Chinese tourists do not stay long in Macao, they spend a lot. Mainland visitors spend consistently much more than the average visitor per capita and those from Hong Kong and Taiwan (Table 5.6). The data show that Mainland visitors spend almost twice the amount spent by an average visitor and triple the amount spent by the visitors from Hong Kong and Taiwan. As Mainland visitors travel to Macao mainly for vacation, shopping is one of their important activities, which makes shopping their major spending component with an average of 60% of their non-gaming expenditure spent in shopping (Table 5.7). In addition to shopping, they spend around 20% of their budget on accommodation and approximately 15% on food and beverage.

TABLE 5.6 Visitor's Non-Gaming Expenditure Per Capita (MOP)

Year	Total	Mainland China	Hong Kong	Taiwan
2002	1,454	2,655	957	984
2003	1,518	2,847	947	1,266
2004	1,633	2,991	969	1,310
2005	1,523	3,078	898	1,336
2006	1,610	3,215	955	1,494
2007	1,637	3,080	1,085	1,447
2008	1,729	3,571	1,109	1,361
2009	1,616	3,040	1,159	1,349
2010	1,518	2,039	811	677
2011	1,619	2,048	916	1,052
2012	1,864	2,385	906	1,356
2013	2,030	2,563	911	1,517

Source: DSEC (2002–2013).

TABLE 5.7 Non-Gaming Expenditure Categories of Mainland Chinese Visitors (%)

Categories	2002	2003	2004	2005	2006	2007	2008	2009	2010	2011	2012	2013
Accommodation	16	16	18	18	17	19	20	19	19	20	21	22
Food and beverage	17	16	16	16	14	14	14	13	15	15	16	17
Transport	3	4	4	4	4	4	4	3	3	3	2	2

Other	4	2	2	1	1	1	1	1	4	2	3	3
Shopping	60	62	60	60	64	63	61	63	63	61	58	59
Clothing	21	20	25	23	22	17	21	21	19	19	19	19
Jewelry and watches	37	34	26	20	18	21	18	19	23	21	21	21
Local food products	16	13	11	11	11	13	13	16	19	19	17	17
Cosmetics and perfume	6	9	11	12	11	11	13	15	13	13	14	14
Shoes, handbags, and wallets	3	5	5	8	8	11	15	14	12	13	16	17
Others	17	19	22	26	30	27	21	16	14	15	12	14

Source: DSEC (2002–2013).

Table 5.7 outlines categories of shopping expenditure of the Mainland Chinese. Mainland visitors spend more on jewelry and watches (from 18%–37%), clothing (17%–25%), local food products as gifts (11%–19%), and cosmetics and perfume (6%–14%). However, it is necessary to point out that the percentage of expenditure on the category of clothing has been declining slightly, whereas the percentage of expenditure on the category of accessories (shoes, handbags, and wallets) has seen a steady increase of spending share. It has been observed that Macao has improved the shopping facilities to attract the tourists.

Since Mainland China is the top source market for Macao, it is important to evaluate the experience of Chinese tourists in Macao. From 2002 to 2013, the satisfaction level of Mainland Chinese visitors has been increasing (Table 5.8). Among all the items evaluated, in general, Chinese visitors were most satisfied with the travel agency, accommodation, and gaming facilities of the city. Satisfaction level on the tourist attractions of the city was ranked the lowest, as only an average of 33% of Chinese visitors stated that there are sufficient tourist attractions in the city, whereas an average of 24% of them stated otherwise. In addition, an average of 27% of them refused to comment when evaluating tourist attractions. Insufficient tourist attractions partly explain about the short duration of stay of visitors in Macao.

TABLE 5.8 Mainland Chinese Visitors' Satisfaction Level of Services in Macao (%)

		2002	2003	2004	2005	2006	2007	2008	2009	2010	2011	2012	2013
Services of travel agencies[a]	1	43	44	56	63	67	62	72	77	76	78	80	91
	2	40	39	28	21	25	26	20	13	16	16	14	8
	3	9	9	9	5	6	9	6	3	5	4	4	1
	4	8	8	7	11	2	3	2	6	3	3	2	0
Services of restaurants[a]	1	58	60	62	58	59	60	65	73	77	77	80	84
	2	33	32	32	34	34	32	30	23	19	19	17	13
	3	7	7	5	5	6	6	4	3	4	3	3	3
	4	2	1	1	2	2	2	1	1	1	1	0	0
Services of hotels[a]	1	59	63	65	69	71	72	79	82	85	87	91	92
	2	31	29	28	25	23	22	18	15	12	10	7	7
	3	7	6	6	5	5	4	3	2	1	1	1	1
	4	2	1	1	2	1	2	1	1	1	1	1	0
Services of shops[a]	1	70	72	69	65	65	69	77	84	87	87	88	89
	2	23	23	27	28	29	25	19	14	11	11	10	9
	3	3	3	3	4	4	4	2	2	1	1	1	2
	4	4	3	2	3	3	2	2	1	1	1	1	1
Services of public transport[a]	1	59	72	70	63	59	57	62	76	79	75	80	81
	2	31	20	22	26	26	27	27	17	13	16	11	10
	3	7	5	6	9	13	14	9	5	6	7	7	8
	4	2	2	2	2	2	2	2	2	2	1	1	1
Envrion-mental hygiene[a]	1	68	68	65	62	61	64	70	77	84	82	87	86
	2	26	27	31	32	34	31	26	20	13	15	11	11
	3	4	4	4	5	4	4	3	2	2	2	1	2
	4	1	1	0	1	1	1	1	1	1	1	1	0
Gaming services and facilities[a]	1	–	–	–	61	70	79	84	84	82	87	89	88
	2	–	–	–	29	23	16	12	13	11	9	8	9
	3	–	–	–	3	3	2	1	1	1	2	2	1
	4	–	–	–	6	5	3	2	2	5	2	1	2

Tourist Attractions[b]	1	22	24	25	25	30	34	36	38	40	39	39	42
	2	25	31	34	28	27	26	27	20	16	21	20	16
	3	26	21	18	19	18	16	17	14	14	12	11	11
	4	28	24	23	27	26	24	20	27	30	28	30	31

Note: – indicates statistics not available.
a1 = Satisfied, 2 = fair, 3 = should be improved, 4 = no comment.
b1 = Sufficient, 2 = fair, 3 = insufficient, 4 = no comment.
Source: DSEC (2002–2013).

Thus, in the past 3 years (2011–2013), there have been some changes in the trip characteristics of Mainland visitors to Macao. First, Chinese tourists spent less time in the city with an average length of stay of 1.1 days compared with 1.3 or more in previous years. Second, more Chinese visitors came to the territory in package tours over the past 2 years, as shown in the percentage of package tours. Specifically, the percentage increased from approximately 30% or less before 2011 to 38.6% in 2012 and 40.0% in 2013. Third, in terms of their purpose of visit, Chinese traveling to Macao for vacation accounted less than 70% in the beginning of 2011, whereas traveling for other purpose increased from less than 10% in 2010 to 24%. Fourth, although Chinese visitors are in number one position in non-gaming expenditure, the amount of the expenditure started to decrease from over MOP 3,000 to MOP 2,039 in 2010. Although there was an increase in expenditure in 2012 and 2013, it still could not reach to the level that was in 2009 and before. Fifth, in their shopping behavior, they spent less in clothing and more in cosmetics and perfume and accessories such as shoes, handbags, and wallets. Finally, Mainland visitors were much more satisfied in every aspect of tourism and hospitality services in Macao in 2012 and 2013 than that in previous years. Even so, they still considered tourist attractions insufficient in the area. These recent changes indicate that the challenge faced by Macao to remain a top destination for Mainland visitors is to entice them to stay longer by providing more non-gaming elements. This issue has been acknowledged, and Macao administration, along with industry, has been addressing this issue by developing itself into a world tourism and leisure center.

5.4 BECOMING A WORLD CENTER OF TOURISM AND LEISURE

With China's central government's full support and the flooding of Mainland Chinese visitors, Macao's economy has been expanding and prosperous over the past decade. However, it is also realized that Macao's economy relies overly on its gaming, which is dependent heavily on tourists from Mainland. This single-industry structure makes the economy of the city rather unhealthy and unsustainable. In other words, Macao's economy is highly concentrated on its gaming, which makes it volatile even with exponential economic growth (Vinnicombe and Sou, 2013). This demands Macao government to diversify its economy to enhance its sustainability and stability (Chan, 2006).

Both central government and Macao government have been making efforts to diversify its economy. In terms of tourism industry, recent policies aim to diversify Macao's tourism products by developing Macao into a "World Center of Tourism and Leisure". However, the new positioning of Macao is challenged largely by the small size of the city. Therefore, support from the central government and cooperation with neighboring regions or areas such as Hong Kong and Guangdong are essential to achieve the objective of becoming the World Center of Tourism and Leisure.

National support has been gained when developing Macao into the World Tourism and Leisure Center was infused into the national development strategy, as the development is part of China's 12th Five-Year Plan (Liu, 2011; Xing and Huang, 2012). Macao's administration has also actively pursued regional cooperation in tourism in the formulation of the Guangdong–Hong Kong–Macao Tourism Cooperation Plan, which urges tourism authorities from Guangdong Province, Hong Kong, and Macao to develop and promote multidestination itineraries, co-organize tourism promotion activities, and participate jointly in local travel exhibitions (Macao Special Adminstrative Region of China, 2014; MGTO, 2014). Recent development has been seen in Hengqin New Area. Hengqin, part of Zhuhai, is adjacent to the city of Macao. Hengqin triples Macao in size, which provides the city of Macao with the necessary space to build facilities that are essential for leisure tourism (Lu, 2012). Of particular relevancy is the construction of the Chime Long Ocean Kingdom. The theme park started its operation in 2014, consisting of entertainment facilities, amusement rides, performances, high-tech experiences, and animal watching, as well as the dolphin-themed hotel with 1,888 guest rooms (Macau Daily Times, 2014a).

Locally, within Macao, the government has been making attempts to improve its tourism product development. While regulating and monitoring the development of gaming industry, Macao government has also actively facilitated the development of integrated tourism and other associated industries (Liu, 2011; Xing and Huang, 2012). Casino resorts are encouraged to reinforce non-gaming elements, such as shopping facilities, shows, kid's centers, spas, and clubs. The highlight to that direction is the family-oriented amusement resort and hotels planned by SJM in Cotai. The complex is designed to consist of one 5-star, four 4-star, and one 3-star hotels with over 6,000 guest rooms, shopping malls, convention facilities, an indoor beach and wave pool, amusement rides, a 4D theatre, an equestrian centre, a horse carriage trail, as well as a water sports performance centre (Macau Daily Times, 2014b). The completion of the resort will afford more leisure opportunities for tourists and residents alike.

Emphasis has also been put on developing the convention and exhibition facilities of the city and cultural and creative tourism (Liu, 2011; Xing and Huang, 2012). For example, Macao government completed the legislative process of the Cultural Heritage Protection Law, bringing success in the protection and continuation of intangible cultural heritage. Moreover, the government made great efforts in the development of local cultural and creative industries by establishing the Cultural and Creative Industry Fund and actively promoting the innovation and continuity of Macao's culture and creativity (Macao Special Adminstrative Region of China, 2014). Macao Government Tourism Office (MGTO) has also been designing and promoting new cultural itineraries for tourists to explore more of the city (MGTO, 2014). Those decisions help transform Macao from a casino gaming city to a more family and business travel destination, thus stepping closer to the objective of being a World Tourism and Leisure Center.

5.5 CONCLUSION[1]

Although the government of Macao has been making efforts to diversify its economy, gaming is still the pillar industry of the city. Gaming industry is so lucrative that new investment keeps pumping into the city. It is expected that 8 more mega-casinos will be added to the territory by 2017 (Table 5.9). Nevertheless, non-gaming tourism resources are given more attention by the government, and more promotion has been seen on the aspects of Macao's culture and heritage and MICE facilities. All these endeavors, hopefully, would help turn Macao into a destination with a wide range of choices from resorts, casinos, shopping, MICE, and entertainment. Although these efforts are not exclusively aiming at Chinese visitors, as a top source market for Macao, Chinese visitors will benefit eventually.

[1]When this manuscript was completed in June, Macao saw its first monthly decline in its gaming revenue by 3.7% since June 2009. After that, Macao has suffered continuous decline up until November, with October witnessing the largest dip (23.2%), followed by 19.6% dive in November. It is predicted that this downward trend will continue in December and coming months in 2015. The negative revenue growth is due to the efforts made by the Chinese government on fighting against the corruption, which has been hurting the high-end market of the casinos. Although there are concerns of this trend, the government of Macao has also seen this as an opportunity to slow down its gaming development and diversify its economy by focusing on the development of other economic activities. In addition, the government announced extended opening hours for 3 entry points effective from December 18, including 24-hour crossing at Hengqin checkpoint; 2-hour extension at Gongbei Border Gate; and opening of Zhuhai-Macao Cross-Border Industrial Park to all residents of Macao. (Currently, it is only open to employees of the industrial park.) It is expected that these policies would influence the profile of Mainland Chinese visitors to Macao.

TABLE 5.9 New Casino Projects in Macao

Construction Name	Total Investment (in billion USD)	Number of Rooms	Number of Gaming Tables	Year of Project Completion
Galaxy phase 2	2.00	3,600	150	2015
Galaxy phases 3 and 4	4.60	5,500	tbc	2018
MGM Cotai	2.60	1,600	500	2016
Wynn Palace	4.00	2,000	500	2016
The Parisian	1.50	3,000	tbc	2015
SJM Cotai	3.22	2,000	tbc	2017
Studio City	2.00	2,000	400	2015

Note: tbc, to be confirmed.

So far, shopping tourism is rather successful in luring Chinese tourists to Macao as evidenced by their shopping expenditure. Hong Kong, the other SAR of China, competes head to head with Macao for shopping dollars from Chinese tourists. Similar to Hong Kong, Macao casino resorts are able to offer tourists clusters of high-end international brands and retailers such as DFS, Louis Vuitton, and Prada. Although Macao has a competitive advantage over Hong Kong in offering lower cost, Chinese visitors still shop more in Hong Kong as Hong Kong provides more shopping facilities, better shopping environment, and better service (Wong, 2013). It is suggested that shopping malls and retailers of Macao need to work hard on these areas.

Efforts in developing cultural and creative tourism are not yet successful. The combination of short length of stay and insufficient interesting tourist attractions has caused unequal distribution of tourist flow among Mainland tourists. When in Macao, in addition to visiting casinos and shopping, most of Mainland tourists only sightsee 2 most famous cultural attractions in the Historic Center, namely Ruins of St. Paul and A-Ma Temple. Thus, these 2 sites are congested with tourists, whereas other heritage sites and squares are underutilized (du Cros, 2009). To ease the congestion problems and provide in-depth cultural experience, in 2013, MGTO proposed and promoted 4 new cultural itineraries to diverge Chinese tourists to other cultural sites (MGTO, 2013). However, these new sites have received limited visitation. The reality still remains that most of Chinese visitors access the territory to gamble and shop, and their awareness of the unique historical heritage of the city is no more than fragmentary (Io, 2011).

Mainland visitors have pushed Macao's economy to a level that has never been imagined. However, the massive amount of Chinese visitors has also posed problems, such as overcrowding, traffic problems, and annoying behaviors, which have negative impacts on the quality of the life of residents of Macao (Wan and Li, 2013; Loi and Pearce, 2012). Macao residents feel that their daily life has been disrupted so much so that some of them start to challenge the IVS (Macau Daily Times, 2013). Macao's capacity of receiving continuously increasing number of Chinese tourists needs to be dealt with instantaneously. Although there are no specific measures considered, MGTO has been emphasizing that Macao aims on quality, not quantity, and future efforts will be made to attract more quality Chinese tourists (Macau Daily Times, 2013).

REFERENCES

Chan, S. S. Rationales and options for economic diversification in Macao. *AMCM Q. Bull.* 2006, 18, 17-36.

DSEC. *2012 Yearbook of Statistics*. 2012. http://www.dsec.gov.mo/Statistic.aspx?NodeGuid=d45bf8ce-2b35-45d9-ab3a-ed645e8af4bb (accessed May 30, 2014).

du Cros, H. Emerging issues for cultural tourism in Macau. *J. Curr. Chin. Aff.* 2009, 38 (1), 73-99.

Gaming Inspection and Coordination Bureau. Gaming Statistics. 2014. http://www.dicj.gov.mo/web/en/information/DadosEstat/2014/content.html#n5 (accessed May 30, 2014).

Io, M.-U. Can the historic center of Macao be a popular tourist attraction? Examining the market appeal from tour marketers' perspective. *J. Qual. Assur. Hospital. Tour.* 2011, 12, 58-72.

Liu, Z. *Empirical Investigation of Moderate Economic Diversification in Macao: Development of Tourism Clusters in Macao*. Macao Association of Economic Sciences: Macao, 2011.

Loi, K. I. and Pearce, P. L. Annoying tourist behaviors: perspectives of hosts and tourists in Macao. *J. China Tour. Res.* 2012, 8 (4), 395-416.

Lu, Y. Macau's Path to Economic Diversification. 2012. http://www.china-briefing.com/news/2012/10/22/macau-way-to-economic-diversification.html.

Macau Daily Times. MGTO head dodges capacity question again. 2013. http://www.macaudailytimes.com.mo/macau/50283-mgto-head-dodges-capacity-question-again.html (accessed May 30, 2014).

Macau Daily Times. Hengqin: Chimelong Ocean Kingdom to open before CNY. 2014a. http://www.macaudailytimes.com.mo/macau/49841-hengqin-chimelong-ocean-kingdom-to-open-before-cny.html (accessed May 30, 2014).

Macau Daily Times. Theme park set to be built in Cotai. 2014b. http://www.macaudailytimes.com.mo/macau/19001-Theme-park-set-built-Cotai.html (accessed May 30, 2014).

Macao Special Adminstrative Region of China. Policy Address for the Fiscal Year 2014 of the Macao Special Administrative Region (MSAR) of the People's Republic of China. 2014. http://www.policyaddress.gov.mo/policy/home.php?lang=en (accessed May 30, 2014)

MGTO. MGTO first launches 4 walking routes themed as "Step Out, Experience Macau's Communities" to divert visitor trends and promote community tourism. 2013. http://industry.macautourism.gov.mo/en/pressroom/index.php?page_id=172&id=2585 (accessed May 30, 2014).

MGTO. Community opinion poll session for new additional walking tour routes "Step out, experience Macau's Communities" MGTO collects opinions broadly. 2014. http://industry.macautourism.gov.mo/en/pressroom/index.php?page_id=172&id=2644 (accessed May 30, 2014).

MGTO. Zhongshan, Zhuhai and Macau jointly promote regional tours in cities along high-speed rail. 2014. http://industry.macautourism.gov.mo/en/pressroom/index.php?page_id=172&id=2647 (accessed May 30, 2014).

Ong, C.-E. and du Cros, H. The post-mao gazes: Chinese backpackers in Macau. *Ann. Tourism Res.* 2012, 39 (2), 735-754.

Riley, C. Macau's gambling industry dwarfs Vegas. 2014. http://money.cnn.com/2014/01/06/news/macau-casino-gambling/ (accessed May 30, 2014).

UNESCO. Historic Centre of Macao. 2014. http://whc.unesco.org/en/list/1110 (accessed May 30, 2014).

Ung, A. and Vong, T. N. Tourist experience of heritage tourism in Macau SAR, China. *J. Heritage Tourism*. 2010, 5 (2), 157-168.

Vinnicombe, T. and Sou, J. P. U. Diversifying the Macao economy: insights from profiling mainland Chinese visitors. *J. China Tourism Res.* 2013, 10 (3), 347-362.

Wan, Y. K. P. Increasing Chinese tourist gamblers in Macao: crucial player characteristics to identify and exploit. *UNLV Gaming Res. Rev. J.* 2012, 15 (1), 51-69.

Wan, Y. K. P. and Li, X. Sustainability of tourism development in Macao, China. *Int. J. Tour. Res.* 2013, 15 (1), 52-65

Wong, I. A. Mainland Chinese shopping preferences and service perceptions in the Asian gaming destination of Macau. *J. Vacation Mark.* 2013, 19 (3), 239-251.

Xing, W. and Huang, T. *Development and Innovation of Industries in Macau*. Economic Science Press: Beijing, China, 2012.

CHAPTER 6

MAINLAND CHINESE OUTBOUND TOURISM TO TAIWAN: RECENT PROGRESS

LI SHEN, CHIA-KUEN CHENG, YANN-JOU LIN, and HSI-LIN LIU

CONTENTS

6.1 INTRODUCTION

The economy of China has been growing explosively in the recent decade and is receiving worldwide attention. China not only has the largest population in the world, but also has continued to be ranked first in international tourism expenditure since 2012 (UNWTO, 2014b). The rapid growth of the Chinese outbound travel market over the past decade has drawn the attention of almost every destination market, especially Taiwan. Mainland China and Taiwan are not only geographically close but also culturally similar. However, the people were separated after the Chinese Civil War in 1940s. After 50 years of political separation and barriers, the people in the two areas experienced an increased mutual sense of mystery and curiosity about the other side, which provided a strong motivation for Mainland Chinese tourists to visit Taiwan. The Mainland China and Taiwan governments finally agreed to launch direct flights across the Taiwan Strait in 2008 (Foundation, 2008), and the daily quota of Chinese tourists allowed to visit Taiwan increased to 9,000 per day in 2014 (Foundation, 2013). However, the complicated political and legal issues have made the development of Cross-Strait tourism very different from other international tourism destinations. With the recent boom in tourism flows between Mainland China and Taiwan, it would be of a timely issue to review the changing tourism markets between the two areas.

6.2 TAIWAN'S TOURISM POLICY TOWARD CHINESE TOURISTS

The Cross-Strait tourism is not a completely open market; rather, it is highly regulated by the two governments. Therefore, tourism development between Mainland China and Taiwan is largely affected by the Cross-Strait political relations. Due to changing international situations, Taiwan's policy toward Mainland China after 1949 can also be divided into two major stages (Lee, 2013).

6.2.1 CROSS-STRAIT MILITARY CONFRONTATION (1949–1975)

Following the Chinese Civil War, the Communist Party of China took full control of the Mainland China in 1949. Nationalist leader Chiang Kai-Shek relocated the government (Kuomintang [KMT]) to Taiwan, which many countries continued to recognize as China's legitimate government. Cross-Strait relations have been hindered by military threats and political and economic pressure, particularly over Taiwan's political status, with both governments officially adhering to a "One-China policy." The Cross-Straits diplomatic competition continued for more than two decades.

6.2.2 EXCHANGE AND MODERATION PERIOD (1976–PRESENT)

In Taiwan, the possibility of retaking Mainland China became increasingly remote in the late 1970s, particularly after the establishment of diplomatic relations between the PRC and the United States. Under the administration of Lee Teng-Hui, the main goal of Taiwan's diplomatic policy shifted to international recognition. Since then, Taiwan's government has sought more opportunities of exchanges in academics, economic and trade cooperation, and technology with Mainland China. The frequent exchanges during this period have resulted in the establishment of one set of Cross-Straits negotiation mechanism by both sides, and the attitude of Taiwan's government softened to promote valuable peaceful communication.

Mainland China and Taiwan have been politically separated since the Civil War in the 1940s, and travel across the Taiwan Straits was first permitted in 1987 (Yu, 1997). The travel flows between Mainland China and Taiwan played an important role in shaping the nature of political relations. Both China and Taiwan have experienced many phases of the tourism industry under political tendency and also formulated many relevant regulations. The evolution of the political division between the Mainland and Taiwan is discussed below.

It was not until 1988 that the Taiwan government decided to allow people from China to attend conferences, news reporting, and religious and cultural events. In December 2000, the Legislative Yuan (Taiwan) passed an *Act Governing Relations between the People of the Taiwan Area and the Mainland Area*, which served as basic law on ushering in a two-way flow of travel between the two sides. After establishing the act, the Executive Yuan passed the *Regulation on the Permission of Chinese Tourists' Traveling in Taiwan* in 2008. According to the regulation, there are three major types of Chinese tourists. The first type is Mainland Chinese citizens visiting Taiwan via Hong Kong and Macau. The second type is Mainland Chinese citizens engaged in business travel to other countries and passing via Taiwan. The third type is Mainland Chinese citizens who live outside of China and have obtained permanent residency in foreign countries. The third type of Chinese visitors was permitted entrance in January 2002, followed by the second type in May 2002. It was not until 2008 that the first type of Chinese visitor was allowed to visit Taiwan.

Before 2008, Chinese business visitors or those with permanent residency in foreign countries were only allowed to visit Taiwan by stopping in a third place before arriving in Taiwan. In 2008, both governments agreed to allow tightly controlled tour groups to visit Taiwan and to set up regulations about tourism managements between China and Taiwan (Fan, 2009). On July 4, 2008, the first direct flight tour group from Mainland China arrived in Taiwan, and Mainland Chinese residents have been allowed to travel to Taiwan in tour groups since then. In the same year, the limit of visitors from China was set at 3,000 persons per day (Table 6.1). China's government only permitted residents of 13 provinces to apply to visit Taiwan in group tours in 2008, and later in 2010, it allowed all the provinces to apply to visit Taiwan (Table 6.1). With a limit of 500 persons per day, the *Free and Independent Travel Policy* for tourists came into effect in June 2011, and the first batch of 273 Free Independent Travelers (FITs) from Beijing arrived in Taiwan in the same year. As for the FIT, China's government just permitted residents of three cities to apply in the first year (2011), and 36 cities were permitted to apply for FIT to Taiwan in 2014. This policy would enable the establishment of a direct dialogue between people on both sides of the Taiwan Strait and also provide a more peaceful opportunity to communicate.

TABLE 6.1 A Chronicle of the Tourism Development Between Mainland China and Taiwan

Year	The Important Events
1949	Since the Chinese Civil War, intense military confrontation resulted in many battles and led to competition between two sides.
1987	First group of Taiwan visitors was permitted to visit Mainland China.
1988	The governments of both sides agreed to have outstanding scientists and scholars from Mainland China visit Taiwan.
1992	Taiwan implemented *Act Governing Relations between the People of the Taiwan Area and the Mainland Area*, which explicitly stipulates the Cross-Straits exchange issues.
1993	Taiwan announced *Measures for Permission of Mainland Chinese citizens to Travel to Taiwan.*
	Taiwan announced *Measures for Permission of Taiwanese People to Travel to Mainland China.*
2001	Taiwan issued *Implementation of the Law on Mini Three Links between Jinmen and Matzu and Mainland China*; however, the effect was not obvious.
	Taiwan issued *Measures for Permission of Chinese Tourists to Travel to Taiwan for Sightseeing* and *Travel program on Permission of Mainlanders to visit Taiwan.*
2002	Taiwan allowed Mainland Chinese residing in other foreign countries to travel to Taiwan for sightseeing.
	Allowed Mainland Chinese outbound travel or business tour traveling to Taiwan for sightseeing.
2003	The planes of Taiwan airline companies flew to Shanghai via Hong Kong and Macao.
2004	The relevant units implemented the "Guan Ping Project," which takes measures for checking the name list of people entering Taiwan, and entry-exit personnel to prevent illegal stays.

2005	Mainland China and Taiwan officials approved allowing both sides' charter flights to cities without stopping at a third place.
2008	The first tour group of Chinese tourists carried by the first Cross-Straits direct charter flight arrived in Taiwan.
	Strait Exchange Foundation of Mainland China and Association for Relations across the Taiwan Straits of Taiwan signed two documents: *Minutes of Talks on Cross-Strait Charters* and *Agreement between Taiwan and Mainland China on Chinese Residents Traveling to Taiwan*. After implementation of the Mini Three Links, self-service traveling was initiated between the two sides via Jinmen, Matzu, and Penghu.
	Taiwan set the limit of Chinese group tours to Taiwan at 3,000 persons per day.
	Mainland China permitted 13 provinces' residents to apply to visit Taiwan.
2009	Mainland China permitted 25 provinces' residents to apply to visit Taiwan.
2010	Mainland China government permitted all (31) provinces' residents to apply to visit Taiwan.
2011	Taiwan increased the limit of Chinese group tours to 4,000 persons per day.
	Taiwan permitted Chinese FIT tourists to visit Taiwan with a limit of 500 persons per day.
	China's government permitted three cities' residents to apply to visit Taiwan as FIT.
2012	Taiwan increased the limit of Chinese FIT to 1,000 persons per day.
	China's government permitted 13 cities' residents to apply to visit Taiwan as FIT.
2013	Taiwan increased the limit of Chinese group tours to 5,000 persons per day.
	Taiwan increased the limit of Chinese FIT to 3,000 persons per day.
	Mainland China permitted 26 cities residents to apply visiting Taiwan by FIT.
2014	Taiwan increased the limit of Chinese FIT to 4,000 persons per day.
	Mainland China permitted 36 cities' residents to apply to visit Taiwan as FIT.
2015	Taiwan's government allowed Mainland Chinese to visit Jinmen, Matzu, and Penghu islands with a visa on arrival at ports of entry.

6.3 CURRENT STATUS OF CHINESE TOURISTS TO TAIWAN

As Mainland China's government gradually relaxed the restrictions on outbound tourism visiting Taiwan, the number of the Mainland Chinese tourists in Taiwan increased greatly (Tourism Bureau, 2014b), largely transforming the island's tourist industry. This section summarizes the studies on the mainlanders traveling to Taiwan and provides insight into the impact of Chinese tourists on Taiwan tourism.

6.3.1 DEMOGRAPHIC PROFILE

The growth of Chinese tourists increased substantially after opening the gate. Approximately, 972,123 Mainland Chinese visited Taiwan in groups in 2009, an increase of 195.35% from 2008. In June 2011, the two sides agreed to allow Mainland Chinese citizens to visit Taiwan independently without tour groups or tour packages. Within just one and half years, the total number of Chinese FITs reached 220,000. In 2012, an average of 7,086 tourists per day from China visited Taiwan, which is 58.8% more than that in 2010 (Fig. 6.1).

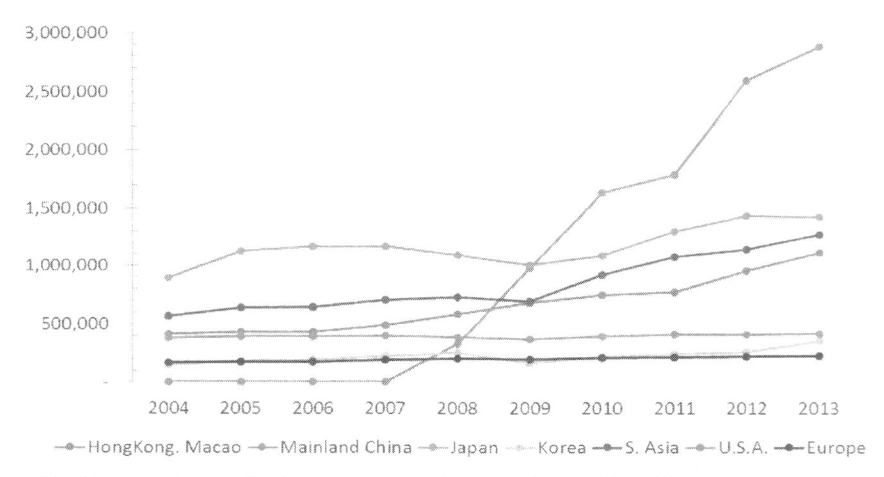

FIGURE 6.1 Taiwan's foreign visitor arrivals by residence (2004–2013). (From Tourism Bureau, 2014b.)

Because there were no official statistics on Chinese tourists' social economic information, especially for independent travelers, several related studies were reviewed to obtain a rough profile of the composition of Chinese visitors, as shown in Table 6.2. Overall, the number of female Chinese visitors was a little higher than that of male visitors (National Immigration Agency, 2014). The Chinese tourists were mainly 30–50 years old and with a university degree or higher. Most of the Chinese tourists were government employees or businessmen, with average monthly income of 2,000–4,000 RMB (Table 6.3). As regulated by the *Free and Independent Travel Policy* in 2014, there is an entry limit of 4,000 individual tourists a day from 36 cities in China. Most Chinese tourists were from the big coastal cities, especially from Southeast China.

TABLE 6.2 Chinese Visitor Arrivals by Age and Gender (2001–2013)

Year	Age							Gender	
	<10	10–19	20–29	30–39	40–49	50–59	>60	Male	Female
2001	6,739	8,377	74,026	56,625	31,081	22,650	11,552	76,199	134,851
2002	6,018	8,219	74,802	65,458	39,724	26,606	12,456	86,916	146,367
2003	5,131	6,972	62,431	63,724	41,101	23,322	10,182	77,746	135,117
2004	5,509	7,073	60,134	64,229	44,741	26,707	11,855	85,943	134,305
2005	2,762	3,654	51,087	59,441	46,660	31,903	15,746	84,290	126,963
2006	3,405	4,710	61,580	82,687	72,535	50,919	23,985	132,928	166,893z
2007	3,677	4,431	66,320	89,393	70,663	48,163	24,129	131,795	174,981
2008	4,500	7,484	75,545	90,840	72,270	47,809	30,756	132,654	196,550
2009	9,844	21,253	136,869	187,805	218,856	187,310	210,186	435,276	536,847
2010	18,060	41,662	200,244	284,404	360,254	328,082	398,029	749,276	881,459
2011	21,728	49,142	235,039	323,426	415,883	354,137	384,830	825,798	958,387
2012	39,127	83,375	354,142	479,636	582,271	510,090	537,787	1,149,839	1,436,589
2013	63,962	109,934	448,714	561,109	571,174	551,964	567,845	1,201,446	1,673,256
Total	190,462	356,286	1,900,933	2,408,777	2,567,213	2,209,662	2,239,338	5,170,106	6,702,565
%	1.60	3.00	16.01	20.29	21.62	18.61	18.86	43.55	56.45

Source: National Immigration Agency (2014).

TABLE 6.3 The Studies of Major Social and Economic Backgrounds of the Chinese Tourists

Source	Occupation	Education	Income
Lin (2003a)	Mostly public officials and technician	Mostly degree of universities	The monthly income was mostly 2,000–4,000 RMB
Lin (2003b)	Mostly administrative directors and managers.	Mostly degree of universities and above	The monthly income was under 7,000 RMB
Chen et al. (2004)	Mostly teachers and then employees of service sector	Mostly degree of universities	The monthly income was mostly over 3,000 RMB, and then 1,500–2,000 RMB
Cheng (2006)	Mostly employees of private enterprises, and then employees of foreign firms	Mostly degree of universities and above	The monthly income was mostly 2,000–4,000 RMB, and then 4,000–6,000 RMB
Yang (2007)	N/A	Mostly degree of senior high school and then colleges and universities	N/A
Hsu and Li (2008)	N/A	Mostly degree of universities	N/A
Lee (2008)	Mostly employees from business and service sectors	Mostly degree of universities and colleges	The monthly income was mostly over 4,001 RMB
Yang (2008)	Mostly public officials	Mostly degree of universities and colleges	The monthly income was mostly over 4,001 RMB
Wu (2008)	Mostly enterprise employees and then public officials	Mostly degree of universities and above	The monthly income was mostly 2,001–3,000 RMB
Teng (2009)	Mostly business people and then public officials	Mostly degree of universities	The monthly income was mostly 1,501–3,000 RMB
Li (2011)	Mostly students and then employees of service sector	Mostly degree of universities	The monthly income was mostly 2,000 RMB
Lin et al. (2013)	Mostly company employees	Mostly degree of universities	N/A
Tung and Chang (2013)	Mostly public officials	Mostly degree of universities	The monthly income was mostly 2,000–3,999 RMB

6.3.2 TRIPOGRAPHIC INFORMATION ON CHINESE VISITORS

According to the 2001–2013 inbound tourism survey (Tourism Bureau, 2014b), the main purposes of Mainland Chinese traveling to Taiwan are sightseeing (65.5%) and visiting relatives (7.7%). After Taiwan's open-door policy for the free and independent Chinese tourists in 2010, the Mainlanders traveling to Taiwan for sightseeing increased sharply (Table 6.4). After the Chinese FITs were permitted in 2012, some Chinese tourists visited Taiwan for medical treatment (Table 6.4), and medical treatment became the second main purpose for Chinese tourists visiting Taiwan. This would become a new and important section of the tourism industry in Taiwan in the next decade, especially for the China tourism market (Hsu et al., 2009).

Several studies have shown that the motivation of Chinese tourists traveling to Taiwan is quite diverse. Because of the separation for half a century, Mainland Chinese citizens had a lot of curiosity about Taiwan (Huang, 2010; Teng, 2009), although they shared the same culture and language. The Chinese visitors wanted to deeply understand the differences between Taiwan and China and experience Taiwan lifestyles (Li, 2005) and Taiwan folk customs (Huang, 2010). The major tourism motivations were visiting the most famous scenic spots and experiencing the rich cultural atmosphere in Taiwan. The Mainlanders also traveled to Taiwan for relaxation, visiting relatives and friends, or medical treatment (Chou, 2012; Li, 2005; Teng, 2009).

Existing studies indicated that most Chinese tourists were first-time visitors to Taiwan for sightseeing and recreation (Table 6.5). Due to the tourism package, they stayed in Taiwan for about 5–11 days, with an average stay of 7 days. Most of them received travel information through travel agencies, relatives or friends, and public media.

Most Chinese tour groups have to prepay tour costs, including airfare, hotel, meals, and transportation. Local expenditures were not prepaid, which could include catering cost outside hotels, transportation, entertainment, and shopping. Chinese tourists spent around US$260 per day in Taiwan (Table 6.6). Over half of their expenditure was spent shopping for jewelry and jade. However, the expenditures of Chinese tourists did not increase from 2008 to 2013; they were lower than the expenditures of average inbound tourists (US$270 per person/per day).

Because FIT travelers to Taiwan were restricted before 2010, most Chinese tourists visited Taiwan by joining tour groups with direct flights. These Chinese tour-group travelers liked to eat the local food and participate in the recreational activities, in which they could experience the local customs of Taiwan. These group travelers also purchased considerable amounts of souvenirs (Chen, 2010).

TABLE 6.4 Chinese Visitor Arrivals by Purpose of Visit (2001–2013)

Year	Business	Visiting Relatives	Conference	Study	Sightseeing	Exhibition[a]	Medical Treatment[a]	Others	Missing
2001	37,371	77,040	5,100	1,682	17,643	0	0	5,217	66,997
2002	39,121	77,107	6,135	3,564	18,735	0	0	7,512	81,109
2003	36,483	67,627	4,171	3,025	23,063	0	0	37,952	40,542
2004	43,465	64,102	4,841	2,062	29,535	0	0	50,687	25,556
2005	21,999	50,600	6,085	926	60,147	0	0	47,078	24,418
2006	30,358	53,338	9,309	1,154	102,782	0	0	55,847	47,033
2007	33,121	56,123	9,501	1,382	84,356	0	0	56,257	66,036
2008	36,621	57,047	13,358	1,216	94,765	0	0	66,935	59,262
2009	69,697	71,341	22,964	3,975	539,106	0	0	109,889	155,151
2010	89,544	104,038	32,843	8,259	1,228,086	0	0	108,024	59,941
2011	125,481	119,074	22,564	9,060	1,290,933	0	0	118,986	98,087
2012	49,185	58,052	3,707	3,366	2,019,757	3,392	55,740	393,229	0
2013	46,560	59,148	2,824	6,644	2,263,635	3,292	95,778	396,821	0
Total	659,006	914,637	143,402	46,315	7,772,543	6,684	151,518	1,454,434	724,132
%	5.55	7.70	1.21	0.39	65.47	0.06	1.28	12.25	6.10

[a]New categories for FIT after 2012.
Source: Tourism Bureau (2014b).

TABLE 6.5 Travel Information of Chinese Tourists

Sources	No. of Trips to Taiwan	Purpose	Average Length of Stay	Tourism Information Sources
Lin (2003)	Mostly first trip	Mostly cultural exchange and then business travel	Most stayed 10 days, and then shorter than 9 days	N/A
Lin (2003)	Mostly first trip	Mostly sightseeing, and then attending the conferences and business travel	Most stayed 5–7 days and then 8–15 days	Mostly from newspapers and then from TV
Chen et al. (2004)	Mostly first trip	Mostly tourism, and then attending conferences and visiting friends	N/A	Mostly from tourism bureau and then from news and advertisement
Yang (2008)	Mostly first trip	Mostly enjoying a vacation and then participating incentive tour	Most stayed 7 days	Mostly from travel agencies and then from relatives or friends
Wu (2008)	Mostly first trip	N/A	N/A	Mostly from internet and travel agencies and then from TV.
Teng (2009)	Mostly first trip	N/A	Most stayed 6–9 days and then 10–14 days	Mostly from travel agencies and then from relatives and friends
Huang (2010)	Mostly first trip	N/A	N/A	Mostly from relatives and friends and then from TV

TABLE 6.6 The Expenditures of Chinese Group Tourists in Taiwan Per Person-day[a]

Year	Expenditure Costs (In Hotels)	Catering Costs (Outside Hotels)	Trans-portation (In Taiwan)	Enter-tainment	Shopping	Others	Total
2008	95.04	9.18	24	34.14	131.36	1.28	295
	32.22%	3.11%	8.13%	11.57%	44.53%	0.43%	100.00%
2009	68.9	14.78	14.62	17.29	115.31	1.21	232.11
	29.68%	6.37%	6.30%	7.45%	49.68%	0.52%	100.00%
2010	47.31	17.05	16.63	21.13	142.37	1.74	246.23
	19.21%	6.93%	6.75%	8.58%	57.82%	0.71%	100.00%
2011	37.9	18.89	15.95	26.26	163.91	3.44	266.35
	14.23%	7.09%	5.99%	9.86%	61.54%	1.29%	100.00%
2012	32.11	23.89	22.57	20.22	164.95	3.58	267.32
	12.01%	8.94%	8.44%	7.56%	61.71%	1.34%	100.00%
2013	27.78	28.4	37.77	7.55	160.59	2.34	264.43
	10.51%	10.74%	14.28%	2.86%	60.73%	0.88%	100.00%

[a]Unit: USD.
Source: Tourism Bureau (2014a).

6.3.3 SATISFACTION AND WILLINGNESS TO REVISIT

Studies indicate that Chinese tourists had really high levels of satisfaction about their visits to Taiwan (Lin, 2003). They were most satisfied with the quality of tourism services, including the tour guides with professional knowledge and high-quality services (Table 6.7). They also liked the hotel service since they always provided the service or resolved problems for the customers as soon as possible (Hsu, 2008; Li, 2005; Teng, 2009). Most Chinese tourists were impressed by Taiwanese peoples' passion and hospitality (Lin, 2003). Since Chinese tourists had high satisfaction with Taiwan, it is not surprising that they would have higher willingness to revisit (Fornell, 1992; Selnes, 1993; Chen et al., 2004; Teng, 2009). However, tourism satisfaction may be affected by motivations of traveling, destination images of Taiwan, or the experiences in Taiwan (Chen, 2010; Huang, 2010; Li, 2011).

Studies have also revealed that when Chinese tourists stayed in Taiwan longer and with higher expenditure, they had lower willingness to revisit (Chen et al., 2004). These Chinese tourists would more likely revisit Taiwan if the quality of lodging, tour package design, restaurant sanitation, and products in stores were improved (Chen, 2010; Hsu, 2008; Tung and Chang, 2013).

TABLE 6.7 The Studies and Conclusions of Chinese Tourists' Satisfaction in Taiwan

Sources	Key Findings
Lin (2003)	The tourists from China most satisfied with "the tour guides' attitude" and "the friendliness of Taiwanese people," especially the first-time visitors.
Chen et al. (2004)	The Chinese tourists were most satisfied with "the hospitality of local residents," "service attitude of local tour guide" and "facilities in hotels."
	When the tourists had higher satisfaction, they would have higher willingness to revisit Taiwan. A better standard of food and beverage quality was provided, and the higher intention of revisiting willingness was shown.
Li (2005)	The Chinese tourists had higher satisfaction with "service attitudes of tour guides," "professional knowledge of tour guides," and "interpretation skills of tour guides."
Hsu (2008)	The tourists from China were most satisfied with "the attitude of the tour guide," "the quality of the interpretation," and "the convenience in Taiwan."
Teng (2009)	The tourists from China were most satisfied with the quality of travel service, such as the politeness, trustfulness, and professionalism. When the tourists had higher satisfaction, they had higher willingness to revisit Taiwan.
Chen (2010)	The Chinese tourists were most satisfied with "the quality of accommodation," "the quality of service," and "the shopping environment." When the tourists had higher satisfaction, they had higher willingness to revisit Taiwan.

6.4 THE IMPACT OF CHINESE TOURISTS

The rapid growth of Chinese tourism since 2008 has caused several political, economic, and cultural problems. Several studies have focused on the impact of Chinese tourism development on Taiwan society. Most of the Chinese tourists provide a welcome boost to Taiwan's economy; they have overtaken the Japanese as the biggest tourist group in Taiwan (Su et al., 2012). The government estimated that Mainland tourists brought in some US$330 million a year by 2013 (Tourism Bureau, 2014a). Although the financial benefits may be considerable, the sheer number of Chinese visitors has given rise to concerns for the tourism industry of Taiwan. It is important to examine the overall benefits and impact on the tourism industry, which would aid policy establishment and management direction and also boost the tourism industry in Taiwan.

6.4.1 ECONOMIC DEVELOPMENT

Currently, there are both positive and negative opinions about the influences of Chinese tourists on Taiwan's economic development (Table 6.8). The optimists think that the Chinese tourists contribute to the demand for transportation, hotels, restaurants, travel agencies, and tickets for different activities (Wu, 2005). These would not only provide more opportunities for businesses and employment, but also promote Taiwan's economic development (Jen, 2010; Tsai, 2011). However, others consider that although Chinese tourists increase economic profit, they might slowly push other countries' visitors out of the tourism market in Taiwan. This may become a hidden worry in the long-term. Furthermore, the tour agents who can handle the Chinese tourists are oligopoly markets in both Taiwan and Mainland China, which neither increases Taiwan's competitiveness in China's tourism market nor increases the economic profits sustainably.

TABLE 6.8 The Studies and Conclusions of the Economic Impacts of Chinese Tourists on Taiwan

Types	Sources	Key Findings
Positive	Wu (2005)	Permitting Chinese tourists to travel to Taiwan is a good policy, which helps domestic airline companies, hotels, transportation, and retail industry.
	Wang and Wen (2010)	Between July 2008 and June 2010, total consumption by Chinese tourists was 59,600 million NTD, GDP increased by 109,200 million NTD, and 52,943 jobs were provided. The main beneficial industries are hotels, entertainment culture, road transport, and catering services.
	Tsai (2011)	The Cross-Straits tourism business can facilitate tourism development, and Chinese tourists can stimulate consumption.
	Chang (2012)	Permission for China tourists to travel to Taiwan has positive effects on Taiwan, such as growth of Taiwan tourists, increased domestic GDP, and providing job opportunities.
	Wang (2013)	Stock prices of tourism and construction industries may have cumulative abnormal returns due to permitting individual Chinese tourists to travel to Taiwan.

Negative	Chin (2004)	The "zero-tour fare" is a poor marketing method to attract Chinese tourists and affects tourism quality a lot.
	Fan (2010a)	The one-stop tour service may bring benefits to Chinese tourism operators; however, profits of local tourism industry may decrease in Taiwan.
	Chen (2011)	Some Taiwan travel agencies are involved in price-cutting competition.
	Chang (2012)	The Chinese tourists may have a crowding out effect on existing tourists and increase costs of tourism in Taiwanese.

Some study results have indicated the positive impacts of Chinese tourists as follows: (1) input of foreign exchange to Taiwan, (2) boosting Taiwan's economy, (3) increased employment opportunities for Taiwan's people, and (4) promoting positive interaction between Taiwan and Mainland China (Fan, 2006). Tourism survey reports show that the annual Chinese tourists' total travel expenditures in Taiwan were higher than the domestic tourists after permitting FIT Mainland tourists to Taiwan (Tourism Bureau, 2014a, 2014b). Among the Mainland tourists' expenditures, shopping expenses were the highest (54.91%), followed by hotel bills (16.58%) and local transportation (13.39%). Thus, the profits of the tourism industry such as hotels, restaurants, and activity tickets increased significantly. According to Chung-Hua Association for Financial and Economic strategies (CAFES) statistics, Taiwan's openness policy for Chinese tourists created 52,943 jobs and contributed 109,200 million NTD to the gross domestic product (GDP) of Taiwan. Many people believe that the openness policy will increase business opportunities for the travel industry and Taiwan's economic development (Chen, 2013; CIER, 2012). Although the considerable amount of Chinese tourists has strengthened the economy of Taiwan, the large amount of tourism expenditures also increased the living expenses in the tourism destinations; this may lead to negative impacts on the local economy (Chang, 2012; Fan, 2010a). Besides, some Taiwanese travel agencies tried to attract Chinese tourists by cutting prices (Chen, 2011). As time passes, these operators may suffer losses and be unable to make ends meet. As a result, they may make loans or fail to conduct sustainable operation, which may deteriorate the tourism service quality (Chin, 2004). The tourism industry cannot conduct sustainable operation and may be affected by serious shock (CIER, 2012; Wang, 2014).

Overall, permitting Chinese tourists to visit Taiwan has positive impacts on the general economy of Taiwan, including more employment opportunities and promoting Taiwan's consumer market (Wang and Wen, 2010; Wu, 2005). The tourism industry is a driver of Taiwan's economic development and service quality improvement. But, there are still several problems related to the policies toward Mainland

Chinese tourists. The current policies cannot make certain that the tourism enterprises gain profit from Chinese tourists; therefore, the travel agencies will try to sacrifice service quality for price cutting. Without satisfactory travel experiences, the Chinese tourists will not revisit Taiwan, which can cause great damage to Taiwan's tourism industries.

6.4.2 SOCIAL IMPACTS

Tourism is regarded as one of the strategies for influencing international affairs (Butler and Mao, 1996). Taiwan has been separated from the Mainland China ever since 1949, which created substantial mistrust and tension between the two sides. Over the past decade, the two governments have started to lift the ban on travel, and the tension between the two sides has gradually been relieved. The increased interaction of the people promoted mutual understanding and business opportunities between the two sides of the Taiwan Straits (Chi, 1995; Fan, 2010b). The agreement about three "direct links" signed by Beijing and Taiwan also achieved a new level of Cross-Strait stability (Chen, 2011; Yu, 2010). Mainland Chinese travelers also help to promote mutual understanding between the two sides.

Because China and Taiwan were separated for half a century, the habits and customs have become considerably different. Nationalism and prejudice also exacerbate the cultural differences. Therefore, Chinese tourists made a great cultural impact in Taiwan at the initial stage of tourism (Hsu et al., 2010). For example, Lee's (2013) study indicates that the queue-jumping behavior of many Chinese tourists has bothered a lot of the local Taiwanese. The boisterous nature of some Chinese tourists is also very different from other foreign tourists as well as local Taiwanese. As Cross-Straits tourism provides contact opportunities between Chinese and Taiwanese, mutual understanding can gradually be promoted (Tsai, 2011; Yu, 2010). Such understanding can further peacefully improve the political relations between the two peoples and governments (Kuo, 2010). China's and Taiwan's governments have also released control of tourism as a low-politics strategy on tension reduction (Yu, 1997).

6.4.3 TOURISM QUALITY

Mainland Chinese tourists have facilitated significant growth in the tourist numbers in Taiwan. To compete for the growing Mainland tourists market, Taiwanese travel agents have engaged in a price war. However, the poor quality of tours and delayed payment by Chinese agents have compressed the profits, even with the increased volume of tourists (Table 6.9). Unfortunately, the price cuts usually came at the expense of service quality (CIER, 2012), and some Chinese tourists have complained about the low service quality and held negative impressions of Taiwan (Yeh, 2009). As for the development of the Chinese tourism market, the tourism industry in Taiwan needs to endeavor to maintain Chinese tourists' satisfied experiences, as well as their willingness to revisit Taiwan (Wu, 2011).

TABLE 6.9 The Studies of the Taiwan Tourism Quality Influenced by Chinese Tourists

Types	Sources	Key Findings
Positive	Kuo et al. (2012)	After opening up Cross-Strait flights, the development of the hospitality industry improved the quality of customer service.
Negative	Wang (2009)	It is considered that the uncivilized behaviors of Chinese tourists may affect the quality of sightseeing spots and crowd out other tourist groups.
	Chen (2011)	Chinese tourists may crowd out tourists from other countries.
	Chen (2013)	The wave of the Chinese tourists caused a negative impact on resources and environment, which is a big concern in Taiwan.

After Taiwan's openness policy for Chinese tourists in 2008, Chinese tourists significantly crowded out Taiwan's international tourists from Japan and the United States (Chen, 2011). Su et al. (2012) modeled the soaring number of Chinese tourists and pointed out that the Chinese tourists would crowd out Taiwan's existing and diverse international tourists. The model indicated that the effect was a decrease of 4,084 tourist arrivals from Japan and 1,449 from the United States per month. The crowding-out effect would result in a loss of Taiwan's tourism revenue by US$5 million from Japan tourists and US$4.9 million from the US tourists (Tourism Bureau, 2014a). However, the openness policy for Chinese tourists increased Taiwan's tourism revenue of approximately US$47.7 million per month. Although the increased number of Chinese tourists brings in large tourism revenue to Taiwan in the short run, the long-term crowding-out effect of Chinese tourists should not be underestimated (Wang, 2009). To maintain the quality of tourism, Taiwan's government should actively enhance either tourism capacity or comprehensive planning; otherwise, the overloaded tourism capacity would disrupt Taiwan's tourism industry and damage Taiwan's tourism reputation in the long run in the global market.

6.5 MANAGEMENT OF CHINESE TOURISTS AND STRATEGIES

Opening up Taiwan to Chinese tourists has had a huge influence on Taiwan's tourism industry. Some researchers have analyzed and studied the current policies and regulations to understand the effects and deficiencies of the management. There are several key issues regarding the increasing tourism flows between Mainland China and Taiwan. The two governments need to cooperate to find the solutions to maintain travel quality for tourists from Mainland China to Taiwan. Taiwan is a small island with relatively limited tourism capacity. With the opening policy, there are 5,000 tourists visiting Taiwan by package tours and 4,000 tourists by FITs at the moment. The number of Chinese tourists has generated tremendous amounts of revenue and pressure in Taiwan. In the management of the exit–entry flow, the research by Ko and Tsai (2010) indicates that government authorities should sign the Cross-Strait exit–entry administration agreement for crime prevention. The governments also should establish a more comprehensive mechanism for the application and inspection of exit–entry administration, which effectively controls illegal activities of tourists on both sides. The exchanges must be carefully handled between the two sides to ensure the security without increasing mistrust relation.

After Taiwan's openness policy for Chinese tourists, Chinese visitors now account for approximately 40% of visitor arrivals to Taiwan. Taiwan's continued openness to Chinese tourists may lead Taiwan's tourism market to excessively rely on China. The study by Cheng (2008) cautioned that overdependence on the China market is risky, even without the complicated Cross-Strait political conditions. Furthermore, the tilt tourism market would be vulnerable to any changes in the Cross-Straits policy; for example, the governments of both sides terminating permission of Chinese tourists to travel to Taiwan may upset the whole tourism market of Taiwan. Some studies also indicate that Taiwan's government should participate in the Economic Cooperation Framework Agreement (ECFA) service negotiations and strive to establish Taiwan's travel agencies in China to achieve mutual benefits (CAFES and CIER, 2013; CIER, 2012). In addition, research has demonstrated that Chinese tourists significantly crowd out Taiwan's international and domestic tourists (Su et al., 2012) and the opening policy may even exacerbate the crowding-out effect. Cheng (2010) considers that the government should not overly rely on Chinese visitors and should monitor the status carefully. Dispersing tourists to different destinations may reduce the tourism pressure and impact, as well as increase tourists' satisfaction with the travel destinations and services.

In response to the increasing demand trends of Chinese tourists, Taiwan's tourism industry needs to coordinate with the relevant authorities for improving the management and service quality and to enhance the image of Taiwan's tourism (Cheng, 2010). The tourism Bureau should inspect scenic sites, hotels, and stores to make certain that tourists from all around the world are fairly treated during their stay in Taiwan. The government also could provide incentives to encourage reports of irrational low tour fares or the behavior of violating contracts by arranging extra activities not listed in the tour schedule. Due to the influx of Chinese tourists, some studies have proposed charging fees for foreign tourists to control tourist numbers at the destinations, as well as to supplement the budgets to manage environment impacts caused by the tourists (CIER, 2012).

Since Taiwan is a small island with limited natural resources, Taiwan's government should actively expand tourism capacity or adjust the pace of openness. Therefore, in addition to the improvement of service quality, local culture and festival resources must be used properly to increase the willingness to revisit Taiwan

(Cheng, 2008). Moreover, attractions could be designed to improve the activities of folk culture (Guo et al., 2006). The contents of the tourism products should be improved, and the types of tourism products should be expanded. Tourism products ought to be developed cooperatively and promoted through joint efforts, such as with film, drama, music, and other pop culture (Timonthy, 1999) to enhance the attractiveness to Chinese tourists.

6.6 FUTURE DEVELOPMENT TRENDS AND SUGGESTIONS

According to the UNWTO (2001), China is expected to become one of the largest outbound tourist generators in the world, representing 6.4% of the total market share or more than 100 million outbound travelers by 2020 (Kim et al., 2005). Chinese tourists caused an increase in international travel expenditure of 16% in the first three-quarters of 2014 (UNWTO, 2014a). To date, the number of Chinese tourists is considerable, and many tourism industries will benefit from this surge as well. Chinese tourists will become increasingly savvy, independent, and demand high-quality experience and service. After Taiwan permitted Chinese tourists to Taiwan in 2008, this policy has significantly impacted Taiwan's economy and Cross-Strait culture exchange.

A recent survey reported that Asian countries will continue to gain the most benefit from the growth of Mainland Chinese tourism (CLSA, 2014). According to the survey, the respondents who plan to travel selected Thailand, Taiwan, South Korea, and Singapore as their top destinations in the next 3 years. In face of the growing number of Chinese tourists, therefore, determining how to lift the bar in current policies and attract more Chinese tourists to visit Taiwan has become a critical topic for Taiwan's government. A continuous effort to improve political relations is the most important approach for developing cross-border tourism. Then, tourism could act as a positive force for stimulating peaceful and steady relations by increasing understanding at the grassroots level.

Taiwan remains a popular neighboring destination for Chinese tourists (for the past 6 years) because of the geographical proximity and cultural similarity. However, such advantage will eventually fade away as the Chinese tourists head off for more exotic destinations. Accordingly, Taiwan's government should endeavor to expand the list of cities for FIT from China and also repackage the tourism resources, which provide the atmosphere of home and mystery for the Chinese tourists (Huang, 2013). The home-like atmosphere and LOHAS (Lifestyles of Health and Sustainability) are appropriate images of Taiwan to be marketed for Mainland China. By emphasizing the home-like aspect of cultural proximity, Taiwan can enhance Chinese tourists' sense of familiarity as well as travel intention.

In the recent half decade, the openness policy for Chinese tourists has driven a new wave of visitors, which also degraded the quality of tourism in Taiwan. Furthermore, the relevant policies and regulations on cutthroat competition of tour fare and tour arrangement, are needed to enhance the quality. The government and tour operators in the private sector have to join hands to upgrade infrastructure and develop new tourist hotspots. The government also has to build a system of supervision at scenic spots, hotels, and stores to make certain the tourism quality in the whole industry. These tour operators should also set up self-discipline to maintain service quality and assistance in long-term healthy development. Taiwan and Mainland China should also try to make agreements to prevent cutthroat competition in operating Chinese tour groups. The pleasant tourism experiences between tourists and the local people would enable a better understanding of each other and lead to an improved relationship between both sides. In the short-term development, this openness policy has attracted fairly high numbers of Chinese tourists to Taiwan, but this caused some considerable long-term impacts (Chang, 2012; Cheng, 2008; Huang, 2013). Since the increase in international tourist arrivals to Taiwan is mainly from China, the decline of tourists from Japan and the United States indicates a potential crowding-out effect (Su et al., 2012). Although increased Chinese tourists bring large revenue to Taiwan, the quality may collapse due to exceeding Taiwan's tourism capacity. Taiwan's government should monitor the composition and demand of international tourists to provide enough service and maintain Taiwan's tourism reputation. With annual surveys and monitoring, Taiwan's government should carefully examine the opening policy for the sustainable development of the tourism industry.

REFERENCES

Butler, R. W. and Mao, B. Conceptual and theoretical implications of tourism between partitioned states. *Asia Pac. J. Tour. Res.* 1996, 1(1), 25–34.

CAFES and CIER. *The Effects of the New Tourism Law on the Mainland Torism Development.* Taipei: Chung-Hua Association for Financial and Economic Strategies, Chung-Hua Institution for Economic Research, 2013.

Chang, M.-W. Mainland Tourists on the Impact of the Development of Taiwan's Tourism. Master, Shih Hsin University, Taipei, 2012

Chen, C.-M. K. L.-J. and Hsiung, H.-L. The study about the effects of direct flight cross-strait on the international tourist hotels in Taiwan. *J. Jia-Da Sport Health Leisure.* 2012, 11(2), 84–93.

Chen, C.-Y. A preliminary study on perceptions of tourism impacts and attitudes toward inbound travel market from China to Taiwan. [A Preliminary Study on Perceptions of Tourism Impacts and Attitudes toward Inbound Travel Market from China to Taiwan]. *J. Island Tour. Res.* 2013, 6 (1), 49–71.

Chen, K.-H.; Yung, C.-Y.; Chen, I.-J. The consumer behavior and revisiting willingness for group package tourists from China to Taiwan. *J. Tour. Stud.* 2004, 10 (2), 95–110.

Chen, W.-C. There are more damages than benefits of opening policy of Chinese tourism. *Formosa Wkly.* 2011, 102, 30–32.

Chen, Y.-J. A Study on Travel Consumption Preferences, Travel Satisfaction and Revisit Intention of Mainland Tourists Traveling to Taiwan. Master, National Penghu University of Science and Technology, Penghu, 2010.

Cheng, C. *Taiwan Tourism Image and Travel Intention from Chinese People.* Master, National Dong Hwa University, Hualien, 2006.

Cheng, L.-P. *Blue Ocean Strategy of Tours to Taiwan by People in the Mainland Area.* Master, Transworld Institute of Technology, Yunlin, 2008.

Cheng, Y.-J. *A Study on the Impact and Security Policies of Mainland TouristsTraveling to Taiwan.* Master, National Penghu University of Science and Technology, Penghu, 2010.

Chi, S. *Taiwan's Economic Role in East ASIA.* Washington, D.C.: The Center for Strtegic and Internationa Studies, 1995.

Chin, T.-C. Taiwan tourism will collapse with no tour prices. *Bus. Wkly.* 2004, 864, 88.

Chou, M.-F. *The reserch of mainland residents traveling to Taiwan for medical tourism.* Master, Tamkang University, New Taipei City. 2012.

CIER. *The Effect of Increasing Mainland Tourists on Taiwan Consumer and Labor Market,* 2012

CLSA. Chinese Tourists – Exploring New Frontiers. Hong Kong: CLSA, 2014.

Fan, S.-P. *A Political-Economic Analysis on the Impacts of the Mainland Chinese Tourists Traveling to Taiwan on the Cross-strait Relations.* Taipei: Showwe Information Co., Ltd., 2010a.

Fan, S.-P. To observe the political meaning of China's policy change on Taiwan affairs from Chinese Tourist arrivals to Taiwan. [To Observe the Political Meaning of China's Policy Change on Taiwan Affairs from Chinese Tourist Arrivals to Taiwan]. *East Asia Stud.* 2010b, 41 (2), 1–40.

Fan, S. P. A study of the legal effects of Mainland China's opening up of trips to Taiwan on the two sides. [A Study of the Legal Effects of Mainland China's Opening up of Trips to Taiwan on the Two Sides]. *Pros. Q,* 2006, 7 (2), 217–267.

Fan, S. P. The study of the development of cross-strait relations after the Mainland tourists visiting Taiwan. *Pros. Explor.* 2009, 7 (1), 60–74.

Fornell, C. A national customer satisfaction barometer: the Swedish experience. *J. Mark,* 1992, 56(1), 6–21.

Foundation, S. E. *Cross-Strait Air Transport Agreement.* Taipei: Straits Exchange Foundation, 2008.

Foundation, S. E. *Cross-Strait Signed between SEF and ARATS Concerning Mainland Tourists Traveling to Taiwan.* Taipei: Straits Exchange Foundation, 2013

Guo, Y.; Kim, S. S.; Timothy, D. J.; Wang, K.-C. Tourism and reconciliation between Mainland China and Taiwan. *Tour. Manag.* 2006, 27 (5), 997–1005.

Hsu, M.-S. A study of the travel satisfaction of Mainland China tourists. *J. Sport Leisure Hospital. Res.* 2008, 3 (4), 22–42.

Hsu, M.-S. and Li, Y.-M. A study of the travel motivation of Mainland China tourists. *Bio Leisure Ind. Res.* 2008, 6(2), 131–142.

Hsu, T.-K.; Wu, R.-L.; Lin, Y.-N. The exploratory study of Taiwanese psychological impacts on tourists from Mainland China. *Fu Jen J. Human Ecol.* 2010, 16 (1), 121–144.

Hsu, T. K.; Tsai, Y. F.; Wu, H. H. The preference analysis for tourist choice of destination: a case study of Taiwan. *Tour. Manag.* 2009, 30, 288–197.

Huang, P.-H. *A study of PRC tour hroup is coming to Taiwan.* Paper presented at the Marketing'10, National Taipei University, New Taipei City, 2010.

Huang, Y.-L. Mainland China Tourists and Its Political and Economic Impacts on Taiwan (2008–2012). Master, Nanhua University, Chiayi, 2013.

Huang, Y.-P. A study of tourism motivations, tourism image, visitor satisfaction and destination loyalty of the cross-straits—sun moon lake scenic area. Master, Chung Hua University, Hsinchu, 2010.

Jen, H.-C. Open Mainland tourists to Taiwan tourism predicament and reflection. *Pros Explor.* 2010, 8 (6), 27–51.

Kim, S. S.; Guo, Y.; Agrusa, J. Preference and positioning analysis of overseas destinations by Mainland Chinese outbound pleasure tourists. *J. Travel Res.* 2005, 44, 212–220.

Ko, Y.-R. and Tsai, J.-J. The current situation and review of China's Mainland tourists to visit taiwan people crowd of border management mechanism. *J Homeland Secur. Border*. 2010, 17, 55–112.

Kuo, A.-M. The effects and prospects of opening Mainland tourists to Taiwan. *J. Hsing Wu Coll.* 2010, 47, 157–172.

Lee, C.-Y. The Research of Mainland Chinese Tourists on Package Tour in Taiwan. Master, Shih Hsin University., Taipei, 2008.

Lee, M.-L. The Profit of Tour Policy after Opening Mainland Tourist in Taiwan. Master, Chinese Culture University, Taipei, 2013.

Li, H.-H. The Study on Tourism Image, Experience Marketing, Consumer Value and Behavior Intention for China Travelers. Master, Chaoyang University of Technology, Taichung, 2011.

Li, J.-S. Personalities, Traveling Motives, and Satisfaction - A Case of Mainland Chinese and Japanese Tourists to Tainwan. Master, National Taiwan Normal University, Taipei, 2005.

Lin, C.-R. The Study of Mainland Professionals' Shopping Behavior in Taiwan. Master, National Dong Hwa University, Hualien, 2003a.

Lin, H.-W. A Study of the Relationships among Travel Participation Pattern, Tourism Image Satisfactions and Revisiting Willingness of China Travelers to Taiwan. Master, Shih Hsin University, Taipei, 2003b.

Lin, Y.-J.; Su, A.-T.; Tzeng, W.-H. Xiamen residents' familiarity with Kinmen and its effects on travel intention. *J. Island Tour. Res.* 2013, 6(2), 1–19.

National Immigration Agency, R. O. C. Entry Persons by Gender, Age, Identification. 2014 http://www.immigration.gov.tw/lp.asp?ctNode=29986&CtUnit=16677&BaseDSD=7&mp=2 (accessed Aug 31, 2014)

Selnes, F. An examination of the effect of product performance on brand reputation, satisfaction and loyalty. *Eur. J. Market.* 1993, 27 (9), 19–35.

Su, Y.-W.; Lin, H.-L.; Liu, L.-M. Chinese tourists in Taiwan: crowding out effects, opening policy and its implications. *Tour. Manag. Perspect.* 2012, 4, 45–55.

Teng, C.-P. A Study of the People Republic of China Travel Motivation Consumer Satisfaction Level and Intention Behavior. Master, National Taiwan Normal University Taipei, 2009.

Timonthy, D. J. Cross-border partnership in tourism resource management: international parks along the US-Canada border. *J. Sustainable Tour.* 1999, 7 (3/4), 182–205.

Tourism Bureau, R. *2013 Annual Survey Report on Visitors Expenditure and Trends in Taiwan.* Taipei: Tourism Bureau, ROC, 2014a.

Tourism Bureau, R. Annual Statistical Report on Tourism 2013. In R. Tourism Bureau, Ed. Taipei: Tourism Bureau, ROC, 2014b

Tsai, L.-C. A Study of Impact on Tourism Internationalization after Lifting the Ban on Mainland Chinese visit to Taiwan. Master, Ming Chuan University, Taipei, 2011.

Tung, J.-J. and Chang, H.-C. The study of the relationship between service recovery and revisit intention: a case of Mainland tourists in Taiwan. *Market. Rev.* 2013, 10 (2), 211–233.

UNWTO. Tourism 2020 Vision. MAdrid, Spain, 2001.

UNWTO. Internation tourism shows continued strength. *UNWTO World Tour. Barometer*. 2014a, 12, 1–5.

UNWTO. UNWTO Tourism Highlights. 2014 Edition. Madrid, Spain, 2014b.

Wang, J.-C. *The Study about the Development Trends of Mainland Tour Groups*. Taipei: Chung-Hua Association for Finacial and Economic Strategies, 2014.

Wang, P.-F. The Effect of the Announcements of Free Independent Travel (FIT) on Taiwan Tourism and Construction Stock Returns. Master, National Chung Cheng University, Chiayi, 2013.

Wang, S.-M. and Wen, P.-C. The economic benefit analysis of open Mainland tourists travel to Taiwan. *Prospect Q*. 2010, 11(3), 133–175.

Wang, Y.-F. The effects of Chinese tourists on Taiwan tourism industry. *New Century Think Tank Forum*. 2009, 45, 74–83.

Wu, H.-Y. A Study of the Taiwan Traveling Image from China Tourists. Master, National Dong Hwa University, Hualien, 2008.

Wu, M.-D. Taiwan's new business opportunities – The Chinese FIT. *Cross Straits Business Monthly* (July), 2011.

Wu, N.-C. The future of the commerce and trade between Taiwan and China under Present Chen, Shui-Bian. *New Taiwan*. 2005, 479.

Yang, C.-H. Applying the ordered probit data transformed method to measure the Chinese tourists' perception of importance degree and the satisfaction of actual experience on Taiwan tourism service quality. *Tour. Manag. Res*. 2007, 7 (1), 1–30.

Yang, M.-H. The Study of Impact of Socio-Demographics and Travel Characteristics of Tourists from Mainland China on their Travel Satisfaction. Master, National Taiwan Normal University Taipei, 2008.

Yeh, G.-P. An interaction of tourism and politics: the study on tourism policy across the Taiwan Straits. *J. Macau Univ. Sci. Technol*. 2009, 3 (1), 66–74.

Yu, L. Travel between politically divided China and Taiwan. *Asia Pacific J. Tour. Res*. 1997, 2 (1), 19–30.

Yu, T.-T. A Study on the Influence of Allowing Mainland Tourists to Visit Taiwan on State Security. Master, Ming Chuan University, Taipei, 2010.

CHAPTER 7

MAINLAND CHINESE OUTBOUND TOURISM TO ASIA: RECENT PROGRESS

HANQIN QIU and LEI FANG

CONTENTS

7.1 INTRODUCTION

With a recorded history of over 5,000 years and a land area of 9,600,000 square kilometers, China is well endowed with its rich cultural and natural resources. However, the great potential for developing tourism was not recognized until 1978 when China's top leaders realized the importance of economic reconstruction and decided to implement open-up policy. As a result of this economic and political reform, tourism industry has been influenced significantly.

Outbound tourism is important not only for the increase of revenue, but also for widening the visions of the Chinese people. More and more Chinese citizens are now willing to travel abroad, which boosted China's outbound tourism industry at a faster rate. In addition to the support from political policy, the main reason contributing to the growth can be the high economic growth rate. Figure 7.1 describes the statistics on growth domestic product (GDP) and GDP per capita from 1994 to 2013. The GDP (GDP per capita) was increased from US$559,000,000,000 (US$469) to US$9,240,000,000,000 (US$6,807) during the last 20 years. The economic boom increased consumption capacity of the Chinese people. The average living standard of the Chinese population was steadily improved, especially for the residents of large cities in coastal provinces and special economic zones such as Shanghai, Beijing, Tianjin, Jiangsu, Zhejiang, and Guangdong, where people are most likely to travel abroad.

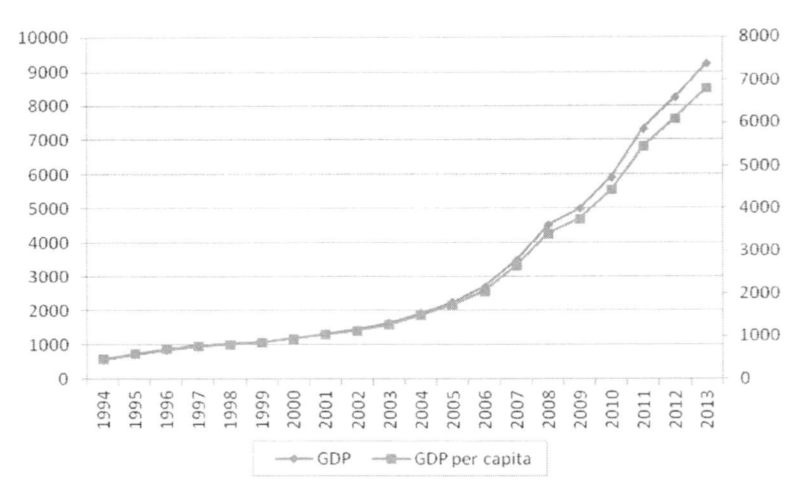

FIGURE 7.1 Statistics on GDP in Billions US$ and GDP per capita in US$ (1994–2013). (From World Bank. GDP. Available at: http://data.worldbank.org/indicator/NY.GDP.MKTP.CD and http://data.worldbank.org/indicator/NY.GDP.PCAP.CD.)

Large population size is another factor that affected the fast development of outbound tourism. A population size of over 1,300,000,000 means that even a very small percentage of population can constitute an important part of outbound market. Table 7.1 describes the China's population size and the number of outbound trips made by Chinese tourists as well as their yearly growth rates from 1994 to 2013. In general, the market size was increasing rapidly with the number of outbound trips increasing from 3,730,000 in 1994 to 98,190,000 in 2013. Even in 2008 and 2009, when the world suffered from global financial crisis and economic recession, China still saw the growing numbers of outbound travelers. In a report by the World Tourism Organization (WTO) released in 1997, it was expected that by 2020, China would be the fourth biggest outbound tourist-generating country in the world (WTO, 1997). But the factor is that China already ranked first in the year of 2013 in terms of both the number of outbound trips and outbound consumption (CNTA, 2014). The achievements suggest huge potential in China's outbound tourism market, with China's opening up to the rest of the world, increase of domestic personal income, appreciation of the RMB, and continuous improvement of the travel environment.

TABLE 7.1 Outbound Trips and Population of China (1994–2013)[a]

Year	Outbound Trips (Growth Rate in %)	Population (Growth Rate in %)
1994	3,730,000 (–)	1,198,500,000 (–)
1995	4,520,500 (21.19)	1,211,200,000 (1.06)
1996	5,060,700 (11.95)	1,223,900,000 (1.05)
1997	5,323,900 (5.20)	1,236,300,000 (1.01)
1998	8,425,600 (58.26)	1,247,600,000 (0.91)
1999	9,232,400 (9.58)	1,257,900,000 (0.83)
2000	10,472,600 (13.43)	1,267,400,000 (0.76)
2001	12,133,100 (15.86)	1,276,300,000 (0.70)
2002	16,602,300 (36.83)	1,284,500,000 (0.64)
2003	20,221,900 (21.80)	1,292,300,000 (0.61)
2004	28,852,900 (42.68)	1,299,900,000 (0.59)
2005	31,026,300 (7.53)	1,307,600,000 (0.59)
2006	34,523,600 (11.27)	1,314,500,000 (0.53)
2007	40,954,000 (18.63)	1,321,300,000 (0.52)
2008	45,840,000 (11.93)	1,328,000,000 (0.51)
2009	47,656,300 (3.96)	1,334,500,000 (0.49)
2010	57,386,500 (20.42)	1,340,900,000 (0.48)
2011	70,250,000 (22.42)	1,347,400,000(0.48)
2012	83,182,700 (18.41)	1,354,000,000 (0.49)
2013	98,190,000 (18.04)	1,360,720,000 (0.50)

[a]National Bureau of Statistics of the People's Republic of China (1994–2013).

Sources: Statistical Bulletin on National Economy and Society Development; China National Tourism Administration. 1994–2013. Statistical Bulletin on China's Tourism Industry.

China's outbound tourism has experienced three phases of development, namely Hong Kong and Macao tours from 1983, border tours from 1987, and overseas tours from1988 (Zhang and Lai, 2009). In November 1983, the outbound tourism first started in Guangdong Province, where residents were allowed to join tours to visit friends and relatives in Hong Kong and Macao. Dating from 1987, border tours began at the border between China and North Korea and were extended to Mongolia, Russia, North Korea, Kazakhstan, Kyrgyzstan, Burma, Laos, and Vietnam. The third phase began in 1988 when Chinese citizens were allowed to visit relatives in Thailand. The Chinese government then announced the Temporary Rules on Mainland Chinese Traveling to Singapore, Malaysia, and Thailand in 1990 and the Philippines in 1992. The development history of China's outbound tourism emphasizes the importance of Asian outbound destinations. Although in recent years, the increase of personal income and prolonged holidays enable Chinese citizens more likely to choose destination countries in America or Oceania, Asia is still positioned in the first place in China's outbound tourism market. In 2012, the outbound travelers to Asian destinations accounted for approximately 70% of total outbound tourists in 2012 (CEN, 2014). In addition to Hong Kong and Macao, there were 14 other Asian countries (region) among the top 25 largest outbound destinations in 2012 (CNTA, 2013). Figure 7.2 describes the geographical location of these 14 destinations. Most of the destinations were located in East and Southeast Asia, including South Korea (2,995,000 trips), Taiwan (2,630,000 trips), Japan (1,962,000 trips), Mongolia (325,000 trips), and North Korea (237,400 trips) in eastern Asia, and Thailand (2,245,000 trips), Cambodia (1,845,000 trips), Malaysia (1,372,000 trips), Vietnam (1,340,000 trips), Singapore (1,167,000 trips), Indonesia (714,000 trips), Myanmar (546,000 trips), and Philippines (272,000 trips) in southeastern Asia. The United Arab Emirates (UAE) was located in the Central and Western Asia. Although none of the destination countries were located in South Asia, Chinese citizens are becoming more interested in "Island tours" in southern Asia such as the Maldives.

In this chapter, the focus is on Asian destination countries, and in the following sections, the progress of China's outbound travel to destinations in East Asia, Southeast Asia, South Asia, and Central and West Asia, respectively, are discussed. It should be noted that although Hong Kong and Macao are two largest outbound tourism markets, these two outbound destinations are excluded in this chapter. Taiwan will also be discussed only briefly. These three markets are discussed in detail in other chapters of this book.

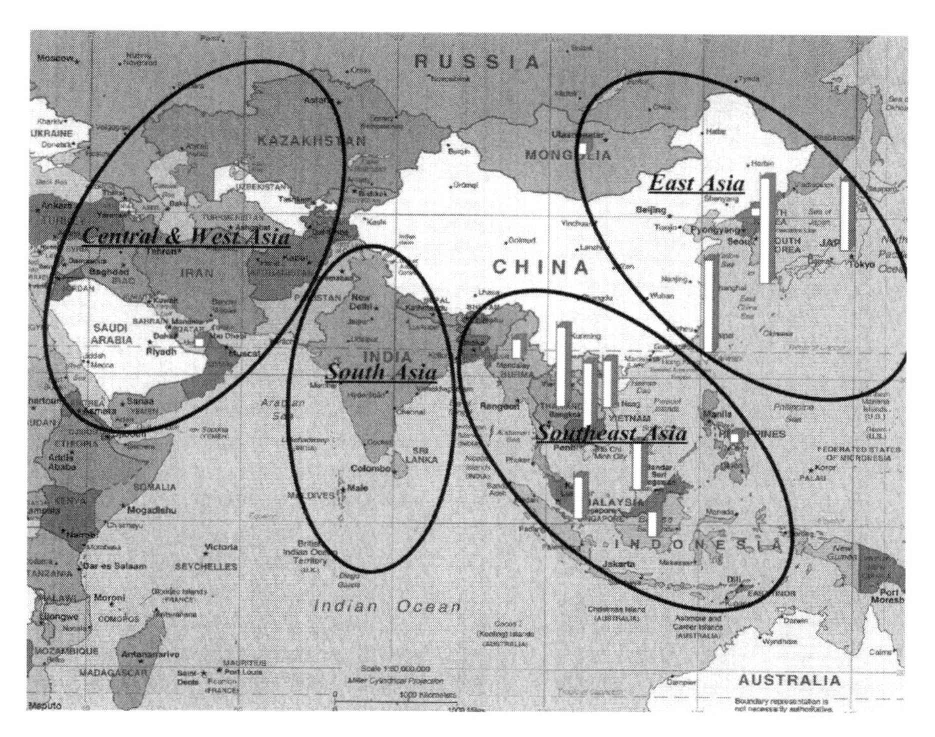

FIGURE 7.2 Top 25 Asian Destinations Receiving Chinese Outbound Tourists.

7.2 EAST ASIA

Figure 7.3 describes yearly statistics on Chinese outbound trips to Asian countries/areas including Japan, Taiwan, Mongolia, Democratic People's Republic of Korea (North Korea), and Republic of Korea (South Korea) from 2000 to 2012. Dating from 1987 when China started the border tours in Dandong City of Liaoning Province, North Korea was the first country as China's outbound tourism destination. At that time, Dandong people were allowed to visit Sinuiju for one day. However, the number of trips for North Korea was relatively small when compared with the other two Eastern neighboring countries, Japan and North Korea. Nonetheless, due to the open policy of North Korea in recent years, the number of Chinese citizen going to North Korea has risen steadily. In 2012, 237,000 trips were made by the Chinese people traveling to North Korea, ranking among the top 25 outbound tourism destinations. Mongolia is famous for the Gobi Desert and Mongolian-Manchurian grassland, as well as its culture such as the traditional Mongolian musical instrument morin khuur and the main national festival Naadam. Although only a small increase in the number of Chinese tourists can be observed in recent years owing to the shortage of tourism infrastructure and services, Mongolia was still among top 25 China's outbound destination markets. Japan, South Korea, and Taiwan, as three of the most important developed neighbors of the Mainland China, are preferred by Chinese citizens as major outbound tourism destinations, partly due to the low travel expense, advanced modern civilization of the countries (region), and similar origins of the cultures between China and these destinations.

As a developed country, Japan is characterized by its beautiful natural and cultural attractions such as Fuji Mountain, Hakone and Kiyomizu Temple, and diversity of commodities. In 2008, consumption per capita of Chinese visitors to Japan reached 160,000 Yen, ranking at the top of Japan's inbound tourists (around double that of visitors from other countries). At the beginning of the new century, digital products were the most favored by Chinese visitors. In recent years, clothing and cosmetics have become more attractive. Due to depreciation of the Yen, the decline of traveling expense in Japan and the reduced cost of purchasing luxury goods attracted more young Chinese citizens. Japanese comics and cartoons play an important role in the daily life of young Chinese during their childhoods. Many young people travel to Japan because of "cartoon pilgrimage". They go to Akiba and Ura-Harajuku for shopping, visit Big Sight for anime fair festival, and visit Mitaka Sum Museum and Osamu Tezuka Museum for saluting Hayao Miyazaki and Osamu Tezuka. In addition to Tokyo and Osaka, which are two major destinations for Chinese visitors, Kansai area, especially Kobe, has recently received more and more attention from Chinese tourists. Moreover, the famous Chinese film *If You Are the One* enables Hokkaido to be more widely known to Chinese people. More Chinese citizens, especially the youth, began to choose Hokkaido as their outbound tourism destination. It should be pointed out that the number of trips made by Chinese tourists to Japan fluctuated by year (see Figure 7.3). The main reason could be the problems of territorial disputes and Japan's attitudes toward Japan's invasion of China during the World War II.

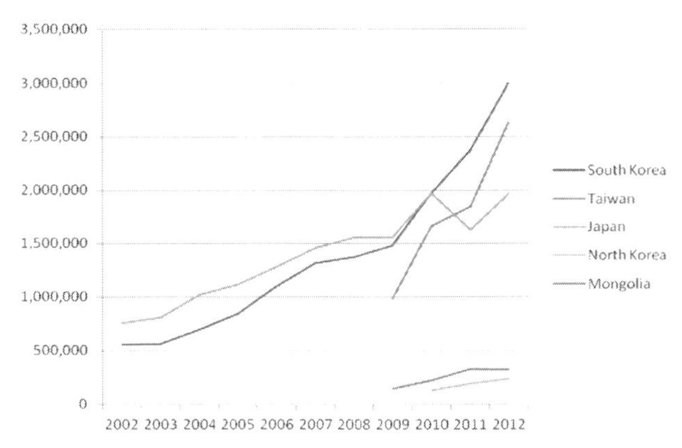

FIGURE 7.3 Statistics on Chinese Outbound Tourists to East Asian Countries (2002–2012).

In 2010, top leaders of China and South Korea made a series of agreements, such as "Tourism Year" and short-term visa-free travel. In addition, more effective measures were taken by relevant sectors of both countries such as increasing air capacity and opening direct flights. These advantages have already made South Korea one of the most popular outbound tourism destinations of China. According to *Travel to South Korea 2014* issued by Ctrip (2014), in 2013, there were nearly 4,000,000 Chinese visitors to South Korea, and the statistics in the first half of 2014 show that the growth rate reached 40% of that in the first half of 2013. It is expected that the number of Chinese visitors to South Korea will become the largest China's outbound tourism market. Seoul, Jeju Island, and Gangwon-do are three major destinations. Among the tourists, 80% were born post-1980s and 90s, and approximately 70% were female. This is partly due to the influences of South Korean TV dramas which are favorites of many Chinese young ladies.

7.3 SOUTHEAST ASIA

Southeast Asian outbound tourism started at the beginning of the third phase of China's outbound tourism (overseas tours). The beautiful natural and cultural attractions, acceptable expense, and adjacency to China attracted a large number of Chinese citizens who were eager to travel abroad. Among all Southeast Asian destinations, Singapore, Malaysia, and Thailand were regarded as "Golden Triangle". The Southeast Asian market reached its heyday at the end of the 20th century. From Figure 7.4, which describes the numbers of China's outbound trips to Southeast Asian countries from 2000 to 2012, one may observe that the numbers were fluctuating violently. The Southeast Asian tourism market was more likely to be influenced by internal or external factors. First, malignant competition led to many problems. For instance, the travel agencies forced customers to purchase products and local merchants swindled the visitors. Second, more countries, particularly European countries, opened their doors for Chinese tourists. This allowed more outbound tourism destination choices for Chinese citizens. With increasing personal income, the Chinese people have larger consumption capacity. More Chinese tourists, especially young people, would choose European or American countries as their destinations. In addition, the tourism perception of Chinese people has changed. "Leisure tourism" has become popular among the Chinese youth. The tight itinerary of the traditional "Golden Triangle" could not meet the requirements of many Chinese tourists. Third, disasters, economic crisis, and political turmoil negatively influenced the tourism market. A typical example is the tsunami at the end of 2004, which stopped the pace of Chinese tourists visiting Southeast Asia. Although only Phuket Island and Maldives were hit severely, the disaster not only negatively impacted the tourism markets of the two countries, but also some other Southeast Asian countries. In addition to natural disasters, the global financial crisis in 2008 also imposed negative influences on Southeast Asian tourism market. It can be observed that in 2009, the numbers of tourists declined for most Southeast destinations. Moreover, political and safety problems led to the fluctuation of China's outbound travel to Southeast Asian countries, such as the disputes on South China Sea between China and some countries in the Association of Southeast Asian Nations, the political turmoil of Thailand, anti-China events in Vietnam, incidents of kidnapping Chinese tourists by Abu Sayyaf Organization, and hijacking Hong Kong citizens by a former police officer in Philippines.

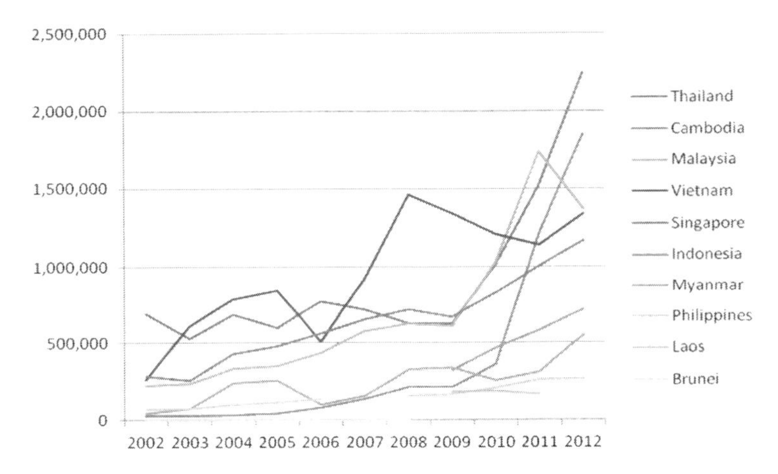

FIGURE 7.4 Statistics on Chinese Outbound Tourists to Southeast Asian Countries (2002–2012)

Facing these problems, Southeast Asian countries started to revive their tourism industry. They adjusted their strategies and developed new tourism products. China's outbound tourism to these countries started to change since the mid-2000s. On the one hand, for Singapore, Malaysia, and Thailand, the focus of the promotion is no longer the low expense of the package but the individual characteristics. The disintegration enables the three countries more concerned with the development and improvement of their local tourism products, which reinforced the recovery of their tourism markets. The most successful winner is Thailand. Since 2010, the number of trips made by Chinese tourists to Thailand has been increasing at a rapid pace (see Figure 7.4). In particular, it benefited from the Chinese Film *Lost in Thailand* released at the end of 2012. By the end of 2013, it had become the most popular overseas tourism destination. With more convenient visa policies, Thailand will continue its growth in Chinese outbound tourism market. Compared with the other two countries, the tourism resources of Singapore are relatively limited. In dealing with this situation, Singapore collaborated with Indonesia and the package of "Singapore and Indonesia" sold well in China's travel agencies, which increased the numbers of Chinese tourists since 2010. Due to incidents such as insulting Chinese customers and hijacking Chinese tourists in Malaysia, the number of Chinese tourists to Malaysia declined after 2-year growth. In 2014, the disappearance of Malaysian flight and crash events significantly and negatively affected the Malaysian tourism market. Nevertheless, Malaysia is still a favorite travel destination for Chinese tourists, particularly Sabah, which is among the most popular resorts to Chinese youth.

On the other hand, the disintegration of "Golden Triangle" brought benefits to tourism markets of other Southeast Asian countries. Chinese young customers' preference for Bali Island accelerated the rise of Chinese visitors to Indonesia. Since 2010, the number of Chinese tourists has increased steadily. According to China News Agency (2014), there were 750,000 Chinese tourists visiting Indonesia in 2013, and it is expected to continue to increase due to good relations between the two countries. During October, 2013, when China's President Jinping Xi visited Indonesia, the two countries signed an agreement on China–Indonesia collaboration on tourism, which laid solid foundation for enlarging Indonesia's tourism market toward Chinese visitors. However, it is also mentioned in the report that the development of the market is constrained by the public infrastructures and limited number of professional translators. The government of Indonesia has realized such problems and made effective countermeasures to attract more Chinese visitors. Chinese young people could hardly resist the lure of Boracay Island of Philippines. However, the image of Philippines as a tourism destination is badly influenced by its chaotic political conditions. Owing to the safety problem, the number of Chinese citizens visiting Philippines kept stable despite attractive resorts. The rapid increase of the number of Chinese tourists traveling to Cambodia was benefited from a range of successful measures targeted at Chinese people. With a closer relationship between China and Cambodia, the tourism units of the two countries cooperated well to promote the tourism image of Cambodia. The promotion not only focused on the well-known Angkor Wat, but also focused on other attractions such as Khmer dance and beach sights. Meanwhile, the Cambodian government offered preferential treatments to Chinese group tours. The implementation of the Thailand–Cambodia single visa regulation also facilitates Chinese tourists to visit the two countries. According to China News (2012), to attract 1 million Chinese tourists by 2020, the Cambodian government will take more measures on direct flights, hotels, restaurants, and public services to improve the quality of tourism products and services. The number of Chinese tourists to Vietnam varies greatly in recent years, largely due to the maritime disputes between the two countries. In particular, the anti-China events in 2014 dramatically narrowed the Chinese market.

Disintegration does not mean disappearance. The "Singapore, Malaysia and Thailand" package will still exist and play an important role in group tours because of the huge potential of middle-or-low-income Chinese outbound tourists who have strong desire to travel abroad but are constrained by traveling expense. Meanwhile, more upmarket tourism products will be designed by Southeast Asian countries to satisfy the needs of big Chinese spenders.

7.4 SOUTH ASIA

Disasters negatively impact tourism markets. On December 26, 2004, following the 2004 Indian Ocean earthquake, the Maldives were devastated by a tsunami. The damage had continuously influenced the tourism market until late 2000s when the tourism industry was recovered with reconstruction of the country. By the end of 2007, there were 92 resorts in Maldives. Visitors to Maldives do not have to apply for a visa pre-arrival, regardless of their country of origin, provided that they have a valid passport, proof of onward travel, and the money to be self-sufficient while in the country. Despite the heavy expense of visiting Maldives, the number of China's outbound trips to Maldives went up at a rapid speed in the last five years (see Figure 7.5). This is mainly because of the increase of consumption capacity of Chinese citizens, particularly Chinese young couples who love island tours for their honeymoons. Thus far, China has become the largest origin of inbound tourism market for Maldives. In 2012, nearly 230,000 trips were made by Chinese tourists visiting Maldives.

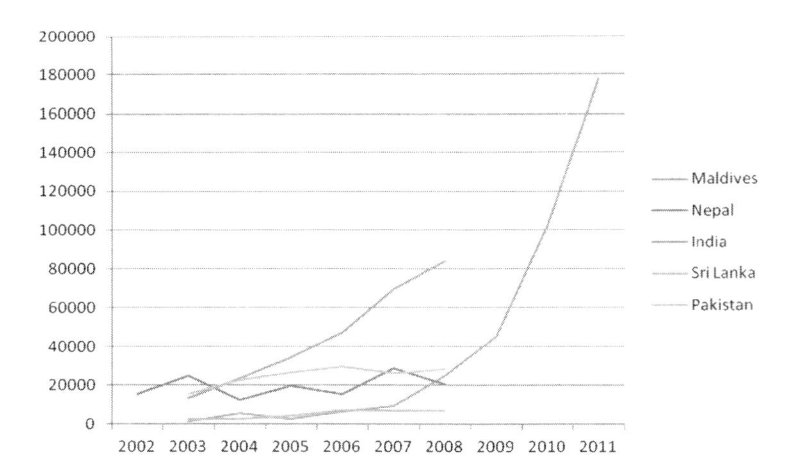

FIGURE 7.5 Statistics on Chinese Outbound Tourists to South Asian Countries (2002–2011).

India is well known for its ancient civilization and Buddhism. Many Chinese people are very much interested in Buddhist culture, and some are even Buddhists. In China, nearly all post-75s, post-80s, and post-90s generations grew up with Chinese TV series "Journey to the West" based on one of the Four Classical Chinese novels. The drama helps the Chinese to better understand India as the origin of Buddhism. In recent years, the number of Chinese tourists to India has risen steadily. This can be attributed to the efforts made by both countries, such as "China-India

Tourism Year", establishment of Chinese Tourism Office in India and Indian Tourism Office in Beijing, and promotions including Indian tourism CDs and maps in Chinese. However, the incidents of rape raised serious concerns about the safety of Indian social environment. This has brought bad effects on the tourism market, especially a negative influence on individual female visitors.

Top-level visits and friendly relationship between China and Nepal, Sri Lanka, and Pakistan promote their local tourism. Among the three countries, the safety problem has negatively influenced the Chinese outbound travel to Pakistan, while Nepal, which is located in the south of Himalaya Range with unique natural scenery, as well as rich cultural and religious heritages, has witnessed a growing trend of the Chinese market. According to Xinhua News (2014), 109,300 Chinese people chose Nepal as their first stop of outbound tours in 2013. During the first half of 2014, the number of Chinese tourists was 22.41% more than that of 2013. The boom of the Chinese market is benefited from the film *Up in the Wind* released in December 2013, and aviation agreement signed in February 2014 increased the weekly number of flights from 14 to 56. It is expected that in the near future, the number of Chinese visitors will increase to 250,000 in each year. Sri Lanka has been among popular "Island Tours" destinations in recent years. CCTV news (2013) reported that there were 16,582 Chinese visitors to Sri Lanka in 6 months of 2014 with a growth rate at 40%. Recent promotion of Sri Lanka's tourism products has focused on the Chinese market such as advertising on public transport in Beijing, Shanghai, Guangzhou, and Chengdu. It is expected that the Chinese visitors to Sri Lanka will continue to grow.

7.5 CENTRAL AND WEST ASIA

The five central Asian countries (Kazakhstan, Uzbekistan, Kyrgyzstan, Turkmenistan, and Tajikistan) went through a hard time at the pioneering stage after independence. In recent years, the economics have started to recover with an average economic growth rate above 8%, which enabled tourism industry to develop at a greater speed. Resulting from closer relationship with China (The Shanghai Cooperation Organization founded in 1996), a growing number of Chinese tourists nowadays visited the continent in the way of overseas tours instead of border tours. Among the five countries, Kazakhstan ranked top in terms of tourism development level accommodated the largest number of Chinese tourists. In 2010 and 2011, the numbers of trips reached 135,300 and 158,500, respectively (see Figure 7.6).

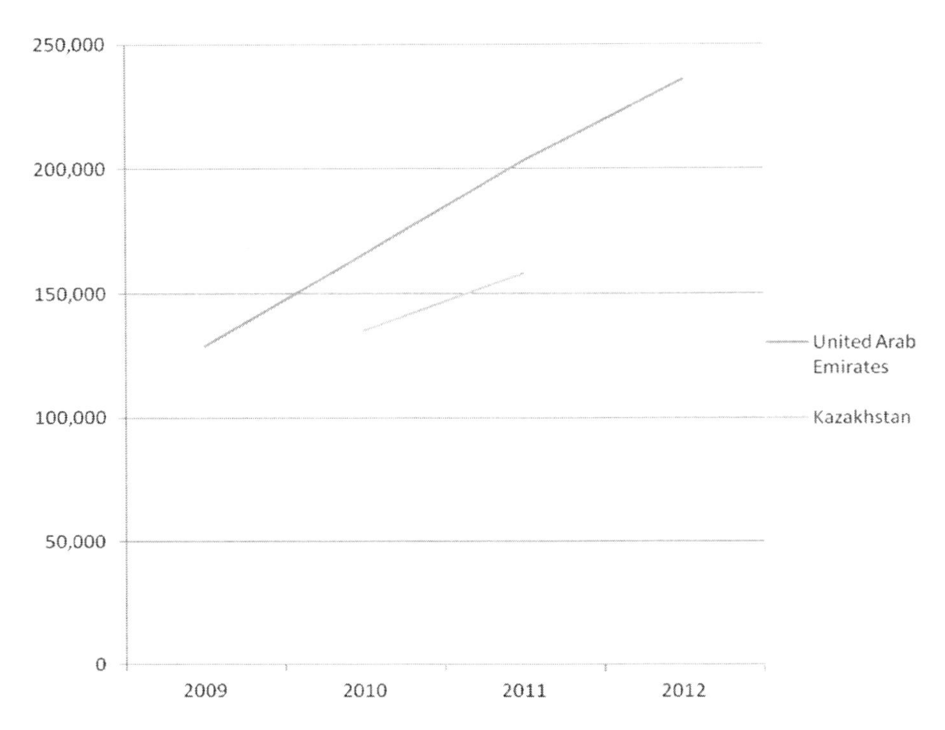

FIGURE 7.6 Statistics on Chinese Outbound Tourists to Central and West Asian Countries (2009–2012).

However, there were still many problems that negatively influenced the Chinese market for the five countries, such as shortage of tourism infrastructures, unprofessional tourism workers, and low service level. Nevertheless, the potential of tourism development in these countries is huge. First, Central Asia stretches from the Caspian Sea in the west to China in the east and from Afghanistan in the south to Russia in the north. The locational advantage enlarges their tourism resources since Chinese tourists can visit Russia, Ukraine, Belarus, Poland, Germany, and other European countries through these five bridges. In addition, the five central Asian countries are well known for their rich natural and cultural tourism resources. For instance, there are many cities with hundreds of architectural monuments from various epochs. By visiting those which have been preserved, it is possible to glimpse pages of history. Water and mountain resources such as Balkhash Lake, Caspian Sea, Communism Peak, and Val R Bbu Valley are suitable for developing special tourism products such as climbing, skiing, hunting, hiking, horseback riding, science exploration, and bicycling tours. Moreover, The Great Silk Road is a rich tapestry of tourism destinations and products based on the unique and outstandingly rich heritage, nature, and traditions of the dozens of distinct histories, people, and cultures all along the timeless route. In June 2014, the United Nations Educational, Scientific and Cultural Organization (UNESCO) designated the route as a World Heritage Site, which is expected to attract more Chinese tourists to visit the 5 countries.

The world's earliest civilizations developed in West Asia, where several major aquifers provide water to large portions of the region. The abundant cultural and natural tourism resources appeal to people all around the world. In recent years, the number of Chinese tourists to West Asian countries has been increasing. Taking Turkey as an example, in 2013, approximately 138,900 trips were made by Chinese visitors, 20% more than that in 2012 (CCDY News, 2014). However, political and social unrests always form a monolithic block for widening the tourism market. Hence, Chinese tourists would rather prefer relatively safe places such as Dubai in the United Arab Emirates (UAE). It can be observed from Figure 7.6 that the number of China's outbound tourists to UAE increased dramatically in recent years. According to Xinhua News (2013), the Tourism Officer of Dubai highlighted the importance of Chinese visitors. The statistics showed that the number of Chinese visitors in 2012 was 26% more than that in 2011. During Spring Festival in 2013, 60% of the customers in Arabia De La Tour Hotel (a well-known 7-star hotel in Dubai) came from China. Spending holidays in super-level hotels on the Palm Jumeirah has become the most fashionable way for the rich people of China. It was expected that Chinese visitors would play a key role in achievement of the goal that Dubai's inbound tourists in 2020 would double the number in 2010.

Like Central Asia, West Asia has advantage in its geographical location. It is located directly south of Eastern Europe. To the north, the region is delimited from Europe by the Caucasus Mountains; to the southwest, it is delimited from Africa by the Isthmus of Suez; while to the east, the region adjoins Central Asia and South Asia. At the moment, the interest of the world is being drawn to the great transcontinental routes of the ancient world. If the safety can be improved in this region, there must be influx of tourists from all over the world.

7.6 CONCLUSIONS

China's outbound tourism market to Asian destination countries is quite huge, but there is great variation by time and location among Asian destinations. The main reasons can be concluded as follows:

(1) Geographical Location

Geographical location is closely related to travel time and travel expense. On one hand, most Chinese tourists are not very rich. The travel expense is still an important factor for choosing travel destinations; on the other hand, although Chinese government has made great efforts to guarantee the annual leave of Chinese people, it is still not easy for Chinese citizens to obtain a long vacation. Hence, adjacent Asian countries have more advantages than those far from Mainland China.

(2) Open Policy and Diplomatic Activities

On the one hand, China signed the ADS agreement with more and more foreign countries so that Chinese people nowadays have more and more choices for outbound tourism destinations. On the other hand, many countries have realized the importance of Chinese market. They simplified the visa application procedure or implemented visa-on-arrival policy or even granted visa-free travels. Diplomatic activities among top leaders of China and other countries suggest good relationship between the two countries. These activities also included a series of agreements on development of bilateral tourism. For instance, "Tourism Years" organized by China and other Asian countries such as Korea and India enlarged the tourism markets in both countries.

(3) Film and Television Contents

Films and TV dramas play an important role in Chinese people's daily lives. Many Chinese citizens are willing to visit those places that look attractive in films or TV drams. Typical examples include *Lost in Thailand* shot in Thailand and *If You Are the One* delineating the beautiful scenery of Hokkaido. The two films positively affected China's outbound travel to Thailand and Japan.

(4) Safety of the Destination

Bad news on tourism such as incidents of kidnapping in Philippines and rape in India traveled fast across the world, which damaged the tourism images of these countries. Local political conflicts such as the political dispute in Thailand, natural disasters like the tsunami of Indian Ocean in 2004, or even air crashes such as Malaysian Airlines could significantly and negatively impact the tourism markets.

(5) Transforming of Chinese Tourists' Perception on Tourism

With the social and economic development of China and improvement of the quality of Chinese tourists, the traditional way which is characterized by tight itinerary has no longer satisfied Chinese tourists. In recent years, the annual growth rate of the numbers of Chinese outbound tourists organized by travel agencies was far smaller than that of independent travelers. It partly reflects that Chinese tourists enjoy the "freedom" tourism nowadays. More Chinese people are likely to choose destinations that can meet their requirements such as high-quality and leisure travel. The transformation of the tourism perception had great negative influences on the destination market such as "Singapore, Malaysia and Thailand", but imposed positive impacts on island destinations such as Maldives and Sri Lanka.

It is believed that China's outbound travel market will continue to grow and China's outbound travelers will become a major source market for many countries in next decade, provided that economic growth in China continues and Chinese government does not limit outbound tourism. The geographical advantages and variety of tourism products developed by Asian countries provide Chinese tourists with a lot of choices. More Chinese citizens will prefer individual visits rather than group tours. While a larger number of Chinese tourists would like to choose destinations among Western countries, Asian countries (regions) will continue their attractiveness to Chinese visitors. The largest Asian markets will still be those in East and Southeast Asia, mostly due to the locational advantages and variety of luxury commodities. In addition, island tours in Southeast and South Asian countries have continuous appeal to Chinese tourists because the Chinese youth are more interested in leisure tours which are characterized by beautiful sunshine and beach.

The extent to which the various Asian markets can be grown depends, as discussed above, on a range of factors. There is still much room for improvement for both China and Asian destinations. For instance, more open policies can be implemented by China to develop the outbound tourism. The destination countries can try their best to deal with political and social problems to guarantee the safety of tourists. The quality of tourism services need to be improved to satisfy the needs of customers. Nevertheless, Asia will continue to be in first place in China's outbound market over a long period.

REFERENCES

China News Agency. Indonesia's desire for Chinese tourists. 2014. http://www.chinanews.com/sh/2014/07-31/6445185.shtml (accessed Aug 4, 2014).

China News. Cambodian tourism target 1 million Chinese tourists in 2020. 2012. http://www.chinanews.com/hr/2012/05-02/3859980.shtml (accessed Aug 4, 2014).

CCTV News. Increasing Chinese tourists to Sri Lanka. 2013. http://travel.cntv.cn/2013/08/08/ARTI1375928691660531.shtml (accessed Aug 4, 2014).

CCDY News. Increasing Chinese tourist to Turkey. 2014. http://www.ccdy.cn/lvyou/zixun/201402/t20140221_872860.htm (accessed Aug 4, 2014).

China National Tourism Administration (CNTA). 2013. http://www.cnta.gov.cn/html/2013-9/2013-9-12-%7B@hur%7D-39-08306.html (accessed Aug 4, 2014).

China National Tourism Administration (CNTA). Release of "Report of Development of China's outbound tourism 2014". 2014. http://www.cnta.gov.cn/html/2014-6/2014-6-11-9-34-96012.html (accessed Aug 20, 2014)

China Economic Net (CEN). China's outbound tourism market in 2014. 2014. http://fashion.ce.cn/news/201407/03/t20140703_3088110.shtml (accessed Aug 21, 2014).

Xinhua News. Increase of Chinese visitors to Nepal. 2014. http://travel.163.com/14/0806/19/A3060T4N00063JSA.html (accessed Aug 21, 2014).

Xinhua News. The importance of Chinese tourists to tourism development of Dubai. 2013. http://news.xinhuanet.com/2013-05/06/c_115647411.htm (accessed Aug 4, 2014).

World Bank. 2014. http://data.worldbank.org/indicator/NY.GDP.MKTP.CD and http://data.worldbank.org/indicator/NY.GDP.PCAP.CD (accessed Aug 4, 2014).

CHAPTER 8

MAINLAND CHINESE OUTBOUND TOURISM TO AUSTRALIA: RECENT PROGRESS

IRIS MAO and SONGSHAN (SAM) HUANG

CONTENTS

8.1 INTRODUCTION

China is Australia's most valuable inbound market. Australia has experienced fast international arrivals and expenditure growth from China. China will continue to be the fastest growing market among all Australia's inbound tourism markets in the next 10 years. On one hand, Australia is enjoying the growing numbers and economic contribution of Chinese visitors, on the other hand, it also faces fierce competition from other destination countries attracting Chinese tourists. In 2013, Australia ranked number 10 among all out-of-region outbound destinations for Chinese visitors, which do not include Northeast Asian destinations (Tourism Australia, 2014a). Australia's share of the Chinese outbound market has not been growing. In the marketplace, Australia is challenged to maintain its preferred destination status among Chinese outbound tourists. In the new era of China outbound tourism, with more countries joining Approved Destination Status (ADS), Chinese outbound tourism to Australia has also been greatly reshuffled. In this chapter, the historic development and update with the current profile of Chinese outbound tourism market to Australia are reviewed, academic research on this topic is analyzed, and some pressing issues on Chinese outbound tourism to Australia are presented.

8.2 MARKET STATUS OF CHINESE OUTBOUND TOURISM TO AUSTRALIA

8.2.1 APPROVED DESTINATION STATUS

The ADS system was first introduced in China in 1995. It is based on bilateral tourism agreements between the Chinese government and a destination to which Chinese tourists are permitted to undertake leisure travel in groups (Austrade, 2014). ADS restricts the destinations that Chinese nationals can travel to for leisure purposes, allowing individuals from 32 cities to travel to Hong Kong and Macau and group leisure tours to travel only to ADS-approved international destinations. For countries with ADS, a quota was established for the number of tourists permitted to visit each of the relevant countries without an invitation letter. Chinese travel agents administer this quota system and obtain passports for travelers (Austrade, 2014).

ADS was extended to Australia by the government of China in 1999. ADS allows Chinese citizens to travel to Australia on private passports for leisure tourism purposes. In 2014, Australia was one of over 140 countries with ADS status. During 2012–2013, 163,894 Chinese visitors were granted visas under the ADS scheme, accounting for 24% of total visitor arrivals from China to Australia. Australia was the first Western destination (along with New Zealand) to be granted ADS in 1999, which allowed Australia to be promoted in China as a leisure holiday destination. From 1999 to 2004, the outbound group travel to Australia only operated in three cities: Shanghai, Beijing, and Guangzhou. In 2005, the ADS system was extended to allow residents living in the nine approved regions or cities (Beijing, Shanghai, Guangdong, Chongqing, Hebei, Jiangsu, Shandong, Tianjin, and Zhejiang) to undertake ADS leisure travel to Australia. People in other regions must apply for other visas to go to Australia. ADS tours must be booked through China National Tourism Administration (CNTA) approved agents. In 2006, there were only 86 Chinese travel agents participating in the ADS scheme. In early 2014, there were approximately 1,400 agents authorized to operate group outbound travel business in China (Tourism Australia, 2014a). While tourism arrivals under ADS arrangements currently make up a large proportion of holiday visitors to Australia, there are also government- and industry-associated delegations, as well as incentive and study groups that travel on non-ADS visas following more flexible itineraries.

Even though no further restrictions are imposed on an individual Chinese passport holder to travel abroad, it is still difficult for an individual to obtain a tourist visa without assistance of travel agencies. Many destination countries require visa applicants to complete lengthy paperwork in English and even request face-to-face interviews. The ADS scheme gives ADS countries advantages to attract Chinese group tourists under ADS arrangements. However, it should be noted that more and more Chinese outbound tourists, especially those with previous outbound travel experiences, would prefer alternative travel mode other than ADS group tours, such as free-independent travel (FIT) or Free Independent Travel (FIT) and semi-FIT, which allows great flexibility in travel arrangements.

8.2.2 HISTORIC DEVELOPMENT OF THE CHINESE TOURISM MARKET TO AUSTRALIA

China is currently Australia's second largest inbound tourist market following New Zealand. The number of visitors from China totaled 709,000 in 2013, an increase of 14.5% from the previous year and over 476 times the 1,480 arrivals in 1980 (Australian Tourism, 2014a). Historically, Chinese visitor arrivals experienced a decline due to the SARS epidemic in 2003 and the economic recession in 2008, but it surged back quickly after 1 year, in 2004 and 2009, respectively. Another reason for the strong growth is that outbound tourism to Australia was allowed to operate in more regions in China than the initial 3 cities in 2005. Since then, there has been a steady growth in the number of Chinese visitors. According to the Tourism Research Australia (2013), 1,355, 000 Chinese visitors were expected to arrive in Australia in the financial year of 2022–2023, enabling an average annual growth of 7% over 10-year period. In addition, China is also Australia's largest source market in terms of total tourist spend. In 2013, Chinese visitors spent AU$4.8 billion on trips to Australia, with an average expenditure of AU$6,770 per trip (Australian Bureau of Statistic, 2014).

The Chinese tourism to Australia market is developing rapidly. The average annual growth rate for inbound visitors from China was 24% over the decade from

1995 to 2004 and 16.2% between 2005 and 2013 (Figure 8.1). Tourism Research Australia (2013) forecasted that over the next 10 years, Australia's top 5 inbound markets (namely, New Zealand, China, the United Kingdom, the United States, and Singapore) were expected to constitute 51% of the additional 30 million arrivals by 2022–2023 (see Table 8.1). China would continue to be the fastest growing inbound tourism market with an average annual growth rate of 7.1% from 2013–2014 to 2022–2023 financial years. The number of Chinese visitors will reach 1,355,000, representing a share of 14.5% of Australia's inbound market, and make China continue its position as the second largest market for Australia in terms of visitor arrivals until 2022–2023. In terms of tourism expenditure, China will continue to be the largest contributor to international visitor spending in Australia, with a projected expenditure of up to 8.2 billion dollars in 2022–2023 (Tourism Research Australia, 2013).

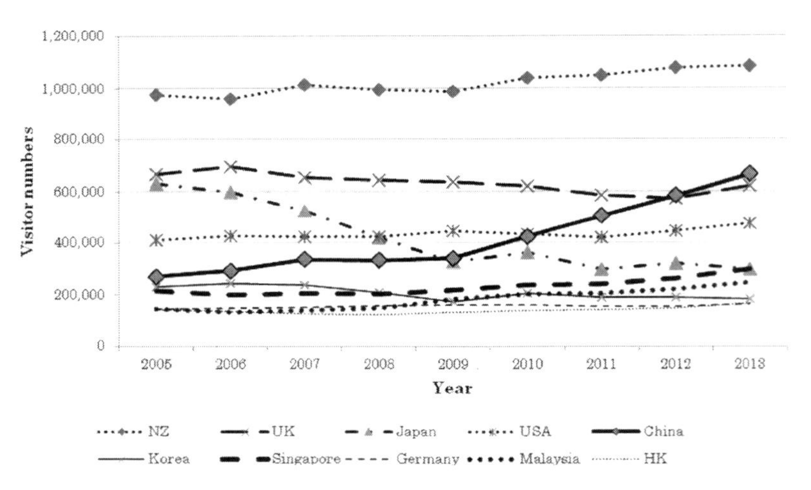

FIGURE 8.1 Australia's Top 10 Inbound Markets in 2005–2013. (From Tourism Australia, 2014a).

TABLE 8.1 Australia's Top 5 Inbound Markets in 2022–2023

Rank	Country	Visitor Arrivals	10-year Average Annual Growth (%)	Share of Total Arrivals (%)
1	NZ	1,513,000	2.4	16.3
2	China	1,355,000	7.1	14.5
3	UK	808,000	2.3	8.7
4	USA	701,000	3.6	7.6
5	Singapore	489,000	3.0	5.3

Source: Tourism Research Australia (2013).

8.2.3 PROFILING CHINESE OUTBOUND TOURISM TO AUSTRALIA

Based on the international visitor survey by Tourism Research Australia, in 2013, there were 17.5% Chinese first-time visitors and 389,950 persons or 55% of all visitors, which was greater than 331,992 persons or 53% of all visitors from China in 2012 (Tourism Australia, 2014a). In 2013, among the total arrivals, 45% were repeat visitors (see Figure 8.2), and 71% of the total arrivals (see Figure 8.2) were for leisure purpose (Tourism Australia, 2014a). In terms of visitor nights, 54% (16,020,688 nights) of the total nights was for education purpose, 25% (7,328,317 nights) was for visiting friends and relatives (VFR), 11% (3,254,446 nights) was for holiday or pleasure purpose, 3% (1,021,718 nights) was on business trip, and 6% (1,906,101 nights) was for employment (Figure 8.3). Compared with leisure travelers for holiday or pleasure purpose, travelers for the purposes of education and VFR would more likely be staying for a significantly longer period. It also explains the high proportion of repeat visitors among the total arrivals. With a large proportion of Chinese visitors traveling to Australia for education who would have higher spending levels and longer stays than other market segments, the average spend of Chinese visitors to Australia amounted $7,194 per trip, and the average length of stay was 45 nights per trip in 2013. The peak travel periods for Chinese tourists to visit Australia were October, December, January, February, July, and August; these are largely determined by school holidays and long public holiday arrangement in China (Tourism Australia, 2014a). The top five accommodation options for Chinese visitors were listed in Table 8.2.

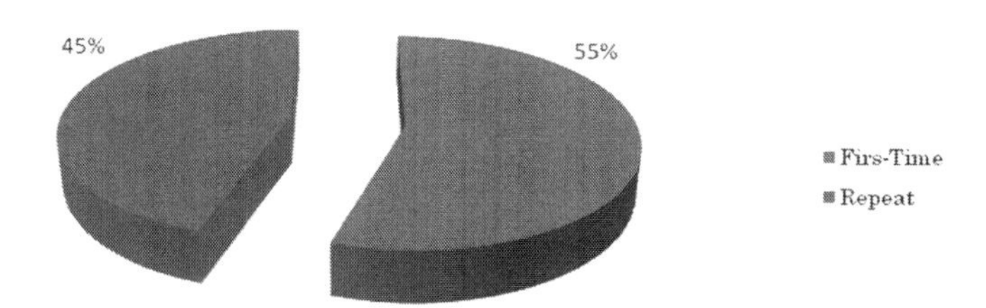

FIGURE 8.2 Composition of First-time and Repeat Visitors from China in 2013. (From Tourism Australia, 2014a).

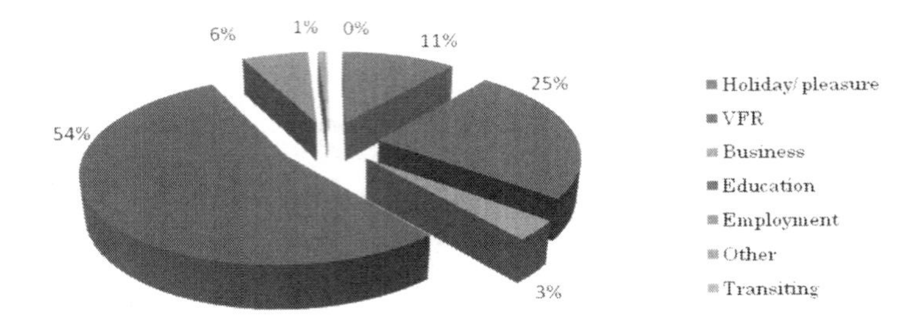

FIGURE 8.3 Composition of Chinese Visitors According to Purpose of Visit in 2013 (From Tourism Australia, 2014a).

TABLE 8.2 Accommodation Type for Chinese Visitor to Australia in 2013

Rank	Accommodations	No. of Visitors	Share %
1	Rented house/apartment/unit/flat	15,790,152	52.9
2	Home of friend or relative (no payment required)	7,444,616	24.9
3	Hotel, resort, motel, motor Inn	2,809,134	9.4
4	Homestay	1,382,162	4.6
5	Own property (e.g., holiday house) (2003 onward)	1,274,709	4.3

8.3 RESEARCH ON CHINESE OUTBOUND TOURISM TO AUSTRALIA

8.3.1 LITERATURE REVIEW ON CHINESE OUTBOUND TOURISM TO AUSTRALIA

A comprehensive literature search was conducted to survey academic journal articles related to the topic of Chinese outbound tourism to Australia at the end of November 2013. A total of 21 academic journal papers published between 2001 and 2013 were located. Based on the analysis of the authors' affiliated institutions, the top three institutions that contributed to the literatures are Griffith University in Australia (10), The Hong Kong Polytechnic University in Hong Kong, China (7), Latrobe University in Australia (6) (see Table 8.3). Of the 47 authors, 32 were from institutions based in Australia, and the rest were from Hong Kong (7), the UK (3), Mainland China (2), Taiwan (2), and New Zealand (1) (see Table 8.4). It is understandable that this body of research was mainly created by academics based in Australia; however, this topic area also attracted reasonable attention from the international research community.

An analysis of the publication outlets reveals that more than half of the papers (66.7%) were published on five journals including *Journal of Travel and Tourism Marketing* (4), *International Journal of Tourism Research* (3), *Tourism Management* (3), *Journal of Hospitality and Leisure Marketing* (2), and *Journal of Vacation Marketing* (2) (see Table 8.5). Judging by the titles of these journals, the articles should have a marketing and management orientation. This somehow reflects the current state of this body of research.

Table 8.6 provides a summary of the research findings from academic journal publications concerning the subject of Chinese outbound tourism to Australia till November 2013. This subject area covered a wide range of topics, from general demand, tourist behavior, tourist experience, marketing issues to government policy. However, Chinese outbound market is continuing to grow and is changing rapidly with new market segmentation, changing tourist behavior, and updated government policy. The current literature provides good reference for future research but still appears far from sufficient. Particular attention is needed for needs and demands of specific segments such as traditional group tourists, backpackers, FIT or semi-FIT travelers, VFR tourists, self-driving travelers.

TABLE 8.3 Institutional Contribution to Research on Chinese Outbound Tourism to Australia

Institution	No.	Institution	No.
Griffith University,	10	University of Surrey	2
The Hong Kong Polytechnic University	7	James Cook University,	2
La Trobe University	6	Victoria University	1
The University of Queensland	4	University of Newcastle	1
Edith Cowan University	2	University of New South Wales	1
Providence University	2	University of Otago	1
University of Wollongong	2	Robert Gordon University	1
University of New South Wales	2	Nielsen Company	1
Monash University	2	Zhejiang Transfar Group	1

TABLE 8.4 Regional Contribution of Research on Chinese Outbound Tourism to Australia

Region	No.	Region	No.
Australia	32	Mainland China	2
Hong Kong	7	Taiwan	2
The United Kingdom	3	New Zealand	1

TABLE 8.5 Publication Sources of Research on China Outbound Tourism to Australia

Journal	No.	Journal	No.
Journal of Travel & Tourism Marketing	4	Annals of Tourism Research	1
International Journal of Tourism Research	3	Journal of Hospitality Marketing & Management	1
Tourism Management	3	Journal of Hospitality and Tourism Management	1
Journal of Hospitality & Leisure Marketing,	2	Annals of Leisure Research	1
Journal of Vacation Marketing	2	Journal of Quality Assurance in Hospitality & Tourism	1
Current Issues in Tourism	1	International Journal of Hospitality & Tourism Administration	1

TABLE 8.6 Summary of Research Findings from Academic Journal Publication on Chinese Outbound Tourism to Australia

Authors	Major Findings
Mao and Zhang (2012)	It is found that word of mouth (WOM) is directly affected by destination satisfaction, whereas destination attachment (DA) is predetermined by the destination preference of tourists before a visit. Thus, the destination experience is the crucial factor in the spread of positive WOM about a destination. Nevertheless, to develop tourists' DA, effective loyalty schemes should be implemented through marketing efforts before they visit a destination.

Authors	Major Findings
Chang et al. (2011)	A total of 15 attributes were identified, which were classified under the following six categories: tourists' own food culture, the contextual factor of the dining experience, variety and diversity of food, perception of the destination, service encounter, and tour guide's performance.
Chow and Murphy (2011)	Four out of five (sightseeing, culture and heritage activity, shopping and dining, and entertainment) and three out of five (sightseeing, culture and heritage activity, and entertainment) travel behaviors were found significant with a combination of demographic and psychographic attributes.
Li et al. (2011)	Language was the top constraint. Four major constraints were categorized: structural (travel as an expensive activity), cultural, information and knowledge. Age and education were the most influential socio-demographic factors to influence travel constraints. Significance was found among 4 clusters regarding destination loyalty.
Chang et al. (2010)	Dominant preference of Chinese tourists toward Chinese food is due to core eating habit, appetizing assurance, and familiar flavor. Their preference to sample local food is mainly for authentic travel experience, prestige and status, and reference group influence. People who are not fastidious about food selection were influenced by their inclination toward group harmony, would compromise in supporting experience and prejudiced advocacy. A typology of Chinese dining behavior is also developed for better understanding of Chinese tourists' dining behavior.
Wang and Davidson (2010a,b)	Chinese holiday makers made greater economic contributions compared with all Chinese tourists, Chinese business travelers, and all international travelers to Australia. Psychological factors were used to verify respondents' spending power.
Wang and Davidson (2010a,b)	The study concluded that Chinese holiday travelers' total and disaggregated expenditures were associated with different sets of sociodemographic trip characteristics and psychological factors. In particular, their total expenditure in Australia was determined by their income, age, place of residence, travel party size, length of stay, and visitation to other destination.
Keating (2009)	The article provided a case study of how the Australian government has responded to concerns about unethical practices in the tourism supply chain from China to Australia.
Sparks and Pan (2009)	Subjective norm and perceived control had significant influence on behavioral intention while attitude had little impact on behavioral intention.

Authors	Major Findings
Zhang and Murphy (2009)	The study identified some significant strategic discrepancies between the travel agents and destination suppliers. It was found that these differences had become a bottleneck problem for the region's international tourism development.
Breakey et al (2008)	The differences between Chinese and Japanese tourism boom in Australia were reviewed. Marketing and cultural issues were discussed. It identified the lack of understanding of specific needs and expectations of the Chinese tourists to ensure their satisfaction and positive experience in Australia
Keating and Kriz (2008)	The authors suggested that new models should be developed to better understand tourists in their planning processing because traditional destination choice models are not adequate enough to capture the dynamics of Chinese tourists.
March (2008)	Empirical evidence was suggested to test the relationship between density of ethnic involvement and the prevalence of unethical practices in certain markets. The Australian government was reluctant to respond to the unethical practices to the local business.
Weiler and Yu (2008)	Cultural and tourist mediation theory was adopted to understand the experiences of the tourists. The findings suggested that their spatial and temporal experiences are not as strong as their memorable experiences which are affective or cognitive. The authors suggested that attention should not only be given to the physical access but also to the interaction, understanding, and appreciation.
Chow and Murphy (2007)	The results showed while this sample's travel activity preferences generally support industry and expert opinions, some differences in travel activity preferences were found, and it was noted that the Chinese market should not be treated as a single homogeneous entity.
Kewk and Lee (2007)	This article identifies current marketing themes in each group. It further elaborates comparative elements in the two markets, providing recommendations to the industry marketers.
King et al. (2006)	The article investigates the online marketing features most influencing travelers in China. Age, area of residence, type of travel website most visited, length of time using the Internet, self-efficacy, domain-specific innovativeness, and perception of the Internet all impact on Chinese lookers becoming bookers.

Authors	Major Findings
Pan et al. (2006)	Information sharing was useful despite the unavailability of some information regarding this market in some regions. There was no mechanism to identify and prioritize information about this market. Issues like lack of knowledge about infrastructure, marketing strategies, understanding and trust of the Chinese operators and extensive managerial experiences, and skills of the Australian operators needed to be highlighted.
Li and Carr (2004)	Satisfaction with a destination's attributes was related to overall satisfaction. Shopping and food were provided low satisfaction among all the attributes. The tourists perceived good value for money.
Pan (2003)	The meaning of Guanxi was broadened with the combination of work and personal relationships. Trust and relationship were the most important factors for the successful management of the Chinese outbound tourism market to Australia.
Pan and Laws (2001)	Provides analysis of the growing demand for outbound tourism from China and describes the special characteristics of the market. It analyses the significance of guanxi and the opportunities for long-haul tour operators to improve the service they provide to visiting Chinese groups.

8.3.2 SATISFACTION AND DEMAND OF CHINESE OUTBOUND TOURISM TO AUSTRALIA

Tourism Australia is the national tourism marketing organization in Australia. Given the significance of the China market, extensive research has been conducted by Tourism Australia on Chinese tourists to inform government policy and decision-making of the industry. Two of the most recent research projects are *Chinese Satisfaction Survey* and *Consumer Demand Project: Understanding the Chinese Consumer*.

Chinese Satisfaction Survey (Tourism Australia, 2014d) was conducted from January 2 to June 30 in 2013. A total of 3,606 Chinese visitors were interviewed from the Sydney, Melbourne, Brisbane, and Gold Coast international airports. It aimed to understand the drivers of Chinese visitor satisfaction in Australia and the causes of their dissatisfaction. Results show that the respondents generally had a high level of overall satisfaction and thought the trip met or exceeded their expectations toward Australia as a holiday destination. The top 3 trip attributes that most positively affected Chinese visitors' travel experience are attractions, value for money, and good shopping. Furthermore, nature-based tourist experiences had also fulfilled their expectations and positively affected their overall experience satisfaction, and thus can be a major drawcard for visiting Australia.

Chinese food experience is important for Chinese visitors. However, the quality of Chinese food, especially that provided to group tourists, needs to be improved. Experiencing quality Western food for lunch or dinner would appeal to some visitors, particularly younger visitors.

Language might not be a significant barrier affecting Chinese visitor's overall satisfaction toward Australia; however, providing Chinese language services signage in Chinese would certainly leverage visitor experience, especially for elderly visitors or FIT travelers, to increase word-of-mouth intentions. Packaged group-tour services apparently need improvement as visitors on a group tour seemed to evaluate their travel experience satisfaction lower than FIT travelers. Tour service providers need to pay particular attention in improving Chinese visitors' experience in areas of nature-based tours, attractions, shopping, and food and beverage.

Understanding the Chinese Consumer (Tourism Australia, 2014b) is part of a major research project undertaken by Tourism Australia to understand international consumer demands on Australia tourism. It investigates how Chinese visitors view Australia as an outbound destination and the factors most likely motivating them to visit the country. The top five destination features for Chinese consumers to visit Australia are (1) world-class beauty and nature environments; (2) good food, wine, local cuisine, and produce; (3) a safe and secure destination; (4) rich history and heritage; and (5) spectacular costal scenery. In terms of these preferred destination features, Australia was ranked among the top three destinations possessing the above destination characteristics except for rich history and heritage. Australia was only second to Hawaii in terms of *world class beauty and natural environments, great swimming beach, spectacular coastal scenery, family friendly destination, value for money*, and *interesting attractions to visit*. Australia was ranked number three after France and Italy in terms of *good food, wine, local cuisine, and produce* and after Hawaii and France as *a romantic destination*. As for *exciting events, local festivals, and celebrations*, Brazil, Hawaii, and France were perceived better than Australia. Australia was perceived to be lacking rich history and heritage and was only ranked number eight among the 15 destinations in the survey. However, apparently, those respondents who had been to Australia ranked Australia higher than those who had not.

8.4 ISSUES AND TRENDS IN CHINESE OUTBOUND TOURISM TO AUSTRALIA

Given the significance of Chinese outbound market, Australia government, in consultation with relevant industry sectors, has developed the *China 2020 Strategic Plan* to prepare Australia tourism toward the China market (Tourism Australia, 2011). The Chinese satisfaction survey (2013) discovered that group tourists are less satisfied with their experience in Australia than FIT travelers. Consumer demand research by Tourism Australia (2014c) identified a perceptual gap between existing and potential Chinese tourists toward Australia as a preferred tourism destination. Therefore, service quality and marketing are major issues facing Australia to attract and retain Chinese outbound visitors.

The issue of unethical business practices within the tourism supply chain from China to Australia has emerged as one of the biggest concerns to both tourism practitioners and authorities in Australia (DITR, 2005; Keating, 2009). According to The CNTA, an analysis of complaints from Chinese nationals visiting Australia revealed a trend of overcharging certain groups and requiring them to pay more upfront fees because they are not expected to spend enough at gift shops (Department of Resource, Energy and Tourism – DRET, 2008; Keating, 2009). A survey conducted by one of China's most influential online travel portals, Ctrip.com, has revealed significant concerns about shopping-subsidized tours, indicating that this practice has impacted negatively on Australia's destination image (DRET, 2008; Keating, 2009). Some academic papers have explored the issue of unethical practices in inbound Chinese group tours to Australia. King et al. (2006) described the unethical practices that had been observed in the context of inbound group tours from China and evaluated the impact of the practices on the growth of the market. Various policies have been implemented to address the issues since this article was published. In December 2005, The N*ational Emerging Markets of China and India* report issued by Australian Government recommended that legislation, as a matter of urgency, should be enacted to address unethical practices that could undermine the future of Australia tourism for the Chinese market (DITR, 2005). However, in February 2007, the Australian government determined that " . . . the issue of eradicating unethical business practices by a small number of tourism operators in the Chinese group tour market should . . . be addressed through industry-based solutions and non-legislative government initiatives" (DITR, 2007).

March (2008) developed a conceptual framework to illustrate the nature and consequences of unethical practices. In his research, he discovered that the reorientation of channel power had occurred in distribution systems for inbound Chinese market to Australia and a variety of unethical practices perpetrated on the innocent Chinese visitors to Australia. The study also found that duty-free stores had obtained some channel power; this phenomenon has not been witnessed with other inbound markets and, as a result, displaced the inbound tour operator as the key intermediary in the Chinese market. March speculated on the reasons for the prevalence of these practices as follows: (a) there is an overreliance on promotion and price-driven, short-term sales strategies; (b) key sectors of Australian inbound tourism are unregulated; (c) unethical industry practices in Australia are regarded simply as "how business is being done around here".

Keating (2009) conducted a case study on the Australian government's response to the concerns about unethical practices in the China–Australia tourism supply chain. Australian government monitors quality on the demand side of tourism supply chain via the Aussie Specialist Program and more recently via the Premier Aussie Specialist Program (Keating, 2009). In 2005, the *ADS Code of Business Standards and Ethics* was promulgated by Australian government. In 2007, Australian government commenced a series of actions in implementing the *Code*, requesting tour operators take immediate duty of care for customer tourists. These changes provide stronger monitoring and more explicitly stated penalties on noncomplying inbound operators. Such a reform also addressed commission shopping, where inbound operators can count on commission income but should provide access to free shopping in designated retail areas prior to leading tourists to commission shopping outlets.

China's tourism law, which came into effect on October 1, 2013, was regarded as a significant external factor influencing inbound tourism from China to Australia. The law contains a number of provisions designed to protect consumers; these include clauses addressing coercive commission-driven shopping, low-price tours, and low-quality tours. The tourism law has the potential to deliver benefits to the Australian tourism industry in the medium and long term, helping the industry adapt to a quality model (Tourism Australia, 2014c); however, in the short term, the industry faces the need to adjust some practices.

Australia has seen an imbalanced tourist flow distribution among its states destinations in receiving Chinese tourists. So far, inbound Chinese tourists, especially package group tourists, are heavily concentrated in New South Wales, Victoria, and Queensland. However, it should be noted that other states, such as South Australia and Western Australia, also treat the China market as the priority market in their future marketing campaigns. For instance, South Australia Tourism Commission published its *Activating China – 2020* tourism development strategy, listing China as one of the most important inbound tourist markets to the state (SATC, 2013). With these efforts made by the state government, it is foreseeable that inbound Chinese tourist flows will further spill over into secondary or second-tier regional destinations in Australia. Unlike Australia's traditional inbound markets (e.g., North America and Europe) and other Asian markets, Chinese tourists were found to be less likely to travel outside Australian capital cities (Jago, personal communication). This trend proves worrying to the industry as even a valuable market, if Chinese tourists tend to stay within cities, they can less likely contribute to regional economies. Australia is a country with a vast land area where most nature-based and wildlife attractions can be found in regions outside capital cities and the countryside. From a sustainable tourism development perspective, it is time for the industry to formulate strategies to encourage Chinese tourists to explore the Australian inlands. On the other hand, food and wine has been identified as the premier resources in Australia that are naturally related to tourism development. Tourism Australia recently launched the Restaurant Australia program (restaurant.australia.com) and a series of collaborative marketing campaigns to promote Australian food and wine.

This seems a unique proposition that Australia can further differentiate itself to Chinese tourists from other ADS destinations.

8.5 CONCLUSION

In conclusion, Chinese inbound tourism to Australia has been undergoing significant changes. Despite the dominant form of group tours, new market segments such as FIT or semi-FIT market are emerging as more important for the industry to cope with. Australia needs to address the quality issue in group tourists' experience to ensure that the overall positive image developed through the years be maintained and further nurtured among Chinese tourists. It also needs to consider providing more information and supporting service to appeal to the new emerging market. Australian government has spent tremendous efforts to ensure its marketing effectiveness and quality assurance of its tourism industry. Australia is one of the most preferred destinations to Chinese outbound tourists. With the transformation of the market structure, Australian government and tourism industry need to take more proactive policies and industry campaigns to guide Chinese tourist flows for sustainable development. The recent focus on developing food and wine resources by Tourism Australia would also bring up opportunities for the China market. The food and wine tourism resources, especially those located in regional Australia, can be further developed and packaged toward Chinese tourists to increase Australia's destination competitiveness.

REFERENCES

Austrade. *China Approved Destination Status (ADS) Scheme*. 2014. http://www.austrade.gov.au/Tourism/Tourism-and-business/ADS

Breakey, N.; Ding, P.; Lee, T. Impact of Chinese outbound tourism to Australia: reviewing the past; implications for the future. *Curr. Issues Tourism*, 2008, 11(6), 587–603.

Chang, R.C. Y.; Kivela, J.; Mak, A.H.N. Attributes that influence the evaluation of travel dining experience: when east meets west. *Tourism Manage.* 2011, 32(2), 307–316.

Chang, R.C.Y.; Kivela, J.; Mak, A.H.N. Food preferences of Chinese tourists. *Ann. Tourism Res.* 2010, 37(4). 989–1011.

Chow, I.; Murphy, P. Predicting intented and actual travel behaviors: an examination of Chinese outbound tourists to Australia. *J. Travel Tourism Mark.* 2011, 28(3), 318–330.

Chow, I.; Murphy, P. Travel activity preferences of Chinese outbound tourists for overseas destinations. *J. Hospit. Leisure Mark.* 2007, 16(1/2), 61–80.

DITR. *National Tourism Emerging Markets Strategy: China and India*; Australian Government Department of Industry, Tourism and Resources: Camberra, 2005.

DITR. *Update on Action to Address Unethical Practices in the Inbound Tourism Industry*; Australian Government Department of Industry, Tourism and Resources: Camberra, 2007. http:/www.aph.gov.au/house/committee/efpa/services/subs/sub053.pdf.

DRET. *ADS Focus Newsletter* 7th *ed.*; Australian Government Department of Resource, Energy and Tourism: Camberra, 2008

Keating, B. Managing ethics in the tourism supply chain: the case of Chinese travel to Australia. *Int. J. Tourism Res.* 2009, 11(4), 403–408.

Keating, B.; Kriz, A. Outbound tourism from China: literature review and research agenda. *J. Hospit. Tourism Manage.* 2008, 15(1), 32–41.

Kewk, A.; Lee, Y.S. Intra-cultural variance of Chinese tourists in destination image project: case of Queensland, Australia. *J. Hospit. Leisure Mark.* 2007, 16(1/2), 105–135.

King, B.; Dwyer, L; Prideaux, B. An evaluation of unethical business practices in Australia's China inbound tourism market. *Int. J. Tourism Res.* 2006, 8(2), 127–142.

Li, M.; Zhang, Q.H.; Mao, I.; Deng, C. Segmenting chinese outbound tourists by perceived constraints. *J. Travel Tourism Mark.* 2011, 28(6), 629–643.

Li, J. W. J.; Carr, N. Visitor satisfaction. *Int. J. Hospit. Tourism Adm.* 2004, 5(3), 31–34.

Mao, I.; Zhang, Q.H. Structural relationships among destination preference, satisfaction and loyalty in Chinese tourists to Australia. *Int. J. Tourism Res.* 2012. DOI: 10.1002/jtr.1919 n - a-n/a.

March, R. Towards a conceptualization of unethical marketing practices in tourism: a case-study of Australia's inbound Chinese travel market. *J. Travel Tourism Mark.* 2008, 24(4) 285–296.

Pan, G.W.; Scott, N.; Laws, E. Understanding and sharing knowledge of new tourism markets. *J. Qual. Assur. Hospit. Tourism*, 2006, 7(1/2), 99–116.

Pan, G. A theoretical framework of business network relationships associated with the Chinese outbound tourism market to Australia. *J. Travel Tourism Mark.* 2003, 14(2), 87–104.

Pan, G.; Laws, E. Tourism marketing opportunities for Australia in China. *J. Vacation Mark.* 2001, 8(1), 38–48.

South Australia Tourism Commission (SATC). *Activating China – 2020.* 2013. http://www.tourism.sa.gov.au/assets/documents/Research%20and%20Reports/satc-activating-china-full-report-2013.pdf

Sparks, B.; Pan, G. W. Chinese outbound tourists: Understanding their attitudes, constraints and use of information sources. *Tourism Manage.* 2009, 30(4), 483–494.

Tourism Australia. *Summary of Tourism Australia's China 2020 Strategic Plan.* 2011. http://www.tourism.australia.com/documents/corporate/TA_China_2020_Strategic_Plan.pdf

Tourism Australia. *China Market Profile.* 2014a. http://www.tourism.australia.com/documents/Markets/MarketProfile_China_May14.pdf

Tourism Australia. *Understanding the Chinese Consumer.* 2014b. http://www.tourism.australia.com/documents/Statistics/Consumer-demand-project-CHINA.pdf

Tourism Australia. *China 2020 Consumer Research.* 2014c. http://www.tourism.australia.com/documents/Statistics/Consumer-demand-project-CHINA.pdf

Tourism Australia. *Chinese Visitor Satisfaction Research.* 2014d. http://www.tra.gov.au/documents/Chinese_Satisfaction_Survey_FULL_REPORT_FINAL_24JAN2014.pdf

Tourism Research Australia. *Tourism Forecasts.* 2013. http://www.tra.gov.au/publications/forecasts-Tourism_Forecasts_Spring_2013.html

Wang, Y.; Davidson, M. Pre- and post-trip perceptions: an insight into Chinese package holiday market to Australia. *J. Vacation Mark.* 2010a, 16(2), 111–123.

Wang, Y.; Davidson, M. Chinese holiday makers' expenditure: implications for marketing and management. *J. Hospit. Mark. Manage.* 2010b, 19(4), 373–396.

Weiler, B.; Yu, X. Case studies of the experiences of Chinese visitors to three tourist attractions in Victoria, Australia. *Ann. Leisure Res.* 2008, 11(1/2), 225–241.

Zhang, Y.; Murphy, P. Supply-chain considerations in marketing underdeveloped regional destinations: a case study of Chinese tourism to the Goldfields region of Victoria. *Tourism Manage.* 2009, 30(2), 278–287.

CHAPTER 9

MAINLAND CHINESE OUTBOUND TOURISM TO EUROPE: RECENT PROGRESS

BERENICE PENDZIALEK

CONTENTS

9.1 INTRODUCTION

Outbound tourism has been developed as a nonexotic activity for Chinese citizens through the years. It was not encouraged during the time of Mao; however, this started to change by the time of China's "reform and opening" policy, led by Deng Xiaoping in 1978, as the country began its transformation as a key global player. Economic prosperity and social stability brought the possibility of recreational activities, first within China and, second, across its borders. With the inclusion of the United States on the Approved Destination Status (ADS) list in 2007 and Canada in 2010, the majority of tourism attractions worldwide are now authorized for Chinese outbound tourists. Furthermore, the ease of visa procedures and increased air connectivity have been favorable for the development of Chinese outbound tourism worldwide.

In the global tourism space available for Chinese, Europe stands out as a "must visit" due to its attractions, the close distance between European countries, and its retail and luxury brands. In this chapter, the specifics of Chinese outbound tourism to Europe and its changes in recent years are presented. For this, I will draw, among others, on the findings from my doctoral research with the topic, "Performing Tourism: Chinese Outbound Organized Mass Tourists on their Travels through German Tourism Stages" (Pendzialek, in press). This research project draws on a conceptual framework based on notions of Goffman's (1959) everyday performances, Schechner's (2002) performance studies, and the performance turn in tourism studies, where the work of Edensor (1998, 2001), Haldrup and Larsen (2010), Rakić and Chambers (2012), among others, is included. My dissertation's discussion focuses on 3 elements. First, I examine the casting and preparation of the performers. Second, I explore the performances on German tourist stages with respect to roles learned, and also highlight the possibility of improvised performances. Finally, I observe the setting of the stages with regard to allowed choreographies, stage direction, and power relations between categories of players.

The research is based on a phenomenological approach, appropriate to this project due to its ethnographic character. Part of the study's field research was conducted largely from 2011 to 2012, both in China and in Germany. In China, information was gathered by using open and semi-structured in-depth expert interviews with eight representatives from the Chinese tourism industry and tourism boards, as well as with tourism academics. In Germany, the ethnographic fieldwork consisted of participant observations, semi-structured interviews, and short questionnaires to 45 Chinese outbound tourist members of three group package tours organized by Chinese–German travel companies. Complementary to tourist contributions is information provided by the groups' tour guides and tour leaders during four semi-structured in-depth interview and "on-the-run" talks. Equally important are my field notes, diaries, photos, and videos. Furthermore, expert interviews were conducted with nine German industry representatives, experts, and academics in the summer of 2013.

9.2 CHINESE OUTBOUND TOURISM FLOWS TO EUROPE

Tourism relations between the European Union and China officially started with the signing of the Memorandum of Understanding for the ADS Agreement on February 12, 2004. Upon its enforcement in May 2004, selected Chinese tour operators benefit from the organization and promotion of leisure tours to the EU Member states part of the Schengen area, whereas the tourists themselves benefit from the simplified group visa or ADS visa application. Other Non-Schengen countries received the ADS status in the following years. For example, the United Kingdom has been an officially approved destination since 2005.

9.2.1 DEVELOPMENT OF CHINESE OUTBOUND TOURISM TO EUROPE

As can be seen in Fig. 9.1, the number of Chinese tourist arrivals to Europe has been growing steadily in the past decade. In 2008 and 2009, the world economic crisis and travel restrictions implemented due to the H1N1 flu pandemic inhibited the growth of arrivals to Europe. However, after 2010, arrivals saw a remarkable growth. Additionally, the share of Europe in the long-haul Chinese outbound market[1] has remained at the 30% mark over the past 3 years. From the total of 24 million tourists traveling from China to long-haul destinations in 2013, Europe represented 31% of the long-haul outbound market (ETC, 2014a), whereas in 2012 and 2011, it represented 33.4% and 32.7%, respectively. As forecast, the European Travel Commission (ETC) reports that travel to Europe from China will grow on an average of 8.4% per year to 2018 (ETC, 2014b).

With regard to Chinese tourists' expenditure abroad, in Europe, they spent an average of 2,418 euros per trip and 373 euros per night in 2012. These figures are above the world's average expenditure of the Chinese, where they spend an average of 1,690 euros per trip and 295 euros per night (GNTB, 2013). A considerable amount of the Chinese expenditure is allocated to shopping. Global Blue reported that in 2012, Chinese tourists spent per tax-free shopping transaction in Europe an average of 813 euros, where the average of other travelers was 485 euros. In Germany, the average is 628 euros (Global Blue, 2013).

---

[1]According to the European Travel Commission (ETC; 2014a), based on Oxford's Tourism Economics (http://www.tourismeconomics.com), 58.9 million tourists traveled from China in 2013. From this amount, 40.9% traveled long-haul. This travel excludes those trips to Northeast Asian countries (Hong Kong SAR, Japan, Rep Korea, Macao SAR, Mongolia, and Taiwan). Similar to other regions worldwide, these figures differ from the official China National Tourism Administration (CNTA) Statistics. A possible explanation was previously stated by the ETC: "Departure figures such as these *(refers to CNTA's)* always differ substantially from arrivals as measured in the individual *European* destinations" (ETC, 2011 [italics mine]).

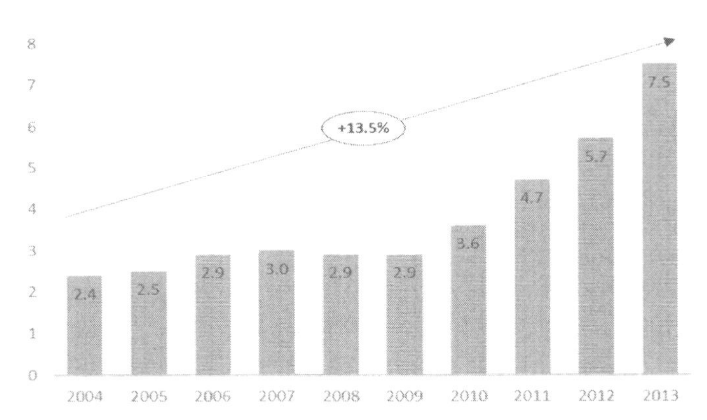

FIGURE 9.1 Chinese Tourists Arrivals to Europe, in Million and CAGR from 2004–2013. **Note:** Including the European part Russian Federation. (Adapted from Tourism Economics via ETC [2014a].)

9.2.2 MAPPING THE ROUTES FOLLOWED IN EUROPE

Based on Chinese tourists' actual visitation data, there are several rankings on the preferred European destinations. Figure 9.2 shows the top destinations as reported by the World Travel Monitor from 2010–2012 (GNTB, 2013). In 2012, Germany, with 16.2% market share, ranked for the first time as the number one destination for Chinese followed by France (14.7%) and the Russian Federation (14.5%). Another ranking comes from the number of Schengen visas issued in European destinations consulates in China (ibid. p. 13). In this respect, the top five destinations in 2012 were France (277,645 visas), Italy (268,538), Germany (236,551), Switzerland (94,368), and Spain (60,666). One more ranking is that of the European cities, where Chinese spend the night. According to the European Cities Marketing (ibid. p.25), the top 10 cities for Chinese tourists in 2012 were Paris, London, Vienna, Florence, Munich, Frankfurt, Prague, Zurich, Berlin, and Venice.

Some of the previously mentioned European destinations are most certainly included in the routes followed by both group tourists and self-organized Chinese outbound travelers. Major sales of tourism products for Chinese mass organized groups in Europe continue to be the typical 9 countries in 14 days. According to a representative of GTA Travel, a popular group package tour sold in Beijing includes at least Italy, France, and Switzerland as main destinations. Besides the many interesting sightseeing points, the route includes shopping in the countries' unique shops. For example, the tour members have the opportunity to buy watches in Switzerland and luxury brands, such as Louis Vuitton, in Paris. Accordingly, self-organized tourists would follow similar routes as group tourists, mostly itineraries that include city destinations. However, they seek to spend more time in every destination (Xiang,

2013). The UNWTO (2012) report of the Chinese blogosphere opinions about Europe reports that the most discussed European attractions by young Chinese online were The Alps, Notre Dame de Paris, the Louvre, the University of Cambridge, Amsterdam's Red-light district, and the Aegean Sea (UNWTO, 2012).

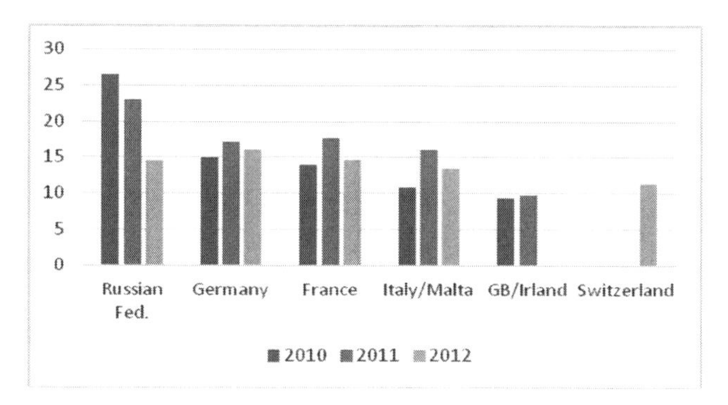

FIGURE 9.2 Top European Destinations for Chinese Tourists, % of Total European Trips 2010–2012.
Note: More than one answer possible. A trip can include more than one country. (Adapted from GNTB, 2010–2013.)

Regarding the preconceptions of Europe by Chinese tourists before traveling, my research on group packaged tours (GPT) shows that a vast majority of tourists regard Europe as one big stage on which to perform tourism. Therefore, they had rather limited knowledge of the continent's geography and differences between countries. From the comments received in my research, Europe was mostly associated with the words new, modern, and civilized. Interestingly, once the tourists arrive in Europe, they might get disappointed because the continent is old and not as modern as expected. With regard to expectations, the young Chinese tourists inside the groups proved to be more informed on Europe's destinations. They associated the attractions with European historical and cultural figures (e.g., Berlin Wall and Goethe's House in Frankfurt), as well as to Chinese own cultural figures (e.g., Cambridge King's College made famous in Xu Zhimo's poem).

9.3 TOURISTS AS PERFORMERS ON EUROPEAN STAGES

Approaching an insider perspective, the following section reports on the Chinese tourists themselves, their demographic aspects, travel behavior, and recent perceived changes in demands.

9.3.1 SOCIODEMOGRAPHIC CHARACTERISTICS

Collected from different sources, Table 9.1 shows the comparison of the sociodemographic details of Chinese tourists worldwide, to Europe and Germany. The predominance of young adults and middle-aged males and females (between 25–60 years old) can be perceived; only the European sample registered a significant number of female tourists. The tourists have a rather high level of education with at least a bachelor degree. Company employees represent the majority of the occupations. Also significant are the students and education-related workers (teachers and professors). In relation to their individual monthly income, the amounts are concentrated in the middle and higher levels.

If the sociodemographic details of Chinese tourists to Europe in 2010 were compared with those visiting Taiwan and Australia, substantial differences can be found with regard to the age group segments. In the case of Taiwan in 2010, the 60 plus-year-old demographic was the largest visitor segment from China, and housewives and retiree were the 2 main occupation groups (CTA, 2012). For the same year, Australia exemplifies the other end, where visitors from 15 to 29 years old represented the majority of the arrivals that year (Tourism Australia, 2013). This can be explained as being due to the significant number of Chinese students in the country. Interestingly, from 2011, the age of Chinese travelers to Australia has shifted to the 45–49 age segment (ibid.), which can be related to the induced visiting friends and relatives (VFR) tourism.

TABLE 9.1 Sociodemographics of Chinese Tourists' Worldwide, Europe and Germany (2011–2012)

Area/ country	Gender	Age	Educational level	Occupation	Individual monthly income
World 2011 (1)	47% female 53% male	0.24%, 0–15	39.66% Bachelor	9.37% Education	RMB per month
		23.10%, 15–24	29.81% College diploma	9.37% Student	13.78%, 1,001–3,000
		44.03%, 25–34	21.8% High school, technical, vocational school	7.90% Manufacturing Ind.	27.31%, 3,001–5,000
		21.00%, 35–44		7.74% Finance	26.44%, 5,001–8,000
		9.57%, 45–59		6.87% IT, Computer Service,	12.07%, 8,001–10,000
		2.06%, 60+	6.63% Master or more	Software Ind. 6.63% Wholesale, retail 5.08% Community services, other services industry	11.51%, 10,001–20,000

Europe 2010 (2)	69% female 31% male	26%, 20–29	66% Bachelor	1% Self-employed	No data
		29%, 30–39	24% Master	45% Managers	
		16%, 40–44	3% Doctorate	33% Non-management	
		19%, 45+	3% Secondary level	16% Professional -technicians	
			4% Primary level	5,5% others.	
Germany 2013 (3)	More male as female	All group levels, up to 55 years. The majority is concentrated between 25 and 44 years old. Average age 39 years	High educational level	NA	High income level

Note: This information can only serve as an indication due to the differences of the samples of each of the studies shown in the table.
Source: CTA (2012), ETC (2011), and GNTB (2013).

9.3.2 PURPOSE OF VISIT AND AVERAGE LENGTH OF STAY IN EUROPE

As can be seen in Table 9.2, the main purpose for Chinese to visit Europe is to participate in leisure trips, followed by business trips. The table also demonstrates that the average trip to Europe lasted four to seven nights. Additionally, the number of trips increased in the 3-year period, although the average stay of travelers has been decreasing. This supports the challenge perceived during my doctoral field research, where the average stays decreased due to Europe's struggle to attract second-time visitors. Recent information on average stay from the ETC (2014d) reports that Chinese trips last on average 8–15 days. Of these, a maximum of 3 nights are spent in each European country.

TABLE 9.2 Purpose of Travel and Average Stay of Chinese Tourists in Europe, Million Trips from 2010–2012

	2010	2011	2012
Trips	3.8	4.0	5.2
Purpose	71% Leisure trips	54% Leisure trips	72% Leisure trips
	27% Business trips	42% Business trips	25% Business trips
	2% VFR/ others	4% VFR/others	3% VFR/others
Stay	71% Long stays	73% Long stays	73% Long stays
	29% Short stays	27% Short stays	27% Short stays
	49% Trips lasted 4–7 nights	43% Trips lasted 4–7 nights	42% Trips lasted 4–7 nights
	8.3 nights in average	7.8 nights in average	6.4 nights in average

Note: Long stays = 4+ nights, short stays = 1–3 nights.
Source: GNTB (2010–2013).

9.3.3 PERCEIVED BEHAVIOR BY MARKET SEGMENT

After more than 15 years of Chinese outbound tourism, it is possible to identify specific market segments. A general segmentation is presented by Li et al. (2013), where they identified three main segments: "Entertainment/Adventure Seekers," "Life-seeing Experience/Culture Explorers," and "Relaxation/Knowledge Seekers." Accordingly, Fugmann and Aceves highlighted two general leisure segments: "mass travelers" and "free independent travelers," as they explained:

The first segment can be characterized mainly as tourists in search of an intensive sightseeing experience, who have almost no travel know-how, are highly price sensitive, have no foreign language skills to confront the destination alone while simultaneously seeking security within a group. On the other hand, there is the "FIT" segment. These tourists have traveled in groups before and are now seeking individual adventures that allow them to stand out from the typical tourism flows. They are increasingly tech-savvy, and most of them have some experience outside of China, often because of their studies (Fugmann and Aceves, 2013).

The segmentation of the market to Europe considerably varies from author to author. For example, the UNWTO (2012) divides the market into five tribes: (a) traditionalists, similar to the mass travelers' characteristics mentioned above; (b) wenyi youth, white-collar Chinese with westernized mentality and bohemian-romantic ideals; (c) experienced-centered, they place more emphasis on the material and less on the emotional travel experience, enjoy the whole travel experience from planning to travel; (d) hedonists, for them luxury shopping, eating well, and having a good time is what counts, scenery is secondary; and (f) connoisseurs, they have concrete knowledge of what they want to visit and taste, with more focus than the experienced-centered tribe. From these 5 tribes, the traditionalists continue to have the major share of the market with 70% of the study's netnography sample.

Regarding the motivations to undertake long-haul travel, my research among group tourists in Germany highlights that tourists travel in order to increase their social capital. As China is a hierarchy-based society, travel to far away destinations increases the tourist's prestige among their peers. Contrarily, the traveling young Chinese highlighted the learning element that travel represents; moreover, they used travel to seek future working-business opportunities. Similar to my findings, main motivation to traditionalists, as reported by the UNWTO, was prestige. For the other tribes, they were freedom and uniqueness for the wenyi youth, togetherness and curiosity for the experienced-centered, pleasure for the hedonists, and aesthetics and knowledge for the connoisseurs.

With regard to planning, in the course of my research, mass group tourists' criteria for the selection of a GPT relied on price, itinerary, accommodation, and the companionship of a Chinese tour guide. Interestingly, just a few tourists commented on the country attributes as a factor in selecting a specific tour. Booking was done directly with a travel agency, either by the tourists themselves or by their family and friends living or studying in Europe. Sources of information before traveling were limited to the agency's information online, recommendations from family and friends, and in the case of young Chinese, online sources (namely blogs). Moreover, the mass group market operations in Europe continue to be an area of expertise for the Chinese diaspora. They give the tourists, especially first-timers, a homelike feeling by providing a (re)making and interpretation of *their European* home (Leung, 2009 [italics mine]). The majority of routes offered for group tourists in Europe continue to promote hectic and superficial consumption of touristic stage, whereby sightseeing of must-see attractions and shopping are the main activities. In general, the mass market continues to be of an offer-driven nature, instead of experience-driven one.

Self-organized travelers are more likely to do intensive research of a destination before traveling to Europe. Xiang (2013) argues that these travelers intensively "do homework" about the destination before traveling in order to "enhance perceived control over strangeness" (Xiang, 2013). Information is gathered mainly from blogs, forums, and social media channels. For Europe, one example is the website www.qyer.com. It started as a forum directed to young Chinese living in Europe where they could exchange experiences and advice about travel around the continent. It was initially a small project by a Chinese student, at the time residing in Hamburg. The online community offers User Generated Content (UGC), as well as free Chinese language eGuides created by the company, supplying the future travel performer with first-hand information collected by previous young travelers. Services are likely to be booked on websites like qyer.com, especially if taking into consideration that self-organized travelers strongly rely on online booking tools.

While traveling to European destinations, self-organized travelers differ from their "Western" counterparts. Their self-control and uncertainty avoidance bring them closer to the group tourist than to the adventurous backpacker or "free walkers" they intend to be. Interestingly, Xiang (2013) suggests that individual Chinese tourists do not follow the beaten track by exploring beyond the tourist bubble, rather they are "within a bigger bubble" (p. 140). Moreover, their introvert behavior and language insecurity stop them from communicating with locals. This could be observed during my participant observation in German destinations.

Also from these observations, it was interesting to recognize predictable and spontaneous behaviors abound with everyday aspects (e.g., Chinese-like banquets, public behavior, casual-loud conversations, and hints of self-importance), influenced by Chinese cultural aspects (collectivism, negotiation, search for harmonious relations, hierarchy, soft "no's" instead of direct answers) and political issues (international relations and conflicts, political ideology).

During their trips, group tourists had an overall positive impression of Europe and Germany. They regarded the people as friendly and warm-hearted, enjoyed the clean cities and environment, and felt safe. On the negative side, the majority of complaints came from the travel agencies' arrangements (e.g., little time at attractions and a long time on the bus, poor information received by the guide, accommodation far from city center, and no activities with local people). With regard to the destination, the tourists highlighted the lack of adaptation to Chinese needs (e.g., lack of free internet connection, food, no information materials in Chinese, and high prices). My findings on group tourists to Germany concur to some extent to those reported by the UNWTO (2012). From the survey, the top positive aspects perceived by Chinese netizens about Europe were peaceful, clean cities, not crowded, good preservation of historic attractions, environmental protection, rich cultural past, and friendly, helpful people. Discontent came from high prices, tedious visa application process, lack of Chinese language information and material, inability to use Chinese debit cards in most places, and bad food in general and bad Chinese food in particular.

9.3.4 RECENT CHANGES: EXPERIENCED, CONSCIOUS, WILLING TO TRADE UP, AND MOBILE

After more than a decade of international travel experience, Chinese outbound travelers to Europe have developed their own travel career. From my research, more than 70% of the respondents had already traveled inside Asia before, whereas more than 50% had traveled to Europe in the past years. The travel career evolution perceived throughout my research is similar to the one presented in Hilton and SOAS (2011) Blue Paper. They reported that first-timers usually take part in group tours during their initial contact with European destinations. Afterward, they might possibly come once again as part of a group, but this time covering different destinations or traveling semi-/self-organized in order to break free from the group rules. Their travel experience allows them to minimize the initial fear of visiting a new destination and makes them aware of quality aspects that need to be included in the touristic services.

Even when sightseeing is regarded as one of the main motivations for traveling abroad, it was interesting to see how tourism is being positioned as an activity that offers relaxation from everyday life. The research shows that the consumption of European and German touristic spaces was highly influenced by the binary "ordinary every day China–extraordinary other." Outbound travel was regarded as a liminal phase, where extraordinary activities such as enjoying blue skies and clean and peaceful lakes and living life at a slower pace were allowed to occur. This reflects the current situation in China, where pollution is a problem in China's major cities and the urban lifestyles are far from Europe's simplicity and balance.

Over the past few years, the Chinese government's influence on the tourist's behavior abroad and in European destinations has also been increasingly recognizable. In order to help their country break away from its "Third World" position (Oakes, 2012), current and future Chinese tourists need to act out exemplary and politically congruent performances worldwide. To achieve this, the government has implemented civilization campaigns[2] for the performer's preparation of their role as a tourist and as "folk ambassador." These campaigns started back in 2006 but were probably not forcefully pursued due to other governmental tourism priorities (one of them, for example, was to control Chinese tourists' mobility by negotiating ADS status worldwide or holding back the status to countries that were not politically aligned with China, such as the United States and Canada). However, allowing the so-called "ugly Chinese tourists" to continue performing without knowing the destination's rules and conventions, which admittedly makes China lose face (e.g., the Egypt graffiti scandal in 2013), is something the government cannot afford if it wants to succeed in its current geopolitical goals. Interestingly, my research findings show how group tourists on several occasions stand ideologically for their country (e.g., avoiding manifestations against China at Brandenburg Gate or reassuring China's position in the world by discussing the country's achievements throughout the years).

Partially visible in my research, but widely commented in other publications, is the conscious budget allocation of Chinese travelers in Europe. The BCG (2013) survey on young affluent travelers shows that the areas where Chinese travelers are willing to "trade up" or allocate a disproportional amount of their budget are lodging, shopping, and dining (p. 6). More specifically about Europe, the UNWTO

[2]Detailed explanation of the campaign can be found on the "Chinese Citizens Civilization" website http://www.xinhuanet.com/travel/wmcjy/index.htm.

(2012) netizens' survey highlights that Chinese travelers would try to spend less on food, accommodation, and transportation and more on shopping and entertainment (p. 10). The latter is possible, thanks to the vast amount of information available online about tricks on how to save money in the destinations. For the majority of group tourists observed, budget was controlled and rationalized throughout the whole trip, rather than allocated. Just a few group members mentioned to have saved on food in order to enjoy shopping, which occurred sporadically and acquired objects were small and not high-priced.

Another current change found in my research was that of young Chinese tourists constantly blurring the lines between market segments in order to maximize time during holidays. Some young tourists mentioned that they take part of a GPT to fill the gaps of their individually planned routes. Others mentioned that it was not easy to find travel companions back home; therefore, they book group tours to enjoy the companionship of other Chinese and also meet new people who might want to join them afterward when continuing their route alone. These examples highlight that tourist were able to leave behind the Chinese travelers versus Chinese group tourists fight in order to take full advantage of their holiday time.

Also recognizable during the touristic performances was the increased influence of technologies, especially of mobile ones. Besides cameras and video cameras, in recent years, tourists also bring along their smartphones, tablets, and GPS devices; however, their use is highly restrained in Europe and German destinations. Even when initially thought that these new objects were only popular among independent and business travelers (UNWTO, 2012), my research showed that group tourists in German destinations are also actively bringing and using them when possible. For the majority of group members, being quite price conscious, the use of the mobile phones was highly restrained due to overseas mobile fees and the willingness to pay them. Amongst the young people, only a minority of participants used their mobile phones, namely, to instantly share photos on Chinese social media channels or navigate with their GPS. Unfortunately, access to Internet connection in Germany is not freely available in public spaces or in hotels. This prevents tourists from staying in touch with people back home over the net and also from instantly sharing or bragging online about the outcomes of their touristic performances. This also prevents Germany from making use of the electronic Word of Mouth (eWOM) generated by these tourists, who are quite dependent on their mobile phones (from China's 618 million Internet users in 2013, 81% were mobile internet users, Hong, 2014) and who constantly visit their social media sites (92% of Chinese Internet users visit their average social media channels at least three times a week ETC, 2014c).

9.4 EUROPEAN TOURISM INDUSTRY'S RECENT CHANGES TOWARD CHINESE OUTBOUND TOURISM

In order to meet the current demands of Chinese group tourists in Europe and Germany, my research partners highlighted the following recent changes in group offers. First, the number of group members has been decreasing in the past 3–4 years. At the beginning, groups used to have at least 40 people; now, they are around the half that size. Moreover, there are the so-called "mini-groups," which mostly consist of families traveling alone. Second, group travel cannot be associated with only cheap, kickback offers and poor organization. The offer spectrum for groups ranges all levels, from low to high-end market. Third, group routes are shrinking. From 10 countries in 14 days to three to four countries in the same amount of time. Mono-destination tours for group tourists are also available (i.e., "11 Days around Germany").

Other recent changes were mentioned by the experts, which apply for both group and self-organized travelers' offers. First, individual demand for thematic leisure experiences lead to customized products. One example is the customized offer for auto fans "The Shanghai-Hamburg New Silk Road Rally" created by a Chinese inbound operator based in Hamburg. Second, the creation of touristic products where learning is the main motivation for travel. Successful themes in Europe are classical music (Germany and Austria) and wine culture (Italy and France), amongst many others. One unique learning experience available in Germany is the possibility to experience an outbound version of "red tourism" by visiting the birthplace of Karl Marx in Trier. As highlighted by Fugmann and Aceves (2013), Chinese tourists pay a visit to this value-laden "red" site mostly to take a photo and fulfill their political duty. It remains uncertain to what extent they seek to learn Marxism, especially when the Karl Marx museum itself presents Marx ideology from several angles, even critical ones (p. 7).

Beyond the new product offers, a vast majority of European destinations and travel-related brands are progressively making use of digital marketing to contact the tech-savvy and highly internet-engaged Chinese customer. A first step to building their digital presence was to present their homepage content in Chinese Mandarin, although there is still much to be done in this area. A digital benchmark study, auditing 45 European National Tourism Organisations (NTO), reveals that only 30% of them have a Chinese language website (DTTT, 2013). Interestingly, those available websites tend to have a one-to-one translation, which is not culturally customized to Chinese needs. In other cases, these Chinese language versions offer a rather limited amount of information in comparison with their native language one. From this restricted content, destinations tend to have available the following information in Chinese: tourist attractions, shopping and transportation, and city guides, whereas hotel bookings online tend to be left behind (Chinavia, 2013).

With regard to the social media presence, the microblog site Sina Weibo is the most used by European national and city tourism organizations. But with rising stars like Tencent's WeChat,[3] Europe tourism stakeholders need to keep up the conversations with Chinese travelers by following the trends on the Chinese social media sites. Another form of digital communication is to engage with Chinese tourists via UGC. An example in this respect is to invite famous bloggers to promote a destination. In 2013, the company Chinese Friendly International invited two famous bloggers to experience a 15-day route around Spain's certified Chinese Friendly Cities: Barcelona, Madrid, La Rioja, Valladolid, Sevilla, Granada, Madrid, and Segovia. Each day, the bloggers posted at least 4 entries in their different social media channels, which included photos, text, and at the same time, started discussions about Spain's touristic offer (For more on Spain and COT refer to Grötsch et al., 2014).

9.5 CONCLUSION: POTENTIALS AND FUTURE EXPECTATIONS

One of the main challenges to visiting Europe continues to be the visa application. My findings present the contradictions inside the Schengen area. On the one hand, European countries wish to receive Chinese tourists, mainly to inject capital and help them out of their current economic recession, however, on the other hand, there is still strong control of these new flows of tourism. Moreover, it can be seen that the Schengen countries, instead of presenting a unified regulation front, compete against each other to attract more Chinese to their destinations. What is also particularly interesting from my analysis is that even when a fear of illegal immigration is latent and prevents further facilitation of a visa procedure, the experts remain positive toward the future developments in the visa granting process.

In the following years, it will be interesting to monitor the strategies adopted for European destinations in order to attract more than one-time visitors, which is the case in countries like France and Germany. A possible solution could be to create awareness of other German and French destinations, away from the typical highlights. For Germany, one example is the work of the Historic Highlights of Germany (HHOG), a co-marketing association that represents 13 cities known for their historical significance and which is an alternative to the popular destinations such as Munich, Frankfurt, and Cologne.

As far as product adaptation is concerned, specific cultural characteristics need to be taken into consideration at the time of engaging with the Chinese market. It is necessary to avoid stereotypes and embrace the diversity that the Chinese

[3]From Q1 2013 to Q1 2014, WeChat was reported as the fastest growing chat application worldwide living behind other Western apps like WhatsApp, Facebook Messenger, and Instagram (Mander, 2014).

market represents. Training seminars for Europe's tourism industry stakeholders are necessary to narrow the current knowledge gap between China and Europe. With regard to marketing strategy, as mentioned elsewhere in this chapter, China is by now a multi-segment market. Therefore, industry practitioners should aim to concentrate on a specific market segment-niche and target their campaigns accordingly. For both main leisure segments, group and self-organized, the tourism industry should seek to include tourists in everyday, authentic experiences with European locals in order to promote multi-sensual participation rather than just gazing from afar.

For infrastructure, it is necessary to include information in Chinese Mandarin language at airports, main tourism attractions, and hotels. In many cases, the information can be read by the Chinese in English, so in this case, its objective is not just to communicate but also to make the Chinese feel "welcome." Beyond information, it is also recommended to have Chinese-speaking staff who could take care of Chinese tourists. Above all, destinations should provide tourists with free WiFi so they can instantly share their experiences online. Free Internet is increasingly becoming a basic need for travelers but is still not so common in Europe.

Even if the present situation in Europe is far from being perfect, experts believe that the market still has a lot of potential and will continue to grow. It can be expected that all kinds of segments, niches, and modes of travel will be further developing in accordance with the evolution and sophistication of the Chinese tourists. In the coming years, it will be exciting to observe whether the industry decides to commit on a long-term basis and cater their products and services toward Chinese outbound tourists' needs, rather than continue on in the current manner, where they expect Chinese tourists to learn to behave as Western tourists and keep operating "as usual."

REFERENCES

BCG. Winning the Next Billion Asian Travelers – Starting with China. 2013. www.bcg.com (accessed July 17, 2014).

Chinavia. Best Practice Study: City Destinations Targeting Chinese Visitors. 2013. http://www.visitcopenhagen.com/copenhagen/about-chinavia-pilot-project (accessed July 17, 2014).

CTA. *Annual Report of China Outbound Tourism Development. 2012;* Tourism Education Press: Beijing, 2012.

DTTT. The European NTO Digital Benchmark. 2013. http://thinkdigital.travel/reports/The_European_NTO_Digital_Benchmark.pdf (accessed July 25, 2014).

Edensor, T. Performing tourism, staging tourism: (re)producing tourist space and practice. *Tourist Studies* 2001, 1(1), 59–81.

Edensor, T. *Tourists at the Taj. Performance and Meaning at a Symbolic Site*; International Library of Sociology; Routledge: London, New York, 1998.

ETC. European Tourism in 2013: Trends & Prospects. Quarterly Report (Q4/2013). 2014a. http://www.etc-corporate.org/reports/tourism-trends (accessed July 14, 2014).

ETC. European Tourism in 2014: Trends & Prospects. Quarterly Report (Q1/2014). 2014b. http://www.etc-corporate.org/reports/tourism-trends (accessed July 14, 2014).

ETC. Market Insights: China. 2011. http://www.etc-corporate.org/images/library/ETCProfile_China-1-2011.pdf

ETC. Market Insights: China. 2014c. http://www.etc-corporate.org/images/library/ETCProfile_China-1-2011.pdf (accessed July 14, 2014).

ETC. Meet the Chinese Travellers Brochure. 2014d. http://www.etc-corporate.org/?page=report&report_id=53 (accessed July 25, 2014).

Fugmann, R.; Aceves, B. Under control: performing Chinese outbound tourism to Germany. *Tourism Plann. Dev.* 2013, 10(2), 159–168.

Global Blue. The Global Blue Review 2012. 2013. http://business.globalblue.com/ch_de/services/market-intelligence/global-blue-review1/ (accessed July 14, 2014).

GNTB. Incoming-Tourismus Deutschland 2014 China, Hong Kong. 2013. http://www.germany.travel/media/pdf/ueber_uns_2/DZT_Incoming_GTM13_de_web.pdf (accessed July 14, 2014).

Goffman, E. *The Presentation of Self in Everyday Life*; Anchor Books; Doubleday: New York, NY, 1959.

Grötsch, K.; Monasterio, M.; Vera, C., Eds. *Libro Blanco del Turismo Chino en España.*: Chinese Friendly Editions: Spain, 2014.

Haldrup, M.; Larsen, J. *Tourism, Performance and the Everyday. Consuming the Orient.* In *Routledge Studies in Contemporary Geographies of Leisure, Tourism, and Mobility 15*; Routledge: London, New York, 2010.

Hilton and SOAS. Blue Paper: How the Rise of Chinese Tourism will Change the Face of the European Travel Industry. 2011. http://news.hilton.com/index.cfm/newsroom/detail/1660 (accessed July 17, 2014).

Hong, K.; China´s Internet Population Hit 618 Million at the End of 2013. 2014. http://thenextweb.com/asia/2014/01/16/chinas-internet-population-numbered-618m-end-2013-81-connecting-via-mobile/ (accessed July 25, 2014).

Leung, M.W.H. Power of boarders and spatiality: a study of Chinese-operated tourism businesses in Europe. *Tijdschr. Econ. Soc. Geogr.* 2009, 100(5), 646–661.

Li, X.; Meng, F.; Uysal, M.; Mihalik, B. Understanding China's long-haul outbound travel market: an overlapped segmentation approach. *J. Bus. Res.* 2013, 66(6), 786–793.

Mander, J.; WeChat Rises to Become Fastest Growing Messaging App in the Last Year. http://blog.globalwebindex.net/WeChat (accessed July 17, 2014).

Oakes, T. Looking out to look in: the use of the periphery in China's geopolitical narratives. *Eur-Asia. Geogr. Econ.* 2012, 53(3), 315–326.

Pendzialek, B. *Performing Tourism: Chinese Outbound Organized Mass Tourists on Their Travels Through German Tourism Stages*; Catholic University of Eichstätt-Ingolstadt, in press.

Rakić, T.; Chambers, D. Rethinking the consumption of places. *Ann. Tourism Res.* 2012, 39(3), 1612–1633.

Schechner, R. *Performance Studies. An Introduction, 1st ed.*; Routledge: London, 2002.

Tourism Australia. Marketing Matters – China. 2013. http://www.tra.gov.au/asiafocus/Marketing-matters.html.

UNWTO. *Understanding Chinese Outbound Tourism. What the Chinese Blogosphere is Saying about EUROPE;* World Tourism Organization; European Travel Commission: Madrid, Brussels, 2012.

Xiang, Y. The characteristics of independent Chinese outbound tourists. *Tourism Plann. Dev.* 2013, 10(2), 134–148.

CHAPTER 10

MAINLAND CHINESE OUTBOUND TOURISM TO THE UNITED STATES: RECENT PROGRESS[1]

HONGBO LIU, XIANG (ROBERT) LI, and SCOTT C. JOHNSON

CONTENTS

[1]The authors would like to gratefully acknowledge Ms. Tiana Vinciguerra for her assistance to the preparation of this chapter.

10.1 INTRODUCTION

The size of the population and spending power have made Chinese tourists widely sought-after today, creating competition among destinations including the United States. With a quicker visa process and a huge market potential, it is becoming more feasible for American businesses to tap into the Chinese market.

The first Chinese organized tour group came to the United States in 1994 (Chen, 1998). However, the United States did not reach the Approved Destination Status (ADS) agreement with China until the end of 2007, which not only allowed Chinese tourists to undertake leisure travel in groups to the United States, but also allowed American destinations and travel service providers to promote or sell leisure trips in China (Tse, 2013). As in other ADS countries, the implementation of ADS scheme has increased the number of Chinese traveling to the United States dramatically (Arita et al., 2011).

Before signing the ADS agreement, the purpose of Mainland Chinese traveling to the United States was for either visiting friends and relatives (VFRs) or business (Jang et al., 2003). Despite Chinese government's discouragement and the US government's visa restrictions, sightseeing had always been an important component of the itinerary of Chinese VFR and business travelers to the United States (Jang et al., 2003), and United States would be many Chinese travelers' top choice if they had total freedom to choose their outbound destinations (Burnett et al., 2008). According to the Office of Travel and Tourism Industries (OTTI), the past 10 years have seen a continuous and significant growth of China's outbound tourism to the United States, especially after the implementation of ADS in 2008 (Fig. 10.1). Most recently, because of a new China–US visa agreement announced in November 2014, it has been predicted that as many as 7.3 million Chinese travelers will visit the United States by 2021, which is a fourfold increase compared with 2013 (The White House, 2014).

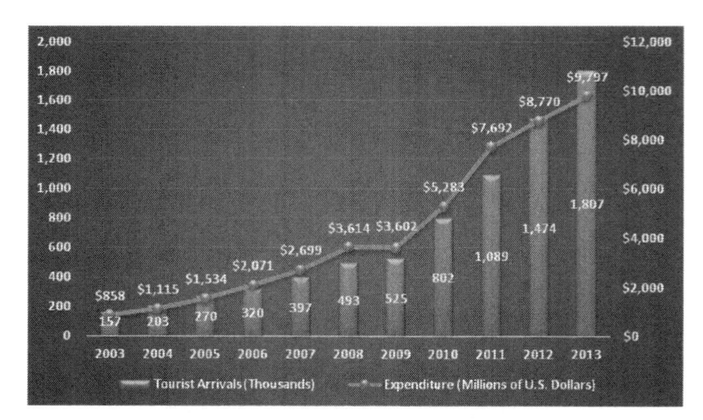

FIGURE 10.1 Tourist Arrivals and Expenditures from Mainland China to the United States (2003–2013).
(From US Department of Commerce, Office of Travel and Tourism Industries [OTTI]).

The purpose of this chapter is to provide an overview of existing research and recent trends of China's outbound tourism to the United States, as well as marketing insights for both academia and industry. There are 4 sections in this chapter: after this brief introduction about China's US-bound tourism market, in the next section, the relevant academic literature is reviewed; the third section provides a discussion about some recent important trends of China's US-bound tourism market, and the conclusion is presented in the last section.

10.2 A REVIEW OF LITERATURE

Despite the rapid growth of Chinese outbound tourists to the United States, studies on this particular market are still relatively scarce, in terms of both quantity and topics covered. It appears that existing studies have mainly covered the following areas: tourists' profiles (Cai et al., 2000; Cai et al., 2001; Jang et al., 2003; Agrusa et al., 2011), travel motivations (Johanson, 2008; Hua and Yoo, 2011; Li et al., 2009; Xu et al., 2010), travel constraints (Lai et al., 2013), tourists' expectations (Li et al., 2011), tourist's preferences (Agrusa et al., 2011), perceived destination image (Li and Stepchenkova, 2012; Stepchenkova and Li, 2012; 2014), market position (Li et al., 2015; Yun and Joppe, 2011), and shopping behaviors (Xu and McGehee, 2012), etc.

10.2.1 TOURIST PROFILE

Understanding the demographic and tripographic characteristics of tourists is critically important for the United States to take a share of the newly emerging Chinese outbound tourism market. Three earlier studies have provided some insights into the profiles of Chinese outbound tourists to the United States in 1990s. Based on the China Outbound Travel Market Survey conducted by the Pacific Asia Travel Association in 1994 (Cai et al., 2000) and the In-Flight survey of International Air Traveler collected in 1996 and 1997 (Cai et al., 2001; Jang et al., 2003), scholars found that in the 1990s, Chinese travelers who visited the United States were mainly middle-aged, relatively wealthy, well educated, married, and male-dominant; most of them worked for the government or state-run institutions and enterprises; most of them traveled for business purpose or VFR; their travel was organized by their employers, and they usually traveled with other company or government delegation members or alone; their length of stay was usually quite long (over 30 nights on average); shops, historical places, theme parks, and museums were attractions that Chinese tourists most like to visit at that time; and main tourism information sources include travel agency, word of mouth, and airline companies.

Nevertheless, Chinese outbound travelers have changed remarkably today. Comparing the latest statistics (OTTI, 2013) with those earlier studies (Cai et al., 2000; Cai et al., 2001; Jang et al., 2003), it is found that Chinese tourists are much younger and wealthier, spend much more during their trip, and there are more female travelers today. Online travel agencies such as Ctrip and Qunar have become important information sources for trip planning. Taking vacation/holiday, rather than business, has become the most frequent trip purpose, with education being another important travel purpose. This is consistent with a recent study in 2011 on the characteristics of Chinese travelers to Hawaii (Agrusa et al., 2011). The significantly high household income and trip expenditure indicate that upper-class tourists are dominating China's US-bound tourism market. In order to seize new opportunities of the emerging Chinese market, it is necessary for future research to identify the distinctive and up-to-date travel needs and characteristics of Chinese tourists to the United States.

10.2.2 TOURISTS' MOTIVATIONS AND CONSTRAINTS

Uncovering motivating factors and traveling barriers could help destination marketers figure out why and why not Chinese tourists travel to the United States. Several studies have tackled the motivations of Chinese tourists to the United States: Johanson's (2008) study identified 4 motivation factors for Chinese tourists traveling to Hawaii from a "pull factor perspective", such as surroundings and environment, culture and history, and so on. Hua and Yoo (2011) extracted 5 motivation factors, including ego enhancement, international exposure, communication opportunities, financial incentives, and destination stimuli. Li et al. (Li et al., 2009; Xu et al., 2010) found 6 similar underlying factors motivating Chinese tourists to visit the United States, including relaxation/escape, prestige, knowledge, job fulfillment, entertainment, and novelty, and 3 market segments were further identified based on the motivations: Enthusiastic, High Brow, and Reluctant travelers. Enthusiastic travelers are the most active and motivated US-bound travelers, who spend much more and are more likely to recommend the United States to others than the other 2 groups; Reluctant travelers are the least motivated tourists who show least enthusiasm for US-bound travel, and High Brow travelers are somewhere in between.

Overall speaking, the motivations of Chinese tourists who travel to the United States were similar to the motivations of Chinese tourists who travel to other Western countries for leisure purposes, which include the following key motivators, suggested by previous studies (Johanson, 2008; Zhou et al., 1998):

(a) Seeing something different
(b) Increasing knowledge about a foreign destination
(c) Rest and relaxation
(d) Being able to share travel experiences after returning home
(f) Experiencing a different lifestyle/culture
(g) Seeing famous attractions

Naturally, some of the motivations of US-bound travelers could be quite different from those of Chinese tourists to Asian destinations. It has been reported that accessibility, safety, and ease of visa application are some reasons for Chinese tourists to choose Asian destinations (Johanson, 2008), yet these factors could in turn be their constraints when traveling to Western countries. Using the Leisure Constraint Model, Lai et al. (2013) conducted a qualitative study on the travel constraints for Chinese tourists traveling to the United States and found that monetary concerns, security concerns, time, visa difficulties, travel inconvenience due to health concerns, having been to the United States before, and negative impression were key factors that prevent Chinese tourists traveling to the United States. Presumably, constraints for Chinese tourists traveling to the United States are largely similar to those who travel to other Western countries.

10.2.3 TOURISTS' EXPECTATIONS AND PREFERENCES

Understanding tourists' expectations and preferences could help enhance tourists' experience and even relieve certain travel constraints to some extent. A study (Li et al., 2011) on the US-bound Chinese tourists' expectations showed that besides the fundamental needs on tourism amenities and services, like cleanliness/hygiene and safety/security of accommodation and food, Chinese tourists have some particular expectations which are related to their cultural habits and lifestyle, such as Chinese-speaking receptionist, hot tea and authentic Chinese food, and so on. The study by Agrusa et al. (2011) reported that Chinese tourists to Hawaii were interested in foreign culture and local community but prefer Chinese food over local food during their stay; they also showed interest in marine sports and shopping for discounted and low-priced products. However, tourists' preferences vary among different market segments of tourists, for example, tourists of different ages. It is found that younger tourists prefer active tourism activities more, whereas elder tourists tend to prefer passive tourism activities (Agrusa et al., 2011). Preferences also vary across destinations based on their distinct destination attributes; therefore, studying the US-bound tourists' preferences at different market segments is highly recommended.

10.2.4 DESTINATION IMAGE

Perceived destination image may affect tourists' visit intention; negative impression on the United States could be a constraint for Chinese outbound tourists, as indicated before (Lai et al., 2013). Based on Echtner and Ritchie's (1993) "stereotypical," "affective," and "uniqueness" destination image model, Li and Stepchenkova (2012) tried to depict the collective mental image of the United States among Chinese outbound travelers. Their findings reveal that the United States is perceived as highly urban, featuring advanced economic development, an open and democratic system, high technology, and big cities; Chinese tourists can only think of those famous cities and attractions in the East and West coasts, whereas many leisure, cultural, and natural attractions have not yet been fully recognized by Chinese tourists, indicating a gap between product offerings and tourists' cognition. Furthermore, the authors (Stepchenkova and Li, 2012) examined the differences of destination image perceptions among 4 groups of past and potential Chinese tourists with different levels of outbound travel experiences. Results show that differences did exist: tourists with more travel experiences (i.e., who have been to the United States or destinations outside Asia) tend to see the United States as an open, democratic, and free society, with a friendly population, while less-experienced tourists (e.g., who have been to countries within Asia or potential overseas travelers) were more likely to view the United States as economically developed, scenic, and beautiful.

10.2.5 MARKET POSITION

Through accessing and comparing the perceived organic images of the United States and its competitor countries, researchers could understand the advantages and weakness of the United States in China's outbound travel market. Positioning analysis of Li et al. (2015) showed that the United States holds a unique position that is different from its 5 key competitors (Canada, Australia, the United Kingdom, France, and Switzerland). In terms of tourist activities, it was considered the best provider of casino visiting/gaming and city sightseeing but relatively weak at dining at local restaurants, visiting an art gallery/museum, and visiting historical/cultural heritage sites. In terms of destination attributes, the United States has relative strengths in terms of information, currency exchange rate, and opportunities to take photos but failed to impress on Chinese tourist for cleanliness, personal safety, and pleasant climate. However, food and safety are considered to be critically important attributes for Chinese outbound tourists (Li et al., 2015; Yun and Joppe, 2011; Agrusa et al., 2011; Li and Song 2013; Lai et al., 2013). A similar comparative study on 7 Chinese long-haul outbound destinations, including the United States, was conducted by Yun and Joppe (2011). Contrary to the former study, this study reported that visiting heritage sites is an area of strength for the United States compared with other destinations (Australia, Canada, France, New Zealand, the United Kingdom, and Germany). Other advantages for the United States include major events, hands-on learning activities, national/state parks, protected areas, and luxury and entertainment products (Yun and Joppe, 2011). From a medical tourism perspective, the United States outperforms its Asian counterparts (Korea, Thailand, India, and Singapore). With its leading medical service quality, Chinese customers prefer the United States for serious medical treatment, despite the differences in culture, language, and political system (Zhang et al., 2013).

10.2.6 SHOPPING BEHAVIOR

Shopping is one of the major travel motivations and most important activity for Chinese outbound tourists (Li et al., 2011; Cripps, 2013). Considered as the most active shoppers globally, Chinese tourists became the world's largest spenders by spending US$102 billion on foreign trips in 2013 (Roberts, 2014). Shopping has been ranked top among all the tourist activities for Chinese outbound travelers – over 80% of them would go shopping during their trips to the United States with an average trip expenditure of over US$7,000 (OTTI, 2013). A qualitative study on shopping behaviors and experiences of Chinese tourists to the United States (Xu and McGehee, 2012) reports that Chinese tourists gave a positive rating on shopping in the United States. Specifically, Chinese tourists were interested in a large variety of the US merchandise, including the US-branded products and luxuries, which were perceived as high-quality and less expensive; overall male tourists tended to be more satisfied with their shopping experiences than female tourists. Provision of Chinese-speaking sales assistants, Chinese shopping guides, and signage would help enhance their shopping experiences.

10.2.7 SUMMARY

Overall, although existing studies on Chinese outbound tourists to the United States provide some useful information for the US destination marketers, they are still limited in terms of the topics covered. There is a lack of studies on pre-trip decision-making process, post-trip evaluations, as well as specific tourists' behaviors during the trip to the United States, like destination choice among different states or cities of the United States, and tourists' preferences in the US Mainland. Further, few studies attempted to explore the underlying sociocultural reasons shaping Chinese tourists' attitudes and behaviors. In addition, almost all of the studies are from consumers' perspective; industrial perspectives warrant more attention. From a methodological perspective, most current studies are descriptive and qualitative; some studies used secondary data, indicating a need for more deep exploration and quantitative studies. Consumer attitudes and behaviors are constantly changing, which is especially true for fast developing economies like China. Thus, it is necessary to provide up-to-date marketing intelligence and monitor the market development. Some recent trends and characteristics of China's US-bound travel market are listed in the next section.

10.3 KEY SHIFTS OF CHINESE TRAVELERS TO THE UNITED STATES

Over 1.8 million Chinese visited the United States in 2013, which was an increase of nearly 23% over the previous year. In 7 of the last 9 years, the US tourism exports to China have grown by over 30% (OTTI, 2013), making China a top international visitor market for the United States (7th in terms of visitor volume and 6th in terms of spending). Although the United States is faced with fierce competition from Asian and European countries, as well as New Zealand and Australia (Grant, 2013), it is clear that China's long-haul travel to the United States is still evolving. In 2013, for the first time since 2007, there were more repeat visitors than first-time visitors from China traveling to the United States. Since 2012, a new trend emerges as more Chinese tourists visit only 1 destination in their trip to the United States, which is a clear departure from the old-fashioned multiple-destination group tours (Johnson, 2014), indicating that Chinese tourists are seeking more in-depth travel experience.

China has been experiencing dramatic socioeconomic transformation, which is reflected by the changing Chinese outbound travel market; key changes emerging recently in the market include the growth of independent travel, increasing need for more in-depth travel, contracted business travel, and new opportunities brought by the visa extension policy. It is necessary to track and take advantage of these important changes. A detailed discussion about 3 trends is presented below.

10.3.1 INDEPENDENT TRAVEL

Traditionally, Chinese tourists prefer all-inclusive group tour, especially in foreign destinations, which is mainly because of their Collectivism culture (Wong and Lau, 2001). However, China is now seeing a new trend in independent travel, which is already reflected in Chinese tourists' travel pattern in the United States. For instance, fewer Chinese tourists to the United States took travel packages or visited multiple destinations in 2013, and their average length of stay in the United States was about 5% shorter than that of 2012 (Johnson, 2014). This, together with the fact that over 55% Chinese tourists in 2013 are repeat visitors to the United States, signals a shifting demand to more "vacation"-focused or recreation-oriented travel (Li et al., 2008) with less structure and more free time during the trip.

Notably, independent travel for China is still not "fully independent" in its traditional sense, and the Chinese tour operators (online and traditional) remain a vital link in the supply chain for the more experienced Chinese leisure travelers. As a result, demand for smaller and less-structured group travel is growing rapidly. However, at the same time, large group travel to the US icon destinations is still in high demand albeit not growing as rapidly as in the past. The shift in demand was supported by the new travel law (came into effect in October 2012), which aims to protect travelers from poor service by tour operators and improve the image of Chinese travelers. The new law, combined with the changing demand, is likely to drive more aspirational and independent travel as reputable Chinese tour operators are getting stronger and adjusting their strategies to adhere to the new law and consumer demand.

10.3.2 BUSINESS TRAVEL

There is a continued downward shift in China's business travel to the United States in recent years: data from the Department of Homeland Security's I-94 form and the Survey of International Air Travelers (SIAT) report showed the percentage of business visas issued to China decreased from 42% in 2007 to approximately 17% in 2013. The dwindling share of business travel could be mainly attributed to leisure travel growth at a disproportionally faster pace – leisure travel by Chinese tourists to the United States has grown from 9% in 2007 to 32% in 2013 (OTTI, 2007, 2013). However, it has also been noted that, starting in 2012 and continuing into 2013, Chinese business travelers' length of stay was reduced, as was the number of destinations visited and spending on dining and beverages and hotels. These changes might be affected by the new anti-corruption campaign by the Chinese government, which establishes very strict regulations for government officials and executives of state-owned companies on their overseas travel.

In the past, many Chinese government's officials and senior-level employees of state-owned companies extended their US business trips to include leisure travel and visit destinations that may not be directly related to their primary purpose of trip. However, anecdotal evidences reported by American inbound operators and receptors indicate the anticorruption campaign restrictions have impacted the number of destinations that can be visited and activities participated in a trip. Some recent trends of business travel from China include the following:

(a) Business travelers (government, professionals) are traveling primarily for business and are less likely to extend their trip for leisure, VFR, or other purposes.
(b) Length of stay shifted to be 10 days or less, including airtime. This has forced a decline in the number of destinations visited from 5–6 destinations in the past to only 2–3 destinations now. Itineraries are focused on West Coast or East Coast rather than combining the 2 coasts.
(c) New caps on spending on lodging, dining, and beverage are now in place, which affected the type of hotels and restaurants used. Alcohol is no longer included as part of travel expenses paid by the state.
(d) Starting in 2012, transportation is only provided for the business portion of a trip, no longer for leisure travel.

10.3.3 VISA EXTENSION POLICY

It is widely acknowledged that visa restrictions have substantive negative impacts on bilateral tourist flow (Neumayer, 2010; Song et al., 2012; Li and Song, 2013). Conversely, visa relaxation and exemption policies could positively affect international tourist flow (Cheng, 2012; Lee et al., 2010). According to Brand USA, visa difficulty is the second biggest obstacle for Chinese tourists traveling to the United States, only following budgetary restraints (Baran, 2014). Most recently, in November 2014, a reciprocal visa extension agreement for tourists, businessmen, and students between China and the United States was reached during the Asia-Pacific Economic Cooperation summit in Beijing, China (Diamond, 2014). The new policy allows Chinese tourists and businessmen to obtain multivisit visas for up to 10 years and students to be able to obtain 5-year visas.

The agreement has substantial and immediate effect in China: Ctrip International in China reported a significant rise in inquiries about travel to the United States, and many Chinese tour operators believe there will be a surge in the number of the US-bound travelers in the near future (Wang, 2014). Considering the money, time, and efforts that will be saved, this visa extension policy is expected to simulate more repeat visitors, potential first-time visitors, and free independent travelers, enabling Chinese tourists to travel to more US cities other than the major icon destinations (Baran, 2014; Huang, 2014). The new visa rules might also boost the US real estate market among Chinese consumers (Wang, 2014), which will further affect their travel patterns: more frequent, less structured, and more in-depth.

10.3.4 INDUSTRY RESPONSES

Faced with the new trends of the Chinese market, American enterprises are making efforts to be China ready. American businesses have noticed the importance of online travel agency as information source of Chinese consumers. Expedia entered China's market through cooperation with eLong, followed by Priceline, who is working on their presence in China by creating a pact with Ctrip International, China's number one travel firm (Tsuruoka, 2014). American hotels also attempt to incorporate Chinese culture, to improve their Chinese clients' experience and satisfaction (Li et al., 2011). Many Starwood hotels are starting to add tea kettles to hotel rooms and have restaurant menus and welcome brochures translated into Chinese, as well as providing Chinese-style rice porridge and noodles (Sanburn, 2013). Hilton also launched its Chinese tourists cultivation program since 2011, "Huanying Program", which includes Mandarin-speaking hotel staff, authentic Chinese breakfast items, Chinese station on guest room TVs, slippers with the images of Chinese dragons, etc. (Murray, 2012). In order to facilitate Chinese tourists' shopping experience, Macy's has incorporated China's domestic payment system, UnionPay, into their payment system at all US locations and arranged Mandarin-speaking assistants at selected stores (Beddor, 2014).

10.4 CONCLUSIONS

In this chapter, the development history of China's outbound tourism to the United States was reviewed, the literature on China's outbound travel to the United States in the past 2 decades or so was scrutinized, and several recent trends and key shifts of this particular market were discussed, through which it is expected to shed some new light on both research and practices relating to this important market.

Prior studies are mainly about tourists' profiles and general tourists' behaviors, with a lack of attention on decision-making process, tourist's behavioral patterns within the United States, and their post-trip behavior and thoughts. Niche market is another overlooked research area. As Chinese tourists are becoming more sophisticated and seeking in-depth travel experiences, they are looking for less "mainstream" products, and clearly, the United States can provide a good variety of options from medical tourism, movie tourism to golf tourism. In addition, most of existing studies are conducted from tourists' perspective; very few studies involved industrial and host communities' perspective.

From a practical perspective, this study discussed some recent trends of China's US-bound travel market. The emerging pattern is that Chinese tourists are visiting fewer destinations and staying in the United States for fewer nights per trip, but there are increasingly more repeat visitors. In the past, both the demand and the products sold focused on the first-time visitors, who preferred the mega multi-destination trips and often combined various purposes in 1 trip. A shift of strategic focus seems to be necessary in light of the current market development. For smaller, lower-tier cities in middle or Western China, first-timers and building brand awareness should still be prioritized, but for tier-one cities like Beijing and Shanghai, it is time to shift the focus to market penetration and improving repeat visitation rate.

To understand the Chinese travel market, one has to closely monitor the country's tourism-related policies and political environment. The new China travel law is likely to drive some structural changes in Chinese travel agencies and help improve travel for the Chinese. It also started to weed out less reputable and low-margin businesses and help reputable operators and agents focus more on designing and promoting better travel products. In the short run, the recent anticorruption campaign in China will have a detrimental impact on the business-related travel by Chinese governmental officials and state-owned enterprises: for instance, because Chinese business travelers will visit fewer destinations, geo-equity among American destination will be reduced. Some secondary destinations, that is, cities that are not key "iconic" business travel destinations, may suffer accordingly (but they might benefit from repeat leisure travelers from China who are looking for in-depth experiences and who take advantage of the new visa extension policy).

Because of the fast development of the Chinese market, market intelligence can get updated easily, and it is hence necessary to track the up-to-date market changes. To ensure that Chinese tourists will continue visiting the United States and enjoy their US travel experiences, American tourism policy makers and travel marketers need to develop a more comprehensive strategic approach. Some recommendations include the following:

(a) Promotion and education of the US products with Chinese trade and consumers.
(b) Need to understand consumer needs and expectations better – by region and segment.
(c) Need to understand Chinese trade needs and business practices, including tour operators, agents, and receptive operators.

(d) Need to understand airline expansion goals to new ports and how the airlines are and will be partners for promotion, information, bookings, and air transport. Destinations need to not only attract more airlifts but also find a way to engage with the airport and airlines to retain that lift over time.

Although the United States is facing great challenges in getting a piece of Chinese market in the highly competitive global environment, undoubtedly, there is still a great potential in this emerging market. Still, both academia and industries need to keep up with the up-to-date trends and shifts of Chinese market. Going forward, the ever-changing Chinese population and their consumption interests will need to be closely observed, researched, and analyzed.

REFERENCES

Agrusa, J.; Kim, S. S.; Wang, K. C. Mainland Chinese tourists to Hawaii: their characteristics and preferences. *J. Travel Tourism Mark.* 2011, 28(3), 261-278.

Arita, S.; Edmonds, C.; La Croix, S.; Mak, J. Impact of approved destination status on Chinese travel abroad: an econometric analysis. *Tourism Econ.* 2011, 17(5), 983-996.

Baran, M. *Industry Lauds Pact with China for Bilateral Extensions of Visas*; Travel Weekly, 2014. http://www.travelweekly.com/Travel-News/Government/Industry-lauds-pact-with-China-for-bilateral-extensions-of-visas/ (Retrieved December 25, 2014).

Beddor, C. *U.S. Spending by Chinese Tourists Fueling Retail Boom*; The American Chamber of Commerce: Shanghai, 2014. http://insight.amcham-shanghai.org/u-s-spending-by-chinese-tourists-fueling-retail-boom/ (Retrieved December 25, 2014).

Burnett, T.; Cook, S.; Li, X. Emerging International Travel Markets: An In-Depth Profile of China's Society, Economy, and Travel Market. Report Published by the Travel Industry Association of America, 2008.

Cai, L. A.; Lehto, X. Y.; O'Leary, J. Profiling the U.S.-bound Chinese travelers by purpose of trip. *J. Hospit. Leisure Mark.* 2001, 7(4), 3-16.

Cai, L. A.; O'Leary, J.; Boger, C. Chinese travellers to the United States – an emerging market. *J. Vacation Mark.* 2000, 6(2), 131-144.

Chen, C. Rising Chinese overseas travel market and potential for the United States. Proceedings of the Third Graduate Education and Graduate Students Research Conference in Hospitality & Tourism, 1998, Houston, TX: University of Houston, 468-478.

Cheng, K. M. Tourism demand in Hong Kong: income, prices, and visa restrictions. *Curr. Issues Tourism*, 2012, 15(3), 167-181.

Cripps, K. *Chinese Travelers the World's Biggest Spenders*; CNN Report, 2013. http://www.cnn.com/2013/04/05/travel/china-tourists-spend/ (Retrieved December 25, 2014).

Diamond, J. *New Visa Policy Elevates U.S.-China Relations*; CNN Report, 2014. http://www.cnn.com/2014/11/10/politics/visa-10-years-obama-announces/ (Retrieved December 25, 2014).

Echtner, C. M.; Ritchie, J. R. B. The measurement of destination image: an empirical assessment. *J. Travel Res.* 1993, 31(4), 3-13.

Grant, M. *Top 25 Most Popular Destinations for Chinese Tourists.* 2013. http://skift.com/2013/09/03/top-25-most-popular-destinations-for-chinese-tourists/ (Retrieved December 25, 2014).

Hua, Y.; Yoo, J. J. Travel motivations of Mainland Chinese travelers to the United States. *J. China Tourism Res.* 2011, 7(4), 355-376.

Huang, J. *New China-US Visa Rules Expected to Boost LA Area Tourism*; Southern California Public Radio, 2014. http://www.scpr.org/blogs/multiamerican/2014/11/13/17557/china-us-visas-los-angeles-tourism/ (Retrieved December 25, 2014).

Jang, S.; Yu, L.; Pearson, T. E.. Chinese travelers to the United States: a comparison of business travel and visiting friends and relatives. *Tourism Geogr.* 2003, 5(1), 87-108.

Johanson, M. M. The outbound Mainland China market to the United States: uncovering motivations for future travel to Hawaii. *J. Hospit. Leisure Mark.* 2008, 16(1-2), 41-59.

Johnson, S. C. China 2013 visits and future path for the United States. *Courier*, 2014, 41(8), 14-15.

Lai, C.; Li, X.; Harrill, R. Chinese outbound tourists' perceived constraints to visiting the United States. *Tourism Manag.* 2013, 37, 136-146.

Lee, C. K.; Song, H. J.; Bendle, L. J. The impact of visa-free entry on outbound tourism: a case study of South Korean travellers visiting Japan. *Tourism Geogr.* 2010, 12(2), 302-323.

Li, S.; Song, H. Economic impacts of visa restrictions on tourism: a case of two events in China. *Ann. Tourism Res.* 2013, 43, 251-271.

Li, X.; Stepchenkova, S. Chinese outbound tourists' destination image of America: part I. *J. Travel Res.* 2012, 51(3), 250-266.

Li, X.; Cheng, C. K.; Kim, H.; Petrick, J. A systematic comparison of first-time and repeat visitors via a two-phase online survey. *Tourism Manag.* 2008, 29(2), 278-293.

Li, X.; Cheng, C. K.; Kim, H.; Li, X. Positioning USA in the Chinese outbound travel market. *J. Hospit. Tourism Res.* 2015, 39(1), 75-104.

Li, X.; Lai, C.; Harrill, R.; Kline, S.; Wang, L. When east meets west: an exploratory study on Chinese outbound tourists' travel expectations. *Tourism Manag.* 2011, 32(4), 741-749.

Li, X.; Xu, Y.; Weaver, P. A. Motivation segmentation of Chinese tourists visiting the US. *Tourism Anal.* 2009, 14(4), 515-520.

Murray, C. E. *Hilton Inculcating Chinese Travelers*; Hotel Interactive, 2012. http://www.hotelinteractive.com/article.aspx?articleID=27061 (Retrieved December 25, 2014).

Neumayer, E. Visa restrictions and bilateral travel. *Prof. Geogr*. 2010, 62(2), 171-181.

Office of Travel & Tourism Industries. *2007 Market Profile: China*; OTTI, 2007. http://travel.trade.gov/outreachpages/download_data_table/2007_China_Market_Profile.pdf (Retrieved December 25, 2014).

Office of Travel & Tourism Industries. *2013 Market Profile: China*; OTTI, 2013. http://travel.trade.gov/outreachpages/download_data_table/2013_China_Market_Profile.pdf (Retrieved December 25, 2014).

Roberts, A. *Chinese Tourists Spend the Most on Duty-Free Goods*; Bloomberg News, 2014. http://o.canada.com/travel/chinese-tourists-spend-the-most-on-duty-free-goods (Retrieved December 25, 2014).

Sanburn, J. *How the U.S. Travel Industry is Adapting to a Growing Wave of Chinese Tourists*; Time, 2013. http://business.time.com/2013/04/09/how-the-u-s-travel-industry-is-adapting-to-a-growing-wave-of-chinese-tourists/ (Retrieved December 25, 2014).

Song, H.; Gartner, W. C.; Tasci, A. D. A. Visa restrictions and their adverse economic and marketing implications-evidence from China. *Tourism Manag.* 2012, 33(2), 397-412.

Stepchenkova, S.; Li, X. Destination image: do top-of-mind associations say it all? *Ann. Tourism Res.* 2014, 45, 46-62.

Stepchenkova, S.; Li, X. Chinese outbound tourists' destination image of America: part II. *J. Travel Res.* 2012, 51(6), 687-703.

The White House. *Fact Sheet: Supporting American Job Growth And Strengthening Ties By Extending U.S./China Visa Validity for Tourists, Business Travelers, and Students*; The White House, 2014. http://www.whitehouse.gov/the-press-office/2014/11/10/fact-sheet-supporting-american-job-growth-and-strengthening-ties-extendi (Retrieved Dec. 29, 2014).

Tse, T. S. M. Chinese outbound tourism as a form of diplomacy. *Tourism Plann. Dev.* 2013, 10(2), 149-158.

Tsuruoka, D. *Expedia, Priceline Upping the Ante in China, Asia*; Investor's Business Daily, 2014. http://news.investors.com/technology/072514-710433-expedia-sticking-with-china-long-term-investment.htm?p=2TTR (Retrieved December 25, 2014).

Wang, Q. *Visa Change May Boost Tourism to the US*; China Daily, 2014. http://usa.chinadaily.com.cn/epaper/2014-11/28/content_18993417.htm (Retrieved December 25, 2014).

Wong, S.; Lau, E. Understanding the behavior of Hong Kong Chinese tourists on group tour packages. *J. Travel Res.* 2001, 40(1), 57-67.

Xu, Y.; McGehee, N. G. Shopping behavior of Chinese tourists visiting the United States: letting the shoppers do the talking. *Tourism Manag.* 2012, 33(2), 427-430.

Xu, Y.; Li, X.; Weaver, P. A. Examining the dimensions of travel behavior: a case of Chinese tourists visiting the United States. *Tourism Anal.* 2010, 15(3), 367-379.

Yun, D.; Joppe, M. Chinese perceptions of seven long-haul holiday destinations: focusing on activities, knowledge, and interest. *J. China Tourism Res.* 2011, 7(4), 459-489.

Zhang, J.; Seo, S.; Lee, H. The impact of psychological distance on Chinese customers when selecting an international healthcare service country. *Tourism Manag.* 2013, 35, 32-40.

Zhou, L.; King, B.; Turner, L. The China outbound market: an evaluation of key constraints and opportunities. *J. Vacation Mark.* 1998, 4(2), 109-119.

SECTION III

CASES & PERSPECTIVES

CHAPTER 11

SERVICE EXPECTATIONS OF CHINESE OUTBOUND TOURISTS

KEVIN KAM FUNG SO, WEI LIU, YING WANG, and BEVERLEY A. SPARKS

CONTENTS

11.1 INTRODUCTION

The China outbound tourist market has become increasingly important to the global tourism industry (Zhang, 2006). In 2013, nearly 100 million Chinese traveled abroad, spending US$102 billion, making China the world's largest source market by both arrivals and expenditure (Xinhuanet, 2014). An increasing disposable income, together with the strong performance of Chinese Renminbi (RMB), more relaxed passport and visa application procedures, and enhanced foreign political and economic collaboration (China Tourism Academy, 2013) contribute to the further expansion of the China outbound market.

The tremendous growth of this market demands service providers to have an enhanced understanding of the service expectations of this market. Described as an essential step toward marketing success in this specific market for all service providers (Lai et al., 2013), this understanding enables better alignment between service provision and customer needs, ultimately achieving greater tourist satisfaction that will sustain the growth from this market in a longer term. This chapter, by drawing on the extant literature on general service expectations, Chinese outbound tourism, and the salient Chinese core cultural values, provides an overview of the relevant research and sets forth a model to describe the sources and determinants of service expectations of Chinese tourists. This model incorporates controllable and uncontrollable factors, as well as several characteristic Chinese cultural values, that are likely to be important in the formation of Chinese tourists' service expectations.

The remainder of this chapter is arranged as follows. The ensuing section presents a brief introduction of how the Chinese outbound tourist market has evolved, followed by a review of the conceptual roots and definitions of service expectations. The subsequent section briefly reviews the general sources of service expectations captured in several dominant expectation models proposed in the literature. Next, the chapter provides a review and discussion of 4 salient Chinese cultural values with relevant conceptual and empirical research that illustrates how each value dimension affects service expectations. The chapter concludes with a conceptual model of influencers of service expectations of Chinese outbound tourists and a discussion of potential future research areas.

11.2 CHINESE OUTBOUND TOURIST MARKET

Chinese outbound tourism can be traced back to the launch of a series of economic reforms, the open door policy, and the approved destination status system (Andreu et al., 2013; Arlt, 2006, 2013; Yun and Joppe, 2011). Some researchers refer to this growth of outbound tourism as "the First Wave" (e.g., Arlt, 2013). Previous research (Arlt, 2006; Keating and Kriz, 2008; Li et al., 2011; Long, 2012; Xu and McGehee, 2012; Yun and Joppe, 2011) suggests that the majority of the First Wave tourists prefer all-inclusive package tours, famous sights in major cities or destinations, and luxury brand products instead of accommodation and food.

Entering the new decade, a Second Wave of Chinese outbound tourism is emerging (Arlt, 2013; Kristensen, 2013; Li et al., 2011; Zeng and Go, 2013), most notably with an increasing number of independent travelers who prefer greater control of their itineraries and travel pace. This new wave of independent travelers has been profiled as relatively young yet experienced tourists (Arlt, 2013), well-educated with good language skills, and consisting of a large proportion of experience seekers and self-challengers (Pearce et al., 2013; Sparks and Pan, 2009; Wu and Pearce, 2014). In a recent study of young Chinese independent travelers to Western Europe, Prayag et al. (2015) found that these travelers have different service expectations from group tourists in many ways, such as a preference for dinning outside of hotels and an expectation for signage and other information to be provided in Chinese, even for those who possess good language skills. Research also indicates that Chinese tourists may have particular motivations, service expectations, and preferences due to several factors, such as cultural and socioeconomic differences, which have not been systematically investigated (Li et al., 2011). A thorough comprehension of the various underlying sources of service expectations necessitates a discussion of its theoretical background.

11.3 THEORETICAL BACKGROUND OF SERVICE EXPECTATION

The term "expectations" has dominated discussion in the customer satisfaction/dissatisfaction (CS/D) and service quality literatures, offering different views from a conceptualization perspective (Boulding et al., 1993; Parasuraman et al., 1988; Zeithaml et al., 1991; Zeithamlet al., 1993). Among various types of expectations discussed, 2 dominant standards of expectations identified in the literature are predicted expectations and normative expectations. In the CS/D literature (e.g., Miller, 1977; Prakash, 1984; Swan and Trawick, 1980), on the one hand, expectations are viewed as a prediction of future event, suggesting that customers form expectations about what *will* happen in their next service encounter with a company. On the other hand, normative expectations are seen as the *ideal* or *desired* standards to measure service quality (e.g., Parasuraman et al, 1991; Parasuraman et al., 1988; Parasuraman et al., 1994; Prakash, 1984; Zeithaml et al., 1993), meaning that customers form their expectations about what *should* happen in their next service encounter with a company. Following this definition, Parasuraman, et al. (1985) propose the gap model in which expectations play a contrast role when customers evaluate the overall service quality based on the comparison between expectations and perceptions of different components of service, highlighting the importance of gauging customer service expectations.

In line with previous research on Chinese tourists' service expectations (Li et al., 2011), our chapter follows the definition of normative expectations. Research on factors influencing Chinese tourists' expectations and service evaluations indicate that Chinese tourists have different travel expectations compared with their Western counterparts. In general, Chinese tourists seek scenic beauty, safety, famous attractions, different cultures, and service in hotels and restaurants during their trips (Yu and Weiler, 2001). Chinese tourists' expectations mainly depend on their travel purposes (Wang et al., 2008), and their satisfaction level may vary based on their gender, education level, as well as travel party (Yu and Weiler, 2001). More recent research (Andreu et al., 2013; Fountain et al., 2010; Li et al., 2011) points out the cultural, social, and economic reasons behind Chinese tourists' expectations and preferences. To provide an overview of how Chinese outbound tourists' service expectations are formed, it is essential to first review the more universal factors and sources that influence service expectations in general.

11.4 GENERAL FACTORS AFFECTING SERVICE EXPECTATIONS

While there is little doubt that expectations play a central role in conceptualizing customer satisfaction and service quality, the sources of shaping expectations are still debatable (Li et al., 2011). In the marketing literature, Zeithaml et al. (1993) developed a conceptual model articulating the determinants of normative customer expectations including uncontrollable factors, such as enduring service intensifiers (e.g., derived expectations and personal service philosophies), personal needs, word-of-mouth communication, and past experience, and controllable factors, such as explicit service promises (e.g., advertising, personal selling, contracts, and other communications), as well as implicit service promises (e.g., tangibles and price). Similarly, Robledo (2001) proposes an expectations management model, which identifies the main sources of expectations, including past experience, reputation or corporate image, formal recommendations (e.g., professional advice provided by travel agents), informal recommendations (e.g., word-of-mouth), personal needs, promotion, and price. Furthermore, as a result of the advent of Internet, electronic word-of-mouth (eWOM) communications, such as online reviews, blogs, destination websites, as well as other online travel information sources, can also influence customer expectations. For example, recent research confirms the influence of web reviews on hotel guests' expectations (Mauri and Minazzi, 2013).

While previous conceptual models contribute significantly to our current understanding of how service expectations are formed in a general sense, Donthu and Yoo (1998) argue that conceptualizations of this kind encapsulate determinants of service expectations relevant primarily in a domestic market. When serving an international market, it is necessary to incorporate the possible impact of cultural differences, due to their influence on service expectations, behavior, and attitudes. Research in both marketing (Donthu and Yoo, 1998) and tourism (Li et al., 2011) offers empirical support for the influence of culture on consumer service expectations. Given that general sources of services expectations are widely documented in the marketing literature (e.g., Robledo, 2001; Zeithaml et al., 1993), the discussion in the remainder of this section focuses on several important Chinese cultural values including face, harmony, interdependence, and group orientation, and how they affect service expectations.

11.4.1 FACE

Face is a central cultural value that has a pervasive influence on interpersonal relations among Chinese (Yau, 1988); it is the public, social, and fluid aspect of the self-concept (Ting-Toomey and Kurogi, 1998). Face may be threatened by lack of respect, feelings of being ignored, or being challenged, often publicly through interpersonal interactions (Chan et al., 2009). According to Mok and DeFranco (2000), to give other people face refers to allowing others to escape the humiliation implicit in not knowing, failing to understand, having been mistaken, or being inferior to others. From this perspective, Chinese service customers would expect service providers to recognize their importance and accord them with the respect or honor that they normally expect, and as such, a face-loss may be experienced when the providers fail to do so (Fox, 2008; Lee et al., 2013). Although face is pervasive in the Japanese and Korean cultures, the Chinese have been described to have higher concern for face in general, and an individual's social standing is more important during interactions with others (Gao, 1998; Oetzel et al., 2001). In a cross-cultural comparison study, Ting-Toomey et al. (1991) found that the Chinese are more likely to maintain others' face than the Japanese and Koreans, and they also have a significantly higher degree of self-face maintenance than the Korean group, underlying the importance of face concern and maintenance in interacting with Chinese travelers.

The value of face is particularly important to Chinese consumers in the service encounter situation (Imrie et al., 2002). Hoare and Butcher (2008) suggest that giving or protecting one's face is seen extremely important in the presence of his family and friends, making this aspect central to the host of a dining group. Thus, it is expected that a service provider would protect or give face to the host of a dining party in front of his/her family, friends, or guests. The authors further suggest that when a customer feels that his status has been enhanced, his face-related expectations are more likely to be fulfilled, and consequently increasing satisfaction with the experience, contributing to forming a long-term relationship when face is present in the service encounter. Similarly, Gilbert and Tsao (2000) also reported that when a hotel consistently provides opportunities for Chinese customers to gain "face", the guests would introduce new clients from their social network.

Customer evaluation of whether face is given during a service encounter is found to be determined by various service interactions between the customer and the service provider, such as service personnel's treatment of guests in a service failure situation, the speed of a response, and tone of voice from wait staff (Hoare et al., 2011). Hoare et al. (2011) suggest that wait staff from non-Chinese ethnic groups need to consider face-sensitive aspects and avoid embarrassing Chinese tourists with regard to their table manners and language ability. In addition, whenever Chinese diners have problems grasping the foreign language, wait staff need to pay close attention to the speed at which they speak and their tone of voice in order to show respect and consideration. Furthermore, in-group wait staff (i.e., Chinese ethnicity) are expected to treat them according to well-known traditions, such as respecting senior citizens and caring for children. Therefore, service employees should be trained to understand Chinese tourists' face-related expectations and how their behaviors in delivering the service offering might affect customer evaluation of the service encounter.

11.4.2 HARMONY

Another salient cultural value that characterizes the Chinese society is harmony, which represents a person's inner balance, as well as the balance between individuals and the natural and social surroundings (Hoare and Butcher, 2008). Although harmony is relevant to several Asian countries, a cross-cultural comparison of Confucian values in China, Korea, Japan, and Taiwan shows that interpersonal harmony is more salient in Chinese consumers. Tourism and hospitality research indicates that harmony dominates and underlines the Chinese tourists' behavior (Kwek and Lee, 2010) and is positively correlated with their satisfaction (Hoare and Butcher, 2008), which is conceptually related to service expectations.

Harmony has profound implications for Chinese consumer behavior and expectations, particularly concerning complaints. In acknowledging such relevance, Mok and DeFranco (2000) and Yau (1988) suggest a lower tendency among Chinese to complain even when they encounter unsatisfactory products or services, because taking public action is often seen as extreme behavior. They are more likely to switch service providers without making it known to the previous providers, unless they are extremely dissatisfied (Mok and DeFranco, 2000). Heung and Lam (2003) also suggest that the Chinese tend to adopt an unassertive style of communication, which often results in avoidance or silence even if they are dissatisfied. The extent to which this conclusion still holds true may seem questionable, particularly with recent anecdotal evidences suggesting that the new wave of Chinese customers are increasingly demanding and often do not hesitate to complain, especially when language barrier is not an issue. However, such a cultural influence is unlikely to have disappeared. While younger generations of Chinese are more likely to voice their discontent than the First Wave of tourists, they are still more conservative in these situations compared with their Western counterparts. Therefore, Chinese travelers are less likely to provide feedback on their satisfaction/dissatisfaction with the service offering, as evidenced in recent studies by Au et al. (2010) and Au et al. (2014). In order to obtain such data, managers need to play a more active role, rather than waiting for the consumer to give feedback (Yau, 1988).

Research demonstrates that the Chinese has a desire to relate to others in a harmonious manner in situations where disputes occur (Lee and Sparks, 2007), indicating that when service fails, harmony may be attained if an aggrieved party shows goodwill, diplomacy, patience, understanding, and tolerance toward the service provider and vice versa. Furthermore, when a service failure occurs and the hotel staff fail to execute a service recovery, Chinese tend to refrain from being too assertive, such as complaining strongly or demanding immediate resolution, because of their desire to deal with others harmoniously and to avoid conflict and confrontation in hospitality service encounter contexts (Lee and Sparks, 2007; Mattila and Patterson, 2004a,b). Thus, when dealing with a service failure, service employees should ensure that the communication leads to the maintenance of harmony that Chinese tourists expect (Lee and Sparks, 2007).

Another aspect of harmony is the unique Chinese dining culture, which embraces the notion of slow eating pace (Hoare and Butcher, 2008). Slow eating is regarded as healthy and elegant, reflecting a harmonious dining atmosphere (Yan, 2000); it offers an atmosphere that is conducive to familial harmony and favorable business outcomes (Hoare and Butcher, 2008). In contrast, quick eating and overindulgence are harmful and can disrupt harmony, adversely affecting satisfaction (Hoare and Butcher, 2008). Furthermore, the notion of harmony is also documented in a recent study in which Chinese travelers expressed that food selection is not their major concern and they are willing to accommodate their family or friends' food preference to maintain group harmony (Chang et al., 2010). Therefore, it is clear that Chinese outbound tourists hold certain expectations relating to their relatively slow eating pace and selection of menu preferred by the group.

11.4.3 INTERDEPENDENCE

Interdependence relates to the fundamental connectedness of human beings to each other (Wong and Ahuvia, 1998), stressing relationships (Mattila and Patterson, 2004a,b). Interdependence has been consistently identified as an important Chinese cultural value that requires special attention from a marketing point of view (Mok and DeFranco, 2000; Yau, 1988). This dimension relates to the flexibility of the Chinese in dealing with interpersonal relations coming from the principle of "doing favors", which signifies one's honor to another (Kindel, 1983) and suggests reciprocity among people as a cause-and-effect relationship (Yau, 1988).

One way to maintain relationship among Chinese is by the presentation of gifts, a symbol of courtesy, respect, appreciation, and friendship (Mok and DeFranco, 2000). When Chinese people visit their relatives or friends, they bring something even if they are not wealthy, because a visitor without a gift is seen as unreasonable or impolite. Yau (1988) further argues that interdependence is especially meaningful to the study of gift-giving behavior of Chinese consumers, who use gift-giving as a way to build relationships.

Research examining the cultural differences between Asian tourist markets and Australian hosts found that gift-giving in Chinese society is an expression of appreciation, gratitude, and remembrance (Reisinger and Turner, 2002). In retail services, when selling products or souvenirs that are normally purchased as gifts, the packaging of these products is extremely important. Service providers should package the items prestigiously and beautifully in red, symbolizing happiness and good luck (Yau, 1988). The prices of product or services offered by a well-known organization or brand can be set higher than competitors' as the Chinese believe in established brands and companies (Yau, 1988). More recent research on service expectations of Chinese outbound tourists by Li et al. (2011) also suggests that most Chinese tourists appreciate traditional collectivism values such as family duty and caring for the children and therefore, when they travel overseas, purchasing gifts for seniors, children, and friends is almost an obligation.

Understanding this interdependence principle is extremely important for anyone who wants to successfully conduct business with Chinese people (Mok and DeFranco, 2000). Instead of aggressive hard-sell methods, the Chinese often expect personal and business relationships to be continuous and broad in scope (e.g., maintaining a long-term relationship and extending such relationship to incorporate other business areas; Mok and DeFranco, 2000). While limited research exists on how interdependence affects service evaluations, research found that people with high interdependent self-construal are highly sensitive to interpersonal treatment during service encounters, and for these customers, service failures led to less favorable service quality ratings than independent customers (Patterson and Mattila, 2008). Such finding underlies the importance of interpersonal service interaction between the customer and the service provider, particularly in cultures where interdependence is highly valued. Therefore, when serving Chinese travelers, a service provider's interpersonal interaction requires further attention.

11.4.4 GROUP ORIENTATION

The cultural aspect of group orientation relates to what Hofstede (1980) describes as collectivism. Group orientation characterizes the Chinese and other Asians and represents the degree to which individuals are integrated into groups. The collectivistic nature of the Chinese is reflected in their family and kinship system serving as a basis for relating to others, meaning continuous and long-lasting human ties, which do not have clearly defined boundaries (Hsu, 1968). The seminal work of Yau (1988) provides particular insight into marketing implications of group orientation that shape service expectations of this unique group of customers. Specifically, first, Chinese consumers often rely on the rumour moiety of the informal channel, such as word-of-mouth communication, rather than what is actually claimed for the product officially (Kindle, 1985). Second, Chinese consumers tend to endeavor to conform to group norms and therefore, expect to purchase the same brand or product other members of the group recommend, rather than deviate from the group norm to switch to a competitive product (Yau, 1988). As such, their satisfaction with a product or service may not be derived solely from one's expectations toward, or disconfirmation with, the offering, but from other members of the family (Yau, 1988). However, the Chinese are only group oriented toward those social units with whom interactions have been established, but appear to be quite suspicious and cold toward strangers (Yau, 1988). Thus, interdependence is more salient within their closer social groups.

The influence of group orientation is also evidenced in several tourism and hospitality studies. Chinese travelers have been found to place emphasis on practising forbearance conforming to the interests of a wider group rather than individual desires (Kwek and Lee, 2010). Hoare and Butcher (2008) also suggest that group conformity is the reason that Chinese consumers tend to stay with the same brand that is recommended by other in-group members. Furthermore, the tourism literature consistently shows that Chinese people prefer to travel in groups rather than individually (Armstrong and Mok, 1995). While recent research shows that the Second Wave of Chinese travelers may include a growing number of independent travelers, package tours would still be the most practical mode for the foreseeable future (Liu et al., 2013). When serving tour groups, instead of catering to individual needs and requests, hotels need to offer special arrangements for groups, such as assigning rooms to tour group members on the same floor or preparing special breakfast items for the groups, helping the hotel exceed customers' expectations (Mok and DeFranco, 2000).

11.5 CONCLUSION AND POTENTIAL FUTURE RESEARCH AREAS

In conclusion, this chapter reviews the extant marketing, tourism, and hospitality literature relating to service expectations and how several Chinese cultural values may affect the formation of these expectations. On the basis of this review, we propose a conceptual model of influencers of service expectations of Chinese outbound tourists (see Fig. 11.1). Building on the work of Zeithaml et al. (1993) and Robledo (2001), this model includes general or universal sources of service expectations. These factors can be broadly divided into predominantly controllable or uncontrollable factors by service providers. In addition, the model takes into consideration the effect of culture on service expectations by incorporating the 4 aforementioned Chinese value dimensions, which are proposed to influence Chinese tourists' service expectations directly. We suggest that these values should be considered when formulating strategies for managing expectations. For instance, decisions on communication channels require the consideration of culture, given that the effectiveness of certain factors, such as word-of-mouth communications, may vary depending on group orientation.

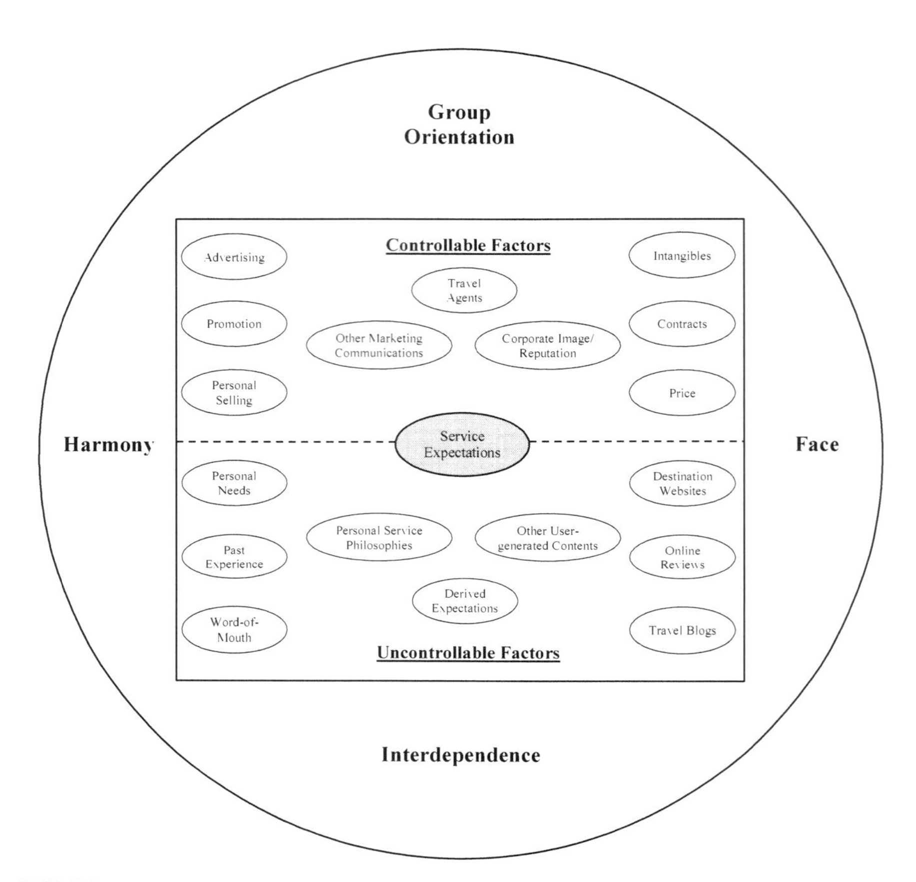

FIGURE 11.1 Model of Influencers of Service Expectations of Chinese Outbound Tourists.

While research supports the importance of these value dimensions in the service settings, it should be noted that their relevance is discussed in general terms based on existing research. There is a lack of empirical research investigating the 4 cultural values collectively in the context of service expectations. It is likely that these cultural values affect service expectations simultaneously. However, as this chapter demonstrates, some values might be more relevant in certain situations or stages of the travel experience. For example, harmony may be more salient in conflict or service failure situations, whereas group orientation would have a strong impact when deciding trip mode, as well as social interactions during the trip. Research suggests that life in contemporary China has undergone significant cultural change as a result of modernization (Faure and Fang, 2008). As such, the new wave of Chinese outbound tourists could have changed substantially from previous generations.

However, the relevance of the cultural values discussed in this chapter, such as harmony and group orientation, should still hold, but their impacts may have weakened as a result of the significant change in the society, as well as people's life. Such an assumption could be evaluated in future research that examines cultural values held by the new wave of Chinese outbound tourists. Future studies need to investigate potential generational differences in these values and how they affect service expectations. Tourists' expectations are constantly revised and updated through various sources, such as customers' experiences and information obtained from tourist information centers during the trip. From this perspective, it would be of practical relevance to examine how Chinese tourists' service expectations change at various stages of their travel experience, including before, during, and after the trip. Another area worthy of investigation is the relative explanatory power of each key determinant included in the model. This could help marketers determine the most effective marketing activities to influence customers' service expectations. In conclusion, this chapter provides an overview of service expectations of Chinese outbound tourists and a conceptual model that serves as an organizational framework that guides future research in this increasingly important topic area.

REFERENCES

Andreu, R.; Claver, E.; Quer, D. Chinese outbound tourism: new challenges for European tourism. *Enlightening Tourism A Pathmaking J.* 2013, 3(1), 44-58.

Arlt, W. G. *China's Outbound Tourism*; Routledge: New York, 2006.

Arlt, W. G. The second wave of Chinese outbound tourism. *Tourism Plann. Dev.* 2013, 10(2), 126-133.

Armstrong, R. W.; Mok, C. Leisure travel destination choice criteria of Hong Kong residents. *J. Travel Tourism Mark.* 1995, 4(1), 99-104.

Au, N.; Buhalis, D.; Law, R. Online complaining behavior in mainland China hotels: the perception of Chinese and non-Chinese customers. *Int. J. Hospit. Tourism Adm.* 2014, 15(3), 248-274.

Au, N.; Law, R.; Buhalis, D. *The Impact of Culture on e-Complaints: Evidence from the Chinese Consumers in Hospitality Organizations*. In *Information and Communication Technologies in Tourism 2010*; Gretzel, U., Law, R., Fuchs M., Eds.; Springer-Verlag: Vienna, Switzerland, 2010; p 285-296.

Boulding, W.; Kalra, A.; Staelin, R.; Zeithaml, V. A. A dynamic process model of service quality: from expectations to behavioral intentions. *J. Mark. Res.* 1993, 30(1), 7-27.

Chan, H.; Wan, L. C.; Sin, L. Y. The contrasting effects of culture on consumer tolerance: interpersonal face and impersonal fate. *J. Consum. Res.* 2009, 36(2), 292-304.

Chang, R. C.; Kivela, J.; Mak, A. H. Food preferences of Chinese tourists. *Ann. Tourism Res.* 2010, 37(4), 989-1011.

China Tourism Academy. *Research Findings of the Blue Books of China Tourism Economy 2013 (No. 5)*. 2013. http://www.ctaweb.org/html/2013-1/2013-1-9-14-45-23752.html

Donthu, N.; Yoo, B. Cultural influences on service quality expectations. *J. Serv. Res.* 1998, 1(2), 178-186.

Faure, G. O.; Fang, T. Changing Chinese values: keeping up with paradoxes. *Int. Bus. Rev.* 2008, 17(2), 194-207.

Fountain, J.; Espiner, S.; Xie, X. A cultural framing of nature: Chinese tourists' motivations for, expectations of, and satisfaction with, their New Zealand tourist experience. *Tourism Rev. Int.* 2010, 14(2-3), 2-3.

Fox, S. China's changing culture and etiquette. *China Bus. Rev.* 2008. https://www.chinabusinessreview.com/public/0807/fox.html (Retrieved March 15, 2012).

Gao, G. An initial analysis of the effects of face and concern for other in Chinese interpersonal communication. *Int. J. Intercult. Relations*, 1998, 22(4), 467-482.

Gilbert, D.; Tsao, J. Exploring Chinese cultural influences and hospitality marketing relationships. *Int. J. Contemp. Hospit. Manage.* 2000, 12(1), 45-54.

Heung, V. C.; Lam, T. Customer complaint behaviour towards hotel restaurant services. *Int. J. Contemp. Hospit. Manage.* 2003, 15(5), 283-289.

Hoare, R. J.; Butcher, K. Do Chinese cultural values affect customer satisfaction/loyalty? *Int. J. Contemp. Hospit. Manage.* 2008, 20(2), 156-171.

Hoare, R. J.; Butcher, K.; O'Brien, D. Understanding Chinese diners in an overseas context: a cultural perspective. *J. Hospit. Tourism Res.* 2011, 35(3), 358-380.

Hofstede, G. *Culture's Consequences: International Differences in Work-Related Values*; Sage: Newbury Park, CA, 1980.

Hsu, F. L. K. Psychological Anthropology: An Essential Defect and its Remedy. Paper Presented at the 1968 Annual Meeting of the American Anthropologist Association: Seattle, Washington, 1968

Imrie, B. C.; Cadogan, J. W.; McNaughton, R. The service quality construct on a global stage. *Managing Serv. Qual.* 2002, 12(1), 10-18.

Keating, B.; Kriz, A. Outbound tourism from China: literature review and research agenda. *J. Hospit. Tourism Manage.* 2008, 15(1), 32-41.

Kindel, T. I. A partial theory of Chinese consumer behavior: marketing strategy implications. *Hong Kong J. Bus. Manage.* 1983, 1(1), 97-109.

Kristensen, A. E. Travel and social media in China: from transit hubs to stardom. *Tourism Plann. Dev.* 2013, 10(2), 169-177.

Kwek, A.; Lee, Y. S. Chinese tourists and Confucianism. *Asia Pac. J. Tourism Res.* 2010, 15(2), 129-141.

Lai, C.; Li, X. R.; Harrill, R. Chinese outbound tourists' perceived constraints to visiting the United States. *Tourism Manage.* 2013, 37, 136-146.

Lee, Y. L.; Sparks, B. Appraising tourism and hospitality service failure events: a Chinese perspective. *J. Hospit. Tourism Res.* 2007, 31(4), 504-529.

Lee, Y. L.; Sparks, B.; Butcher, K. Service encounters and face loss: issues of failures, fairness, and context. *Int. J. Hospit. Manage.* 2013, 34, 384-393.

Li, X.; Lai, C.; Harrill, R.; Kline, S.; Wang, L. When east meets west: an exploratory study on Chinese outbound tourists' travel expectations. *Tourism Manage.* 2011, 32(4), 741-749.

Liu, P.; Lin, Q.; Zhou, L.; Chandnani, R. Preferences and attitudes of Chinese outbound travelers: the hotel industry welcomes a growing market segment. *Cornell Hospit. Rep.* 2013, 13(4), 1-16.

Long, H. An Exploratory Study of Chinese Tourists' Expectations and Preferences of New Zealand as a Travel Destination. Master Dissertation, AUT University, New Zealand, 2012.

Mattila, A. S.; Patterson, P. G. The impact of culture on consumers' perceptions of service recovery efforts. *J. Retailing*, 2004a, 80(3), 196-206.

Mattila, A. S.; Patterson, P. G. Service recovery and fairness perceptions in collectivist and individualist contexts. *J. Serv. Res.* 2004b, 6(4), 336-346.

Mauri, A. G.; Minazzi, R. Web reviews influence on expectations and purchasing intentions of hotel potential customers. *Int. J. Hospit. Manage.* 2013, 34, 99-107.

Miller, J. A. *Studying Satisfaction, Modifying Models, Eliciting Expectations, Posing Problems, and Making Meaningful Measurements.* In *Conceptualization and Measurement of Consumer Satisfaction and Dissatisfaction*; Hunt, H. K., Ed.; School of Business, Indiana University: Bloomington, 1977; p 72-91.

Mok, C.; DeFranco, A. L. Chinese cultural values: their implications for travel and tourism marketing. *J. Travel Tourism Mark.* 2000, 8(2), 99-114.

Oetzel, J.; Ting-Toomey, S.; Masumoto, T.; Yokochi, Y.; Pan, X.; Takai, J.; Wilcox, R. Face and facework in conflict: a cross-cultural comparison of China, Germany, Japan, and the United States. *Commun. Monogr.* 2001, 68(3), 235-258.

Parasuraman, A.; Berry, L. L.; Zeithaml, V. A. Refinement and reassessment of the SERVQUAL scale. *J. Retailing*, 1991, 67(4), 420-450.

Parasuraman, A.; Zeithaml, V. A.; Berry, L. L. A conceptual model of service quality and its implications for future research. *J. Mark.* 1985, 49(4), 41-50.

Parasuraman, A.; Zeithaml, V. A.; Berry, L. L. SERVQUAL: a multiple-item scale for measuring consumer perceptions of service quality. *J. Retailing*, 1988, 64(1), 12-37.

Parasuraman, A.; Zeithaml, V. A.; Berry, L. L. Reassessment of expectations as a comparison standard in measuring service quality: implications for further research. *J. Mark.* 1994, 58(1), 111-124.

Patterson, P. G.; Mattila, A. S. An examination of the impact of cultural orientation and familiarity in service encounter evaluations. *Int. J. Serv. Ind. Manage.* 2008, 19(5), 662-681.

Pearce, P. L.; Wu, M. Y.; De Carlo, M.; Rossi, A. Contemporary experiences of Chinese tourists in Italy: an on-site analysis in Milan. *Tourism Manage. Perspect.* 2013, 7, 34-37.

Prakash, V. Validity and reliability of the confirmation of expectations paradigm as a determinant of consumer satisfaction. *J. Acad. Mark. Sci.* 1984, 12(4), 63-76.

Prayag, G.; Cohen, S. A.; Yan, H. Potential Chinese travellers to Western Europe: segmenting motivations and service expectations. *Curr. Issues Tourism*, 2015, 18(8), 725-743.

Reisinger, Y.; Turner, L. W. Cultural differences between Asian tourist markets and Australian hosts, part 1. *J. Travel Res.* 2002, 40(3), 295-315.

Robledo, M. A. Measuring and managing service quality: integrating customer expectations. *Managing Serv. Qual.* 2001, 11(1), 22-31.

Sparks, B. A.; Pan, G. W. Chinese outbound tourists: understanding their attitudes, constraints and use of information sources. *Tourism Manage.* 2009, 30(4), 483-494.

Swan, J. E.; Trawick, I. F. *Satisfaction Related to Predictive vs. Desired Expectations.* In *Refining Concepts and Measures of Consumer Satisfaction and Complaining Behavior*; Hunt H. K., Day R. L., Eds.; Indiana University: Bloomington, 1980;p 7-12.

Ting-Toomey, S.; Gao, G.; Trubisky, P.; Yang, Z.; Kim, H. S.; Lin, S. L.; Nishida, T. Culture, face maintenance, and styles of handling interpersonal conflict: a study in five cultures. *Int. J. Confl. Manage.* 1991, 2(4), 275-296.

Ting-Toomey, S.; Kurogi, A. Facework competence in intercultural conflict: an updated face-negotiation theory. *Int. J. Intercult. Relations*, 1998, 22(2), 187-225.

Wang, Y.; Vela, M. R.; Tyler, K. Cultural perspectives: Chinese perceptions of UK hotel service quality. *Int. J. Cult. Tourism Hospit. Res.* 2008, 2(4), 312-329.

Wong, N. Y.; Ahuvia, A. C. Personal taste and family face: luxury consumption in confucian and western societies. *Psychol. Mark.* 1998, 15(5), 423-441.

Wu, M. Y.; Pearce, P. L. Chinese recreational vehicle users in Australia: a netnographic study of tourist motivation. *Tourism Manage.* 2014, 43, 22-35.

Xinhuanet. *China's Outbound Tourist Number and Expenditure Become World Number One.* 2014. http://news.sina.com.cn/c/2014-04-11/183929914261.shtml (Retrieved April 18, 2014)

Xu, Y.; McGehee, N. G. Shopping behavior of Chinese tourists visiting the United States: letting the shoppers do the talking. *Tourism Manage.* 2012, 33(2), 427-430.

Yan, Y. *Of Hamburger and Social Space: Consuming McDonald's in Beijing.* In *The Consumer Revolution in Urban China*; Davis D. S., Ed.; University of California Press: Berkeley, CA, 2000 p 201-205.

Yau, O. H. Chinese cultural values: their dimensions and marketing implications. *Eur. J. Mark.* 1988, 22(5), 44-57.

Yu, X.; Weiler, B. Mainland Chinese pleasure travelers to Australia: a leisure behavior analysis. *Tourism Cult. Commun.* 2001, 3(2), 81-91.

Yun, D.; Joppe, M. Chinese perceptions of seven long-haul holiday destinations: focusing on activities, knowledge, and interest. *J. China Tourism Res.* 2011, 7(4), 459-489.

Zeithaml, V. A.; Berry, L. L.; Parasuraman, A. Understanding customer expectations of service. *Sloan Manage. Rev* . 1991, 32(3), 39-48.

Zeithaml, V. A.; Berry, L. L.; Parasuraman, A. The nature and determinants of customer expectations of service. *J. Acad. Mark. Sci.* 1993, 21(1), 1-12.

Zeng, G.; Go, F. Evolution of middle-class Chinese outbound travel preferences: an international perspective. *Tourism Econ.* 2013, 19(2), 231-243.

Zhang, G. China's Outbound Tourism: An Overview. Paper presented at the WTM-China Contact Conference: Beijing, 2006.

CHAPTER 12

CHINESE TOURISTS' ON-SITE EXPERIENCE IN FLORENCE: APPLYING THE ORCHESTRA MODEL

PHILIP L. PEARCE and MAO-YING WU

CONTENTS

12.1 INTRODUCTION

Abundant evidence exists that Chinese tourists are traveling in ever-increasing numbers outside of Asia (China Tourism Academy, 2014). In particular, substantial numbers of both independent and group tourists are now visiting Europe (Arlt, 2013; Lai et al., 2013; Wu and Pearce, 2014). Italy has become a prominent destination for these new waves of visitors. Remarkable growth has occurred in 4 regions: Lazio (where the capital city of Rome is located), Lombardia (with Milan as its central city), the Veneto (where Venice is the popular city destination), and Tuscany (with Florence as its feature city). In 2013, these 4 regions hosted almost 50% of Italy's 538,000 Chinese visitors (CaixinOnline, 2014). For Tuscany, with Florence as its capital, China is now the fifth most important non-European market after the United States, Japan, Canada, and Australia (Ministero Affari Esteri–Agenzia Nazionale del Turismo [MAE–ENIT], 2012). As research on Chinese tourists grows in the Western academic literature, it becomes important to provide detailed information on how the rapidly growing Chinese market engages with pivotal destinations (cf. De Carlo et al., 2009; Woodside et al., 2007). There is a major need to understand the rich reactions of Chinese tourists to the key locations they visit since such studies address tourists' well-being and offer guidelines for destination managers.

The study reported here describes the experiences of Chinese tourists to the center of Florence. The research reaches beyond the mere documentation of demographic facts and the recording of sites visited. Instead, it seeks to use insights from studies of the experience economy to explore in subtle detail the visitors' on-site involvement with the core of the city. In pursuing the specifics of how experiences are formed and the factors which operate when tourists visit key attractions, specific empirical evidence of what Chinese tourists visiting Florence do, see, feel, and what they think of the setting and the people with whom they interact are collected and reviewed. The analysis of on-site experiences can be seen as an informative accompaniment to work on expectations, post-travel surveys, and the kinds of reflection on experience offered in travel blogs (Pearce, 2010; Rossiter, 2011).

The guiding purpose of the study lies in providing new insights and fresh access for researchers into the minds and psychological world of the Chinese visitors. In time, this kind of research can offer a more complete understanding for those who promote destinations, manage attractions, and design tours, as well as for those who provide services to visitors.

12.2 THE ANALYSIS OF EXPERIENCE

Successful and sustainable tourism is always partly defined by rewarding and positive experiences for tourists (De Botton, 2002; Morgan et al., 2010; Ryan, 2000). It is these positive experiential outcomes which support businesses and communities through expenditure, general recommendations, and potential repeat visits (Bowen and Clarke, 2009; Ryan, 1995). The theme of the importance of experiences was identified in early tourism scholarship (Cohen, 1979; Krippendorf, 1987; Pearce, 1988; Ryan, 1995) and has been supported by the United States–based business research of Pine and Gilmore (1999) and the European work of Schulze (1995).

The recent interest in tourists' experiences has been dominated by detailed conceptual analyses of the issues (Jennings et al., 2009; Ryan, 2010; Uriely, 2005). These reviews contain several cautionary remarks. There are some challenging issues in measuring and assessing experiences. These concerns include recognizing cognitive and perceptual cross-cultural variability among tourists (Li and Sofield, 2009; Ryan et al., 2010), appraising experiences over time (Pearce, 2011), avoiding simple positivistic measures such as checklists, and acknowledging that experiences can be mundane, even disappointing, rather than always special or uplifting (Caru and Cova, 2003; Tung and Ritchie, 2011). The analysis of tourists' experience can function at several geographical scales but appears to be well suited to quite specific locations and incidents, where tourists can recall the details of their encounters and environmental interactions (Stewart and Hull, 1996).

For the participants, experience tends to be a holistic and integrated flow of reactions and emotions as they move through space and time. By way of contrast, it is clear that researchers must put in considerable effort to gain the thick description, which is immediately available to any participant. The questions which researchers have to use to access their understanding of someone else's world may be atomistic, demanding, and possibly seem repetitive to the individual whose privileged access to their own conscious thoughts is automatic. This study builds on some earlier efforts to use a conceptual model of experience, which recognizes both the subjective integration of experience and the researcher-driven need to understand it through close inquiry. In this analysis and following precedents from the insights of other researchers (cf. Cutler and Carmichael, 2010; Murphy et al., 2011; Pearce, 2011), tourists' experiences will be understood as analogous to listening to an orchestra. There are multiple contributing parts, but the overall effect is a fusion of the elements. The contributing elements to be assessed are the sensory inputs, the affective reactions, the cognitive mechanisms used to think about and understand the setting, the behaviors available, and the relevant relationships that define the participant's world. The approach is built on many influences but reflects strongly the ideas of Schmitt (2003) and the contributions offered by Baerenholdt et al. (2004). In a research sense, these components may be studied separately, but in the lives of the tourists, they are interacting and fused facets of the tourists' lived world.

For the purposes of this study, the orchestrated approach to viewing tourists' experiences directs attention to specific measurable experiential factors. The time when the experiences are assessed is also a point of concern, and in this study, the Chinese visitors will be surveyed while they are in the central core of the city. While this approach places little burden on the respondents in terms of remembering the way they feel and think, it necessarily misses out on their longer term and cumulative reflections on their visit. Figure 12.1 depicts the approach taken to develop an understanding of Chinese tourists' appreciation of the central heart of Florence.

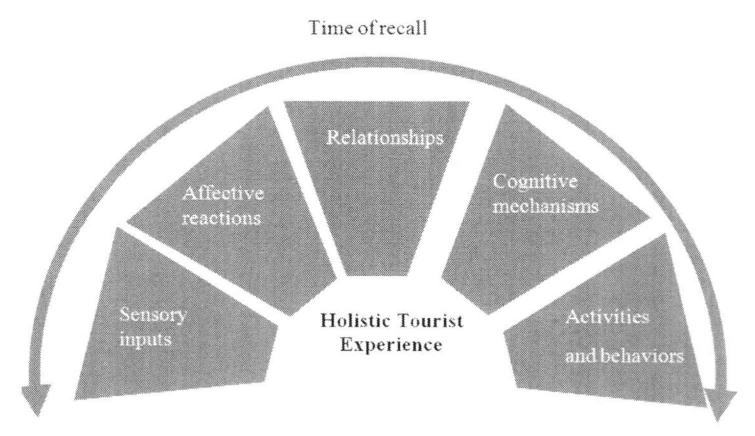

FIGURE 12.1 The orchestra model of on-site experience.

The aims of the study may be specified in terms of these experiential components. The on-site research concerning Chinese tourists in Florence seeks to document the principal sensory, emotional, cognitive, activity-based, and relationship-influenced components, which define Chinese tourists' immediate experience of the setting. In addition to this documentation of their experience, the study seeks to draw out the implications of these accounts for the presentation and management of the attraction space, which is at the very core of tourism in Florence.

12.3 METHODOLOGY

12.3.1 ON-SITE DATA COLLECTION

Preliminary Material

Florence is the capital city of the region of Tuscany, and its rich historical, artistic, and cultural heritage make it one of the main tourists' destinations in Italy and Europe. The principle site studied is the Cattedrale Santa Maria del Fiore, which is usually simply known to tourists as the Duomo and its precinct.

Some initial familiarization and pre-survey work was conducted to tailor the conceptual approach to the study site. Seven informal interviews with Chinese tourists were conducted by one of the researchers who is a native Mandarin speaker. These informal interviewers, undertaken on different days across 2 weeks in May 2012, sought the views of 4 independent travelers and 3 group travelers concerning their experiences in Florence. The conversations helped the research team understand Chinese tourists' immediate on-site behavior and perceptions of their time in the location. These emic voices from informal interviews were of assistance to the researchers in constructing the questionnaire-based survey. Additionally, photographs taken by the interviewees were inspected, and on-site notes were recorded. These informal processes helped formulate the plans to survey the Chinese tourists and provided some tips in terms of how to phrase the questions. The material collected also helped to support the subsequent interpretation of the survey results.

The Major Questionnaire

The questionnaire-based survey was designed in English by the research team and then translated by the Mandarin-speaking researcher. Using the skills of a Mandarin-speaking colleague, back-translation technique was used to correct any expressions lost in the initial translation process. This check on the survey work also served to improve the accessibility of the questions. The survey included 22 questions, 2 of which were open-ended items. The questions covered the 5 aspects of tourist on-site experience (e.g., sensory, affective, cognitive, activity-based responses, and relationships). The design of this material followed the approach reported in the literature review. To support the understanding of the on-site experience, information was also sought on the tourists' motivations for visiting Florence, the length of time they spent at the site, a benchmarked comparison of the Florence central area with other sites in Europe, and the identification of the most impressive features of the setting. Basic demographic information was also collected including travel style, travel experience, and the kinds of travel companions.

On-Site Survey Distribution

The surveys were conducted near the Duomo and in the popular adjacent market area by 2 research assistants from a local university in Milan in June and July of 2012. The assistants were advised about a number of good face-to-face surveying practices by the researchers. They were asked to approach those tourists whom they identified as likely to be Chinese. They were instructed to confirm the tourists' nationality before handing out the survey. To build a diverse sample of Chinese respondents, the assistants delivered a copy of the survey to any small travel group where the party size was less than 5. For tour groups, a maximum of 4 surveys was presented to the group, which usually consisted of 15–30 people. These approaches were designed to ensure the final sample was not dominated by easily accessible group tourists. In particular, a desire to collect a healthy percentage of independent tourists motivated this approach so that the diversity and heterogeneity of the travel experience could be sampled rather than relying on the more uniform patterns of those traveling in larger cohorts. In all, 220 copies of the questionnaire were delivered, with 179 valid copies received.

12.3.2 THE SAMPLE

The core demographic information about the sample is presented in Table 12.1. Cross-tabulation of some of the key demographic variables analyzed with Chi-square established that there was a significant relationships between travel style and travel experience ($\chi^2 = 19.15$, df = 3, $P < .001$) and the time spent on-site ($\chi^2 = 12.94$, df = 4, $P = .012$). Independent tourists were considerably more experienced travelers to Western countries than those participating in the group tours, and they tended to spend considerably more time at the Duomo and its surrounding areas.

TABLE 12.1 The Demographic Background of the Respondents

Demographic Factors		**Frequency (N = 179)**	**Percentages**
Gender	Male	92	51.6
	Female	87	48.4
Travel style	Independent travel	121	67.6
	Tour group	58	32.4
Length of stay	Less than 1 day	16	12.3
	1 day	46	35.4
	1–2 days	34	26.2
	3 days	9	6.9
	4 days and more	25	19.2
Travel companion	alone	13	7.3
	with family	68	38.0
	with close friends	89	49.7
	with others (e.g., online friends and new contacts)	9	5.0
Travel experience in Western countries	None	33	18.4
	1–2 times	74	41.3
	3–5 times	34	19.0
	6 times or more	38	21.2

12.4 RESULTS

The main aims of this study were to identify Chinese tourists' on-site experiences utilizing the comprehensive key dimensions of experience proposed in the Introduction and Literature Review sections. As an initial frame to place these findings in context, it is useful to identify the reported motivations of the Chinese tourists for their visit to Florence.

12.4.1 MOTIVATIONS TO VISIT FLORENCE

Three broad motivating destination features (or pulls in the push-pull account of motivation) were highlighted in the Chinese tourists' responses when they were asked about their motivations for visiting Florence. Multiple responses were permitted for the question. Motivations to visit Florence were mostly leisure-oriented. The leading responses were visiting heritage sites (73.2%), experiencing Italian culture and lifestyle (70.3%), and being attracted by shopping and fashion (45.8%). Very few surveyed participants were involved in business trips (8.4%).

12.4.2 THE ON-SITE EMOTIONAL (AFFECTIVE) EXPERIENCES

The reporting of the component parts of experience to Florence considers in turn the tourists' emotional reactions, their sensory experiences, the influence of relationships on their experience, on-site activities and behaviors, and their appraisal of the site in relation to other European sites. Building on the work of Richins (1997), the Chinese respondents were asked to consider a broad set of possible emotional responses drawn from a model of consumption emotions. The range of responses listed permitted the researchers to detect an array of possible feeling states varying from disgusted and annoyed to happy and relaxed. Table 12.2 presents the major emotions that Chinese tourists experienced at the site. The findings indicated that both independent and tour group Chinese tourists had mostly positive experience in Florence.

TABLE 12.2 Emotional States Reported by Chinese Tourists in the Florence Duomo Area

Emotional State Selected	Overall Percentage (N = 179)	Percent of Group Tourists Reporting This Emotion (N = 58)	Percent of Independent Tourists Reporting This Emotion (N = 121)
Relaxed	46.4	29/50	57/47.1
Happy	48	25/43.1	58/47.9
Romantic	29.6	20/34.5	33/27.3
Surprised	21.8	11/18.9	28/23.1
Peaceful	20.7	18/31.0	19/15.7
Energetic	15.1	11/18.9	16/13.2

In addition to the principal percentages reported in Table 12.2, the emotions of unsafe, stressed, annoyed, puzzled, and disgust were reported by between 1 and 10 respondents. Although the percentages in Table 12.2 appear to indicate trends in the data, the emotional reactions of the travelers in different travel styles were not significant at the.05 level for any of the specific items as assessed by Chi square.

12.4.3 THE ON-SITE SENSORY EXPERIENCES

The sensory experiences of the Chinese tourists were assessed by asking them to consider their responses to a range of inputs. In the same questionnaire, the tourists were asked to offer a number of observations about their sensory experiences. They were asked to rate how pleasant the setting was in terms of its visual appeal, its tastes, its scents and smells, their reactions to the space itself, and their view of the weather conditions. The scale used for the ratings was very pleasant (5) to not pleasant (1). In this section, those who did not respond to any of the sensory perspectives were considered as invalid cases. Thus, 147 cases were considered as valid. Repeated measures one-way ANOVA was adopted to examine how these 5 sensory elements were experienced and whether or not they were perceived as significantly different (Kerlinger and Lee, 1992). Mauchly's Test of Sphericity ($\chi^2 = 38.70$, $P = 0.000$) and Tests of Within-Subjects Effects ($F = 31.93$; $P < 0.001$; $\eta p^2 = 0.237$) indicate that the 5 sensory experiences were perceived significantly differently in terms of how pleasant they were. A post hoc analysis (through Pairwise Comparisons) suggests that Chinese tourists enjoyed the visual components of their sensory experience most ($\overline{X} = 1.35$). Their experiences of the smell ($\overline{X} = 1.79$) and taste ($\overline{X} = 1.96$) were pleasant but considerably less enjoyable than the visual experience. The space ($\overline{X} = 2.47$) and weather ($\overline{X} = 2.42$) experience in the Duomo precinct was perceived as the least pleasant among the 5 sets of sensory experience. Figure 12.2 presents the information visually. The attributes falling in the same dashed circle were considered at the same level using the pairwise post hoc comparison tests and a probability level of 95%.

A mixed-model factorial ANOVA testing for the differences among the 5 sets of sensory experiences and the demographic factors was also conducted. The results for the variables of gender, travel companions, the amount of previous travel experience, and travel style were not significant.

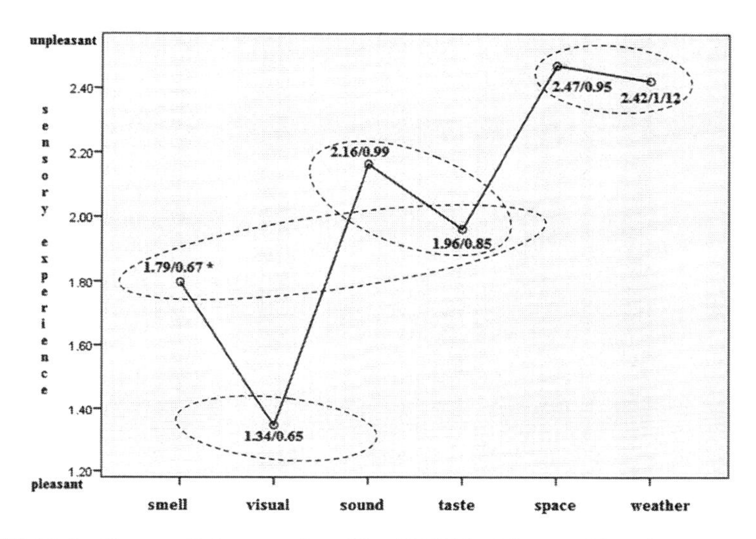

FIGURE 12.2 Repeated Measures One-Way ANOVA of Sensory Experiences.

Note: *Represents the mean value and standard deviation of each dimension.

12.4.4 THE ON-SITE RELATIONSHIP EXPERIENCES

This section of the study presents how the different forms of the tourists' social contacts contributed to their on-site enjoyment. These contacts included relationships with travel companions, contacts with the local service staff and vendors, and interaction with other on-site tourists. A five-point Likert scale with ratings for levels of enjoyment with a not applicable option was used.

One hundred and sixty-four of the survey participants offered specific answers to all of these questions, and they were regarded as valid cases. Descriptive analysis suggested that the Chinese tourists in Florence's Duomo area considered that their travel companions ($\overline{X}$ = 1.65, SD = 0.92) made their trip to this special site much more enjoyable than the interaction with other tourists ($\overline{X}$ = 2.68, SD = 0.94) and local service staff and vendors ($\overline{X}$ = 2.28, SD = 0.93). Tests through Mauchly's Test of Sphericity (χ^2 = 13.52, P = 0.001) and Tests of Within-Subjects Effects (F = 47.73; P = 0.000; η_p^2 = 0.317) through the repeated measures one-way ANOVA indicated that there was a relatively large effect because it explained 31.7% of the reported differences. Pairwise comparisons confirmed that travel companions were viewed significantly more positively, while interactions with local service staff and other tourists only enhance their travel experience to a small degree. A mixed-model factorial ANOVA between the social contacts' dimensions and the demographic factors was conducted. No significant differences among different demographic groups were detected.

12.4.5 THE ON-SITE BEHAVIORAL EXPERIENCES

The 7 most popular activities in the Duomo precinct were elicited from the on-site observation and 7 informal interviews. These activities were exploring the interior part of the site (reading signs, interpretation, appreciate presentations), photography, shopping in the nearby market, observing other tourists and tourism vendors, eating and drinking, and observing the exterior environment of the site.

The on-site Chinese tourists survey revealed that the most popular activities, that is those favored by more than 30% of the respondents, were photography (98/54.7%), observing the exterior environment of the site (73/40.8%), eating and drinking (61/34.1%), and exploring the interior part of the site (reading signs, interpretation, and appreciate presentations) (58/32.4%).

12.4.6 THE ON-SITE COGNITIVE EXPERIENCES

The assessment of on-site cognitive experiences of Chinese tourists in the Duomo site was achieved by asking one benchmarking question and 2 open-ended questions. These questions sought to document tourists' overall perspective of the on-site experiences, consider issues for improvement, and identify the tourists' most memorable on-site experience. The participants were firstly asked to compare their experiences at the chosen site in Florence with their experience at other European sites. One hundred and seventy-one tourists responded to this question. Almost two-thirds of the tourists thought that their experiences at Florence's Duomo area were better or even much better than experiences at other sites. Another 35% of the respondents considered the Florence site to be the equal of many others in Europe.

The second question asked in this section was in an open-ended format. The question was "What kind of facilities and services do you think businesses/managers of the Duomo area should improve to provide a better experience for you?" Fifty-seven out of 179 survey participants commented on this issue. Content analysis of these unstructured answers from the Chinese tourists was used to assess these answers. Interpretation emerged as a key issue. As many as 11 tourists would like to have Mandarin interpetation services, for example, maps, signs, (e) tour guides, and brochures. One-third of the answers identified facilities to make the trip physically easy, for example, more toilets (4), tap water (4), sitting areas (4), shade (3), wifi access (1), mailbox (1), bike service (1), and an elevator to upstairs (1). Other issues noted were the cleanliess of the setting (6) and the limited service hours (6). In addition, confusion at the site caused by the crowding (6) and the safety concerns resulted from the traffic (4) worried quite a few Chinese tourists. Further, some Chinese tourists would prefer a diversity of food, rather than Italian food only.

The last question concerning the Chinese tourists' cognitive experience at the Florence site was to report on the site's the most memorable feature. Ninety-two respondents participated in this open-ended question. They were mostly impressed with the magnificent historical Cathedral (58/63.0%). Ten of the tourists highlighted the on-site Italian food experiences. In the dry and hot weather, ice cream was especially favored. In addition, 9 tourists thought it was amazing to see such a diversity of people on the site. An interviewee suggested that, "I like the diversity of people walking by. It offers me the atmosphere of being in a Western country and reminds me I am in an international setting." This preference was confirmed by another group tourist and an independent traveler as well. Chinese tourists do like to visit places with an international atmosphere and where there are not many Chinese tourists on-site. If there are a large number of other Chinese tourists, it is likely that they feel they have paid a lot of money, but the experience is more like that of traveling domestically.

A small number of tourists (5) emphasized that both the views at the top of the Duomo and the night view of the center of Florence were very peaceful and beautiful. Other impressive features about the site include its bright color (5), the number of pigeons (5), the friendly people they encountered (5), the paintings in the cathedral (4), and the relaxing atmosphere (3).

As mentioned earlier, the lack of order and safety concerns at the site caused by the crowding and traffic worried a number of tourists. These negative experiences were recorded by 5 tourists as their most memorable experience.

12.5 DISCUSSION AND CONCLUSION

The main aim of this study was to document the principal sensory, emotional, cognitive, activity-based, and relationship-influenced components which define Chinese tourists' experience of the Duomo area as a central attraction space in Florence. This aim serves as a specific illustration of the desire to understand how the new waves of Chinese tourists, unused to Western destinations and imbued with their distinctive cultural gazes, appraise some of the world's leading attractions (Pearce et al., 2013). It was reported from the on-site survey work that visiting heritage sites (73.2%), experiencing Italian culture and lifestyle (70.3%), and shopping and fashion (45.8%) were the dominant features attracting visitors to Florence. The activities that Chinese tourists were interested in undertaking in the Florence Duomo area were consistent with their perceived images of and the attractive activities in other Italian cities (Bellini et al., 2012; Corigliano, 2011). The study highlighted that for the Chinese tourists' the most memorable feature of Duomo area is the magnificent historical cathedral.

An overview of the Chinese tourists' reactions to the study site was provided by highlighting that when compared with other European sites, 65% thought it was better or much better, while 35% thought it was of an equivalent standard. These positive appraisals do not necessarily mean that improvements and suggestions for change are unnecessary, but they do frame the findings of the Chinese tourists' experience in comparative terms.

For the sensory components, the overwhelming and significantly different component of this facet of the Chinese tourists' experience was the visual impact of the site. This finding was reinforced by the open-ended responses documenting the most memorable feature of the whole experience where the spectacle of the Duomo was the dominant theme. Broadly positive reactions were also reported for the smell and aromas of the site and for taste, but slightly less positive ratings were recorded for the noisiness of the site and the management of the space. The rating of auditory experience was not so pleasant and supports a recent finding that when visiting well-known sites, Chinese tourists appreciate natural sounds; however, they are not so fond of human-induced noise (Zhao, 2009). Weather conditions were also rated less positively, but these findings need to be placed in the context of the survey being conducted in a notably hot European summer and an awareness that the central area of the site consists of a large amount of exposed concrete.

The ratings by the Chinese tourists of their emotional, relationship, and activity components of their on-site experience, coupled with their open-ended comments, provide a cumulative picture with dominant and subordinate experiential themes. A primary view emerges that most Chinese tourists are emotionally relaxed and happy with the site and report its romantic and international atmosphere. Specific comments relating to activities and memorable features begin to identify the distinctive Chinese view of the attraction space. The Chinese tourists show a willingness to observe, to seek to understand and find out about the site's themes and stories, although at times a lack of interpretation frustrates this effort. There is a consistency here with the reported identification within Chinese tourism and cultural frames of mind with the quest to learn and enjoy the integrating and dominant stories and themes which define places (Li, 2008). Additionally, they react positively to several everyday features of the setting and Italian culture, including an appreciation of the novelty of the food, ice cream, and the abundance of the pigeons. They are active in photographing the site.

It can be argued that these behaviors are underpinned by several facets of the concept of harmony where there is a quest to see the successful conjunction between people and the environments in which they live. In this conception of the world balance, moderation and a fit between the well-being of people and the environment all matter (Nyíri, 2006). These somewhat elevated ideals are expressed in views that people should be well catered for at tourist sites, and this theme finds explicit expression in a concern for more facilities including more items for physical comfort such as water, toilets, and cooling and requests for better translation and information services (cf. Li and Sofield, 2009). These concerns provide directions for those who will manage rising numbers of Chinese visitors in the ensuing seasons.

There are both marketing and managerial implications, which can be drawn from this study. The orchestra model of experience and the specific instances of content unearthed by assessing all these components are that distinctive nuances in marketing content can be developed for the large emerging Chinese market. The study revealed that the central marketing piece for Florence is the Duomo. Preparing visitors with succinct interpretive materials that define the historic role of this significant European building could build a greater understanding and appreciation of the setting. This informative material could be a productive direction for marketing tours and promoting individual travel to the destination. Further, a range of supportive and supplementary components of the Italian experience in this city could be emphasized as additionally attractive features. That is, for the Chinese visitors, the cosmopolitan atmosphere and the sense of the destination being globally popular can be subtly included in the promotional push. Additionally, the smells and tastes of Italy are features to promote. These positive sensory and experiential components captured the attention of both the group and individual visitors. These experiential resources, while perhaps not so novel to European visitors, could form a further part of the marketing of Tuscany and its globally recognizable city.

There is also a material in the study that highlights the need for some enhanced management of the core area of Florence. Previous studies have suggested that the Chinese place safety as their leading concern when selecting their overseas tourism destination (Kim et al., 2005; Tourism Australia, 2012). There is an undercurrent in the data that at least some Chinese tourists are troubled by issues of safety and report degrees of emotional insecurity and tension. The safety concern supports Corigliano's (2011) appeal to build a positive and safe destination image for all the emerging Chinese markets. The central issue identified in the commentary sections highlighted threats from crowds and chaotic traffic. The safety concerns were exacerbated by being in an unfamiliar environment without proper interpretation and sometimes even a lack of physical comfort. For both the dominant view and the less common but not insignificant minor representation of their on-site Italian experiences, there were some indications in the data that the more independent and experienced Chinese tourists were more critical of the setting. This direction in the finding, which was sometimes only at the level of a weak trend in the data, is of concern to those who manage the future image and destination appeal of Florence as well as other destinations.

It is clear from the global figures of Chinese movements outside their own country that many citizens are now increasing not just the number of countries they visit but also their confidence in being able to travel internationally (Arlt, 2013; UNWTO, 2011). An important managerial implication of the rising confidence of Chinese visitors and China's continued economic development is that future Chinese tourists are likely to be more demanding and more critical of service situations which trouble them or which do not embellish and enhance their experience. The data in this study provide some of these criticisms. The managerial thrust of this perspective is that controlling issues of safety and security and attending with more thoroughness to the educational and interpretive needs of Chinese tourists are items for attention, not just in central Florence but throughout Italy and for all those places seeking to host the expanding Chinese market. A specific and further message from this study is that using the full orchestrated model of tourists' experience provides a multifaceted and comprehensive view of tourists' experiences and can be a future tool in other settings to build site and regional analyses of how Chinese (and other) visitors react to sites and destinations and thus highlight suggestions for improvement in these heavily visited spaces.

REFERENCES

Arlt, W. G. The second wave of Chinese outbound tourism. *Tourism Plann. Dev.* 2013, 10(2), 126–133.

Baerenholdt, J.; Haldrup, M.; Larsen, J.; Urry, J. *Performing Tourist Places*; Ashgate: Aldershot, Hants, 2004.

Bellini, N.; Loffredo, A.; Rovai, S. Chinese tourism: a sustainable opportunity for local economies? Paper presented at the Regional Studies Association Global Conference, Beijing, 2012.

Bowen, D.; Clarke, J. *Contemporary Tourist Behaviour*; CABI: Wallingford, Oxon, 2009.

CaixinOnline. *Want Chinese Tourists? Be Friendly to Them*. 2014. htttp//english.caixin.com/2014-02-17/100639696.html (accessed May 28, 2014).

Caru, A.; Cova, B. Revisiting consumption experience: a more humble but complete view of the concept. *Mark. Theory* 2003, 3(2), 267–286.

China Tourism Academy. *China Outbound Tourism Annual Report 2012-2013*; China Tourism Academy: Beijing, 2014.

Cohen, E. A phenomenology of tourists' experiences. *Sociology* 1979, 13(2), 179–201.

Corigliano, M. A. The outbound Chinese tourism to Italy: the new graduates' generation, *J. China Tourism Res.* 2011, 7(4), 396–410.

Cutler, S. Q.; Carmichael, B. A. The dimensions of the tourist experience. In *The Tourism and Leisure Experience*; Morgan, M.; Lugosi, P.; Brent Ritchie, J. R., Eds.; Channel View: Bristol, 2010; pp 3–26.

De Botton, A. *The Art of Travel*; Penguin: London, 2002.

De Carlo, M.; Canali, S.; Pritchard, A.; Morgan, N. Moving Milan towards Expo 2015: designing culture into a city brand. *J. Place Manage. Dev.* 2009, 2(1), 8–22.

Jennings, G.; Lee, Y-S.; Ayling, A.; Bunny, B.; Cater, C.; Ollenburg, C. Quality tourism experiences: reviews reflections research agendas. *J. Hosp. Mark. Manage.* 2009, 18(2–3), 294–310.

Kerlinger, F. N.; Lee, H. B. *Foundations of Behavioral Research*, 4th ed.; Cengage Learning: Belmont, CA, 1992.

Kim, S. S.; Guo, Y.; Agrusa, J. Preference and positioning analyses of overseas destinations by Mainland Chinese outbound pleasure tourists. *J. Travel Res.* 2005, 44(2), 212–220.

Krippendorf, J. *The Holiday Makers: Understanding the Impact of Leisure and Travel*; William Heinemann: London, 1987.

Lai, C.; Li, X.; Harrill, R. Chinese outbound tourists' perceived constraints to visiting the United States. *Tourism Manage.* 2013, 37, 136–146.

Li, F. M. S. Culture as a major determinant in tourism development of China. *Curr. Issues Tourism* 2008, 11(6), 492–513.

Li, F. M. S.; Sofield, T. Huangshan (Yellow Mountain), China: the meaning of harmonious relationships. In *Tourism in China: Destinations, Cultures and Communities*; Ryan, C.; Huimin, G., Eds.; Routledge: New York, 2009, pp 157–168.

Ministero Affari Esteri–Agenzia Nazionale del Turismo (MAE–ENIT). *Rapporto Congiunto Ambasciate/Consolati/ENIT 2012*; Repubblica Popolare Cinese: Hong Kong, Rome, 2012.

Morgan, M.; Lugosi, P.; Brent Ritchie, J. R., Eds. *The Tourism and Leisure Experience: Consumer and Managerial Perspectives*; Channel View: Bristol, 2010.

Murphy, L.; Benckendorff, P.; Moscardo, G.; Pearce, P. L. *Tourist Shopping Villages: Forms and Functions*; Routledge: New York, 2011.

Nyíri, P. *Scenic Spots: Chinese Tourism, the State and Cultural Authority*; University of Washington Press: Seattle, 2006.

Pearce, P. L. *The Ulysses Factor: Evaluating Visitors in Tourist Settings*; Springer-Verlag: New York, 1988.

Pearce, P. L. New directions for considering tourists' attitudes towards others. *Tourism Recreat. Res.* 2010, 35(3), 251–258.

Pearce, P. L. *Tourist Behaviour and the Contemporary World*; Channel View: Bristol, 2011.

Pearce, P. L.; Wu, M-Y.; Osmond A. Puzzles in understanding Chinese tourist behaviour towards a triple-C gaze. *Tourism Recreat. Res.* 2013, 38(2), 145–158

Pine, B. J. II; Gilmore, J. *The Experience Economy*; Harvard Business School Press: Boston, 1999.

Richins, M. L. Measuring emotions in the consumption experience. *J. Consum. Res.* 1997, 24(2), 127–146.

Rossiter, J. *Measurement for the Social Sciences – The C-OAR-SE Method and Why it Must Replace Psychometrics*; Springer: New York, 2011.

Ryan, C. *Researching Tourist Satisfaction: Issues, Concepts, Problems*; Routledge: London, 1995.

Ryan, C. From the psychometrics of SERVQUAL to sex: measurements of tourist satisfaction. In *Consumer Behavior in Travel and Tourism*; Pizam, A.; Mansfeld, Y., Eds.; The Haworth Hospitality Press: New York, 2000, pp 267–286.

Ryan, C. Tourism experience: a review of the literature. *Tourism Recreat. Res.* 2010, 35(1), 37–46.

Schmitt, B. H. *Customer Experience Management*; John Wiley & Sons: Hoboken, 2003.

Schulze, G. *The Experience Society*; Sage: London, 1995.

Stewart, W. P.; Hull, R. B. Capturing the moments: concerns of *in situ* leisure research. *J. Travel Tourism Mark.* 1996, 5(1–2), 3–20.

Tourism Australia. *2020 China ... Building the Foundations – Knowing the Customer*; Tourism Australia: Canberra, 2012.

Tung, V. W. S.; Ritchie, J. R. B. Exploring the essence of memorable tourism experiences. *Ann. Tourism Res.* 2011, 38(4), 1367–1386.

UNWTO. *Tourism Towards 2030: Global Overview*; UNWTO: Gyeongju, 2011.

Uriely, N. The tourist experience: conceptual developments. *Ann. Tourism Res.* 2005, 32(1), 199–216.

Woodside, A. G.; Cruickshank, B. F.; Ning, D. Stories visitors tell about Italian cities as destination icons. *Tourism Manage.* 2007, 28(1), 162–174.

Wu, M-Y.; Pearce, P. L. Chinese recreational vehicle users in Australia: a netnographic study of tourist motivation. *Tourism Manage*. 2014, 43, 22–35.

Zhao, X. A quantification analysis on acoustic landscapes of waterfront scenic areas: a case study of Hangzhou City, China. *J. Asian Archit. Build. Eng.* 2009, 8(2), 379–385.

CHAPTER 13

CHINESE GAMBLING PREFERENCES AND THE EMERGENCE OF CASINO TOURISM

IPKIN ANTHONY WONG

CONTENTS

13.1 INTRODUCTION

With China's economy barreling ahead of the world with 8%–10%increase in gross domestic product (GDP) for more than 2 decades, Chinese are getting affluent and hence, are increasingly demanding to travel outside of the country, like never before. Although there are signs that the country's economy is slowing down in the past year or so to about 7.5%, it does not seem to affect Chinese travel behaviors significantly. Because Chinese are known to enjoy gambling, while the Mainland Chinese Central government prohibits all forms of gambling except lottery, it creates a huge gap between supply and demand. The imbalance between the 2 sides compels Chinese to travel outbound and fulfill an array of their travel needs in gaming destinations. In fact, to fulfill the unmet needs for gambling and casino travel, many countries, especially those in the Asia Pacific region, are legalizing gambling or expanding their presence in gambling in an effort to take advantage of this unprecedented golden opportunity. The world gambling capital, Macau, for example, has seen its economy take off since its liberalization of the casino industry in the early 2000s, and it generated US$45.2 billion gaming revenue in 2013 (7 times that of the entire gaming revenue of Las Vegas). It welcomed about 30 million tourists, over 90% of them being Chinese. Although only around 8% of tourists in the city reported that their primary travel purpose is gambling (Macau Statistics and Census Service, 2013), the author's recent studies suggest that a huge number of tourists flock to Macau to experience casinos—much like other attractions, with gambling just a part of this travel experience (Wong and Rosenbaum, 2012).

In this chapter, the author presents details about why Chinese desire to visit casinos and engage in gambling. Specific coverage will be focused on the motivational factors that underpin casino tourism, which refers to tourists' desire and behaviors in visiting casinos as tourism attractions. Details about Chinese gamblers' preferences and behaviors are also presented in the chapter. Due to their preference for casinos, destinations such as Singapore, Las Vegas, and especially Macau are enjoying competitive advantages over other destinations that do not have casinos. This may explain why destinations such as the Philippines, Taiwan, and Japan have opened or are opening up casinos in order to attract Chinese tourists. Casinos are such important attractions for tourists as they offer tourists an assortment of services and surprises that could hardly be mimicked in other places. Thus, this chapter also details how casinos provide tourism experiences as well as the specific type of experience tourists seek. This chapter concludes with insights into future development of casino tourism.

13.2 CHINESE GAMBLING MOTIVATIONS AND BEHAVIORS

The literature consistently indicates that people gamble for 3 major reasons: economic (e.g., to win money), symbolic (e.g., to take risks and receive a sense of control), and hedonic (e.g., to enhance self-esteem and self-image and to enjoy the excitement from playing) (Cotte, 1997). Chinese are not different; yet, they are particularly fond of financial appeals and taking the risk of gambling while craving symbolic control in the gambling process (Ozorio and Fong, 2004). However, they are more likely to be frequent and hardcore players who have a higher tendency to develop problems and even pathological gambling behaviors.

Chinese gambling preferences show marked differences from those of gamblers from other cultures. The desire of Chinese for personal control during gambling is manifested in the type of games they prefer, as they perceive a loss of such control when playing slot machines. Chinese also consider table games more exciting and entertaining than slots. They believe table games are easier to understand and perceive that they have a better chance to win compared with slot games (Liu and Wan, 2011). Because Chinese are people of a strong collectivistic culture, gaming has connotative meanings as a public display of one's achievement. They like to be accompanied by others rather than to be alone while gambling or watching others gambling. Gambling also serves as a means to deepen friendship and to provide business networking opportunities (Wan et al., 2013). Mainland Chinese gamblers also like smoking in casinos whether they are wagering or not. They tend to speak loud, cause disruptive behaviors to other patrons, and are perceived to be untidy.

Chinese gamblers' preferences for table games are further exemplified by their behaviors as hardcore gamblers who like to shout out in public while peeling cards when they play baccarat (Lam, 2007). Such a behavior reflects Chinese gamblers' strong desire to control their luck and a belief in luck through personal worship, rituals, gesture, and customs. For example, Chinese prefer golden and reddish colors because they symbolize fortune and good luck. Some Chinese wear special clothes (e.g., red underwear) while gambling because they believe that they represent power and luck. They do not like to be touched (especially their shoulders) or to hear mention of books (as the word sounds like "lose" in Chinese) because it would ruin their luck. These superstitious beliefs are related to Feng Shui, and their beliefs about it further manifest in their lifestyle and daily routine. For example, they prefer to gamble in casinos that bring them luck, while avoiding casinos that are deemed to have Feng Shui in favor of the house. They even customize their hotel room in their own way in an effort to enhance their luck. For example, some gamblers like their rooms to face specific directions and organize their personal belongings and room amenities in certain ways. Chinese gamblers also like to pray to the gods and make offers to the gods when they receive a big win. Furthermore, Chinese are very superstitious about numbers. They avoid any number that includes 4 because the word sounds similar to "die" in Chinese. On the contrary, they embrace 3 and 8 because these pronunciations signify propensity/life and fortune, respectively.

Chinese gamblers also prefer baccarat more than other table games because they think that it is easier to understand, to have better control, and to rely more on their skills and hence, that it gives them a higher chance to win against the house. They also hop from one table to another to search for winning opportunities. Chinese gamblers like to cluster together because they believe that if many people gather at a table, the table will bring luck to them. As a result, gamblers often support each other to win against the house. Some gamblers even ask other gamblers to peel the card for them if they are considered lucky people. Chinese gamblers like to talk with each other; some also like to talk to the table dealers. They stay together at a table, record the win–loss pattern (sequence), and keep betting until the sequence of win–loss pattern is broken (Lam, 2007). These behaviors are fairly different from those of other countries' gamblers, which may be attributed to the Chinese's strong presence of collectivism.

Chinese are superstitious about their destiny and luck. Gambling is often perceived as the best means to test their luck and gain control over one's destiny. A win from the table signifies luck while a loss signifies otherwise. The author's ongoing research suggests that testing luck is one of the most important factors to gamblers other than wining money and seeking excitement (see also Papineau, 2005). Luck also determines the fate of Chinese; in that, if they perceive themselves to be lucky, they could enjoy success in the future because it is predetermined by their fate and destiny. People who perceive that they possess luck are happier, enjoy their life more, and have higher self-efficacy. This may explain why Chinese like gambling since it could signify whether their future will be bright or not. In fact, Chinese culture is future-oriented, and they score the highest on Hofstede's long-term orientation dimension.

Gambling is widely acknowledged to have serious negative social effects such as pathological gambling, taking bribes, embezzling public funds, fraud, money laundering, and other issues (Zeng and Forrest, 2009). As a result, gambling and media coverage of gambling is strictly prohibited in the Mainland China. The Chinese government also prohibits government officials from gambling while they travel overseas. Yet, this policy only has a minor adverse effect on gaming destinations like Macau because the general and VIP gamblers dominate the market, while these officials only account for a small portion of the casino clientele. Also, casino marketing efforts have attracted an influx of gamblers who are not government officials, as detailed in the sections that follow. In addition, China's new government has initiated policies to crack down on graft and corruption as well as to tighten expenditures, which have had a trickle-down effect on citizens to focus on austerity. The casino gambling market was fairly robust and bulletproof early on, in that such policies did not have any significant adverse effect on gamblers and casino visitors; gambling revenue in Macau, for example, remains strong, with double-digit growth (19% in 2013), despite a slowing growth rate of tourist arrivals (only 4% in 2013 over the previous year). Yet, negative publicity on gambling and the Chinese Central government's endeavors for thrift and anti-corruption have signaled an early alarm to the sustainability of the casino gaming business as gambling revenue in Macau and other gaming destinations have been slowing down since the second quarter of 2014 primarily due to the decline of the VIP segment. In fact, savvy casino operators in Macau, for example, have already realized that the tightening expenditure policy will continue to hurt the VIP market, which contributes about 60%–70% of the total gaming revenue generated in Macau, so they started to focus on other people by attracting more mass market gamblers and offering more leisure entertainment options for them. In fact, most contemporary casinos are built like palaces and are integrated with virtually all forms of leisure entertainment options. These places not only serve as gamblers' playgrounds but also act as major attractions for tourists. This leads to a new travel trend—casino tourism—in that tourists sojourn casinos much like ordinary attractions. Details about casino tourism are presented in the next section.

13.3 CASINO TOURISM AND TRAVEL MOTIVATIONS

Today's contemporary casinos have changed tourists' perceptions of and motivations for visiting casinos. Many of these establishments are large in scale with an integrated assortment of leisure entertainment options including spa, kids and family playrooms, outdoor entertainment options, live shows, convention and meeting venues, luxury accommodations, shopping outlets, and causal and fine dining restaurants. The gambling floor usually is situated in the middle of the casino and can be accessed from anywhere inside the casino complex since it is the most important cash cow for the casino. These casino complexes are built aesthetically to signify a sense of harmony, beauty, and elegance. Such palace-like environments are excellent attractions for tourists, especially Chinese tourists, because they are not only unique to tourists, but they also embrace virtually all entertainment facilities that tourists need. As a result, many tourists flock to gaming destinations such as Macau to pay visits to casinos. Some tourists have even mentioned that their travel experience is not complete without sojourning to casinos (Wong, 2013; Wong and Rosenbaum, 2012).

The emergence of casino tourism is indeed fused by a huge demand from the Mainland Chinese tourists. Recent empirical studies show that mainland tourists flock to casinos for 5 major motivational factors (Wong and Rosenbaum, 2012). *Entertainment and novelty-seeking* refers to tourists' desire to experience casinos, gambling, and nightlife entertainment and to learn more about the gaming destination. *Casino sightseeing* describes tourists' desire to enjoy casino facilities and to sightsee casinos, similar to sightseeing other types of attractions. *Leisure activity* refers to tourists' preference for high-quality entertainment, shopping, and accommodation offerings. The other 2 motives—*escape* and *socialization*—refer to tourists' desire to utilize casinos as places for relaxation, escaping from mundane life, and socialization opportunities with friends and family members.

The findings suggest that while gambling is still the most important reason for Chinese to visit casinos, many of them perceive gambling as an entertainment option. They spend a limited amount of their travel expenditure on gambling, as it is part of the experience in visiting casinos and gaming destinations. In this sense, a majority of Chinese tourists who partake of casinos go beyond mere hardcore gambling, to enjoy a diverse array of service offerings.

Some discrepancies among Chinese casino tourists are also recognized (Wong and Rosenbaum, 2012). Younger and better-educated tourists enjoy entertainment and socialization opportunities more than their older and less-educated counterparts, who tend to enjoy casino sightseeing and escape alternatives. The younger and better-educated tourists are also more likely to travel with their family and friends, and they match the demographic profile of the Chinese middle class. First-time tourists are more inclined to experience gambling, seek different kinds of casino facilities, and try different nightlife entertainments (Wong, 2012). While short-stay day-trippers are more likely to seek novelty experience from gambling and the gaming destination, as well as socialization opportunities with their family, longer-stay tourists are more inclined to utilize casinos as a place for escape and relaxation.

Furthermore, an ongoing study conducted by the authors in Macau suggests that for-gambling and regular (i.e., not-for-gambling) tourists differ in several ways. Comparing the 2 groups, for-gambling tourists are more likely to be males, be older, have lower education level, come from the Mainland China, have better income, travel with others, and visit the destination/casino more often (i.e., with more travel experience) (Wong and Li, 2014; Wong and Rosenbaum, 2012). These findings are similar to Chinese visiting casinos in other Asian destinations (Kim, 2004). The profile of Chinese casino visitors is also similar to casino patrons in Western gaming destinations such as those in Las Vegas (Shinnar et al., 2004). Furthermore, Chinese tourists are also different in regard to the service perceptions and travel experience. For example, although for-gambling tourists rate gaming facilities higher than not-for-gambling tourists do, they seek entertainment, accommodation, and dining facilities to enhance their travel experience (Wong and Li, 2014).

Casinos also play a major role in fulfilling Chinese tourists' shopping needs. Chinese shopping preferences have been well documented in the literature. Chinese tourists purchase merchandise that varies greatly in product category, brand assortment, and price range (Wong, 2013). Astute casino operators also acknowledge this important business opportunity. Hence, most major casinos contain large shopping outlets that cater to the various shopping needs of Chinese tourists. Although most casinos target high-end shoppers and offer exclusively luxury brands, some casinos have realized that many Chinese do shop for low-end to mid-range brands. Having an assortment of brands at different levels also helps to attract different types of tourists. The Venetian Macau complex, for example, has 2 adjacent malls: the Grand Canal shopping mall which offers mid- to low-end stores such as Zara, Esprit, Nike, and Mango. The Four Seasons shopping mall hosts exclusively designer and premium luxury boutiques such as Louis Vuitton, Dior, Cartier, Gucci, Prada, and Rolex. In fact, all 4 Louis Vuitton boutiques in Macau are located inside casinos. Shopping inside casinos has become a prevalent trend (Wong, 2013; Wong and Rosenbaum, 2012). Hence, mall design becomes a key factor in attracting gamblers and tourists alike. The Grand Canal shopping mall is crafted to mimic a romantic travel theme from Venice with bridges, open squares, and a gondola ride. This replica of Venice is overwhelmingly popular, attracting more than 20 million shoppers to the mall each year. Indeed, shopping and window shopping at casino outlets have become common itineraries for Chinese tourists.

13.4 CASINO SERVICE STRATEGY AND TOURIST EXPERIENCE

Themed casinos and their artistically designed service encounters are highly appealing to Chinese tourists. Many tourists have even commented that elegant and grandiose design of casinos allows them to enjoy a splendid indulgence in such places. They have observed that the scenery of casinos is unique, with a perfectly crafted palazzo-like environment. The author's ongoing research suggests that for many tourists, a casino is more than a place for gambling, a restorative place perceived to help them to get away from ordinary life. Visitors are fascinated by and feel compatible with a casino's rapturous scenery, facilities, and atmospheric attributes. That is, a casino could prompt a healing process for people who seek a better quality of life, much like other service providers (Rosenbaum, 2009). In fact much research suggests that the service environment plays a major role in attracting gamblers to visit a casino, to stay in a casino longer, and to gamble more (Wong and Fong, 2010). These environmental attributes include a casino's background music and noise, lighting, color scheme, smell, cleanliness and comfort of the atmosphere, interior decor, architectural design, furnishing and seating comfort, signage, physical facilities, and more. These very attributes are environmental stimuli that are used to arouse people's 5 senses and to further promote a pleasant service experience. In fact, the author's recent research shows that the casino service environment plays a significant role on gambling impulsivity. That is, people are motivated to gamble while being engulfed within a palace-like enchanting environment. Yet, there are some discrepancies of the service perceptions among Chinese gamblers. For example, male and hardcore players are more likely to perceive casino services favorably and hence, are more satisfied with a casino than female and leisure players (Wong et al., 2012). These differences are due to gamblers' involvement and the intensity of the casino, which suggests that casinos' customer engagement programs and facilities will make a difference to patrons.

In addition, Chinese tourists like to take photos and tour around casinos. Many package tour groups have casino visits on their travel itinerary. Savvy tour guides even charge each tourist about US$20 just to take a casino tour. As a result, many contemporary casinos welcome hundreds of thousands of tourists on a daily basis. That may explain why some of the newly built casinos are so aesthetically crafted in order to attract tourists.

For example, the Venetian is constructed to manifest a virtual experience of being in Venice with a man-made canal and lakes, as well as luxury accommodations and irresistible dinning and leisure entertainment shows. Tourists often find surprises while touring the complex, as staff provide live performances with European customs, thus, making this replica of Venice more realistic. The Galaxy resort is coined as the New Palace of Asia due to its grandiose and breathtaking architectural design. It has the world's largest rooftop wave pool and a huge man-made beach, and they are adjacent to a beautiful indoor garden with lush foliage that they combine to offer tourists an oasis of fun and excitement. Its fortune diamond show and the *laserama* performance further add the "wow" factor to tourists. MGM puts the wow factor at a new high. It has turned its Grande Praca (the central square of the casino) into an aquatic wonderland where tourists can immerse themselves in a breathtaking underwater palace with fishes and corals surrounding the place. Tourists can also enjoy exhibitions, mermaid shows, and "streetmosphere" in the Grande Praca. The cylindrical water-sky aquarium in the center of the Grande Praca allows tourists to view different rare species of live fishes (which further signify luck and fortune).

The Wynn casino offers another form of attraction to tourists. Its renowned Performance Lake and Fountain show exudes a sense of grandiose fascination. Its exaggerated Dragon of Fortune show is not only breathtaking, but the rising dragon also symbolizes vitality, good fortune, and well-being. Resort World Sentosa of Singapore offers another unique experience to tourists. It is not only simply a place for casino gambling but also a home for endless leisure entertainment attractions including the Universal Studio, the Marine Life Park, SEA Aquarium, Adventure Cove Water Park, Dolphin Island, and Maritime Experiential Museum. It also hosts a large array of hotel accommodations, restaurants, and entertainments and shows including the spectacular Crane Dance and Lake of Dreams shows. These facilities jointly have attracted tourists in their millions to the property and allow it to command a leading position in the casino gaming industry in Singapore.

The aforementioned cases are only some exemplars that illustrate casino operators' endeavors in differentiating their casino properties from other casinos. Each casino is carefully designed to implement "wow" factors to surprise and delight its customers. These very esoteric products not only attract tourists to visit casinos, but the constellation of leisure entertainment offerings also helps promote a destination's image. Empirical studies further suggest that themed casinos enjoy competitive advantages over traditional ones. One of the major reasons is that these casinos have an excellent physical environment, which is ranked as the most important service experiential attribute for the Chinese (Wong, 2013; Wong and Wu, 2013).

Wong and colleague (Wong, 2013; Wong and Wu, 2013) further explain that casino service experience embraces a variety of service attributes including the tangible settings, employee services, service convenience and value, hedonic services, brand experience, and unique service offerings. Although experience engendered from the service environment remains the most critical factor in enticing customer satisfaction, especially among Chinese casino patrons, hedonic services (e.g., live shows, entertainments, and streetmosphere) are increasingly important. Similar to the service environment, hedonic and other unique service offerings provide casinos with distinctive advantages over rivals because they improve a casino's brand image and overall brand equity. Tourists often praise the services and experiences they enjoy in casinos such as the Venetian, Wynn Resort, MGM Grand, City of Dreams, and Galaxy. All of them are sprawling integrated casinos that have strived tirelessly to provide a splendid oasis and a memorable experience to gamblers and tourists. Indeed, these very properties rank higher in respect to experiential positioning and hence, enjoy a higher level of customer equity and loyalty (Wong, 2013; Wong and Wu, 2013).

Casinos also utilize other means to attract and retain gamblers and tourists alike. Because accessibility is a key factor in luring tourists, most casino operators have invested heavily in building their transportation system such as shuttle bus and limousine services. Some casinos even have their own charter airplanes and ferries to serve a variety of customers. Competition has driven casino operators to hire young and beautiful hosts to welcome and escort patrons. Casinos also provide complimentary food and drinks for gamblers. Many gamblers have commented that they enjoy casino loyalty programs (Benston, 2011) because such programs allow them to receive perks and recognitions such as free meals, show tickets, rooms, and transportation services; luxury gifts; discounts on shopping; service upgrades; and special privileges. As a result, these offerings help to engage gamblers and keep them loyal to a casino even if they lose money.

Gamblers can be segmented by casinos into tiers based on their gambling dollars spent. Higher tiers allow them to earn rewards more rapidly and hence, to better enjoy the comps, discounts, upgrades, privileges, and casino recognitions. Yet, some high-end players have expressed that it is important that luxury casinos should give exclusive access and provide VIP treatments only to high-end customers, as this gives them a high social status and honor (or *mianzi* in Chinese). In fact, these high-end Chinese patrons will pay a premium just for a chance to gain *mianzi* (Chen et al., 2014).

To better cater to the surging demands of Chinese tourists, casino operators are adopting the Chinese culture and seeking means to better fulfill their needs. In the case of Macau, for example, casinos are recruiting Mandarin-speaking staff and training more staff to speak Mandarin. The appreciation of the Chinese language is best manifested by an example from the Venetian, where gondola riders are now singing Mandarin songs for Chinese customers. In respect to food, there are numerous Chinese restaurants inside each casino. In Macau, for example, all 11 Michelin star-rated restaurants are located in casinos, while 7 of them offer Chinese food. Some casinos even renovate existing restaurants to offer northern Chinese food in order to better cater to the taste of Chinese customers. These initiatives are rewarding, as Chinese tourists visiting Macau and its casinos are on the rise despite tightening policies from the Chinese Central government.

13.5 CONCLUSIONS

In essence, casinos have become "must go" attractions for Chinese tourists. They have blossomed into places where tourists seek endless streams of entertainment options. Contemporary integrated casinos have emerged like theme parks to provide tourists with a variety of exciting service experiences that continue to attract and retain Chinese tourists to gaming destinations. The assortment of esoteric products offered by casinos not only attracts tourists to visit these places, but the constellation of such offerings also helps promote a destination's image. In fact, the concentration of casinos and their unique service offerings have elevated Macau and Las Vegas as world-renowned gaming destinations. Macau, for example, welcomed 30 million tourists in 2013, and over 90% of them were Chinese (for details about Macau, please refer Chapter 5 Mainland Chinese Outbound Tourism to Macao: Recent Progress by Xiangping Li). Casino infrastructures have turned Las Vegas from a desert into a hotbed of endless shows and entertainments. Likewise, similar casinos have turned Macau from a fishing village into a place with glitzy attractions awash with Chinese tourists.

Casino operators are focusing more on the mass market in response to the Chinese central government's efforts to tighten expenditures. In fact, although the VIP market still is dominant (it generated about 60%–70% of the total gaming revenue in Macau, for example), its share has gradually declined. Some casino operators such as the Venetian foresaw the changes in the market and shifted focus primarily to the mass market. Even casinos that traditionally focused on the luxury VIP market, such as Stanley Ho's Sociedade de Jogos de Macau and Wynn, have been shifting emphasis more to the mass market by having more mass-market table games and cutting VIP table games. To promote higher spending from the non-VIP players, many casinos introduced the premium mass-market table games with a minimum bet limit in the hundreds. These strategic moves have helped casino operators remain lucrative, as gamblers in Macau laid down an average of US$1,354 in casinos during their trips in 2012 (or about 7 times the outlay of those visiting Las Vegas) (Lee, 2013). In fact, gambler expenditure has been on the rise in Macau, as gaming revenue has recorded an average increase of over 20% since 2005. In addition, casinos have been utilizing more technologies by offering electronic table games without human dealers. This strategy helps to cut costs while maintaining standardized services.

Another market trend is the focus of diversified hospitality services that go beyond hardcore gambling. This is indeed a global trend in response to Chinese tourists' travel needs for uniqueness and service variety. Meanwhile, casinos have tried hard to customize services based on Chinese gambler preferences—for example, providing ashtrays and better ventilation systems due to heavy smoking habits of the Chinese. The disruptive behaviors and untidiness of the Chinese may also require floors and seats to be cleaned more regularly. Their belief in Feng Shui and other Chinese culture-specific habits should be understood and respected through better employee trainings and service customization.

Due to the lucrative casino business and massive demand of Chinese for gambling, it is clear that gaming destinations and casinos will continue to attract Chinese tourists in their millions and keep the casino business afloat. Macau has overtaken Las Vegas and blossomed into the world gaming metropolis and capital. Other Asian destinations such as Singapore, Philippines, Taiwan, South Korea, and Japan are following Macau's lead by developing their casino infrastructures in order to siphon off some of the market share of global gaming market. Casinos are available in many countries near China, so popular destinations besides the aforementioned regions include South Korea, Cambodia, Malaysia, Vietnam, and India. In fact, casinos are prevalent around the world. Yet, geographic location of a destination does matter, and it is one of the key factors in attracting Chinese tourists; as evidence suggests a curvilinear relationship between geographic convenience and (re)visit intention among Chinese tourists, in that tourists from nearby regions are significantly more likely to visit a gaming destination (Wong and Zhau, 2014). In other words, destinations that are close to China will enjoy competitive advantage over other destinations in luring Chinese tourists to their casinos.

In addition, concentration of casinos and large-scale facilities are also factors that could attract Chinese tourists. Due to the low level of foreign language ability, Chinese also prefer Chinese-speaking destinations and service encounters. This may explain why Macau is one of the most preferable destinations for Chinese (Xola Consulting, 2008), while destinations such as Las Vegas and Monte Carol are less likely to enjoy growth from Chinese tourists as some of the Asian destinations do.

In conclusion, the surging demand among Chinese for outbound travel, gambling, and visiting casinos will continue to provide practitioners and destinations with numerous business opportunities. Casino tourism will be increasingly popular due to the development of the casino industry in respect to both the number of new glittering integrated casinos and an assortment of leisure entertainment options available in these sprawling establishments. For example, a number of multi-billion-dollar mega casino projects are underway in Macau, while Japan and Taiwan are planning to enter the market with large-scale casino projects to attract Chinese tourists.

It is fairly clear that competition for the lucrative market segment of Chinese clientele has intensified and reached a global level. Yet, only gaming destinations and casinos that can fulfill the needs, satisfy the preferences, and understand the culture of the increasingly demanding Chinese tourists will be able to take lead in this "game of fortune." Thus destination authorities and casino operators should work together to create a gestalt effect to capitalize on the golden opportunities of casino tourism offered to Chinese.

REFERENCES

Benston, L. MGM Loyalty Program will Reward Customers with Perks not Imagined Previously; Las Vegas Sun, 2011.

Chen, K.; Sun, L.; Chen, L.; Wong, I.A. In The Interactive Relationship Management Between Managers, Employees and Customers in Macau's Hotels, 12th APacCHRIE Conference, Kuala Lumpur, Malaysia, May 20–24, 2014.

Cotte, J. Chances, trances, and lots of slots: gambling motives and consumption experiences. J. Leisure Res. 1997, 29(4), 380-406.

Kim, W.G. Implications of Chinese casino visitor characteristics in South Korea. J. Qual. Assur. Hospit. Tourism 2004, 5(1), 27-42.

Lam, D. An observation study of Chinese baccarat players. UNLV Gaming Res. Rev. J. 2007, 11(2), 63-73.

Lee, B. What Happens in Vegas... is Nothing Compared to Macau: The Tiny CHINESE Peninsula that is the Real Global Home of Gambling. 2013. http://www.thisismoney.co.uk/money/news/article-2413731/Macau-casino-mecca-world-revenue.html (accessed June 17).

Liu, X.R.; Wan, Y.K.P. An examination of factors that discourage slot play in Macau casinos. Int. J. Hospit. Manag. 2011, 30(1), 167-177.

Macau Statistics and Census Service 2013 Ref. Tourism Statistics. http://www.dsec.gov.mo/ (accessed June 16).

Ozorio, B.; Fong, D.K-C. Chinese casino gambling behaviors: risk taking in casinos vs. investments. UNLV Gaming Res. Rev. J. 2004, 8(2), 27-38.

Papineau, E. Pathological gambling in Montreal's Chinese community: an anthropological perspective. J. Gambl. Stud. 2005, 21(2), 157-178.

Rosenbaum, M.S. Restorative servicescapes: restoring directed attention in third places. J. Serv. Manag. 2009, 20(2), 173-191.

Shinnar, R.S.; Young, C.A.; Corsun, D.L. Las Vegas locals as gamblers and hosts to visiting friends and family: characteristics and gaming behavior. UNLV Gaming Res. Rev. J. 2004, 8(2), 39-48.

Wan, P.Y.K.; Kim, S.S.; Elliot, S. Behavioral differences in gaming patterns among Chinese subcultures as perceived by Macao casino staff. Cornell Hospit. Q. 2013, 54(4), 358-369.

Wong, A.I. Casino travel motivations of Chinese tourists: differences in visitation attributes. Tourism Rev. Int. 2012, 16(3-4), 217-226.

Wong, I.A. Exploring customer equity and the role of service experience in the casino service encounter. Int. J. Hospit. Manag. 2013, 32, 91-101.

Wong, I.A. Mainland Chinese shopping preferences and service perceptions in the Asian gaming destination of Macau. J. Vacat. Market. 2013, 19(3), 239-251.

Wong, I.A.; Fong, H.I.V.; Liu, M.T. Understanding perceived casino service difference among casino players. Int. J. Contemp. Hospit. Manag. 2012, 24(5), 753-773.

Wong, I.A.; Fong, V.H.I. Examining casino service quality in the Asian Las Vegas: an alternative approach. J. Hospit. Market. Manag. 2010, 19(8), 842-865; (b) Johnson, L.; Mayer, K.J.; Champaner, E. Casino atmospherics from a customer's perspective: a re-examination. UNLV Gaming Res. Rev. J. 2004, 8(2), 1-10.

Wong, A.I.; Li, X. Destination services and travel experience in the gambling mecca: the moderating role of gambling travel purpose among Chinese tourists. J. Trav. Tourism Market. 2014, 31.

Wong, I.A.; Rosenbaum, M.S. Beyond hardcore gambling: understanding why mainland Chinese visit casinos in Macau. J. Hospit. Tourism Res. 2012, 36(1), 32-51.

Wong, I.A.; Wu, J.S. Understanding casino experiential attributes: an application to market positioning. Int. J. Hospit. Manag. 2013, 35, 214-224.

Wong, I.A.; Zhau, W. Exploring the effect of geographic convenience on repeat visitation and tourist spending: the moderating role of novelty seeking. Curr. Issues Tourism 2014. DOI: 10.1080/13683500.2013.870538.

Xola Consulting. Chinese Travelers: Trends for Adventure Companies and Destinations. 2008. http://www.xolaconsulting.com/chinese_tourism_trends.pdf (accessed August 31).

Zeng, Z.; Forrest, D. High rollers from Mainland China: a profile based on 99 cases. UNLV Gaming Res. Rev. J. 2009, 13(1), 29-43.

CHAPTER 14

CHINESE OUTBOUND TOURISTS' SHOPPING BEHAVIOR

FANG MENG and PEI ZHANG

14.1 INTRODUCTION

Nowadays, due to the increasingly leisure-oriented setting for shopping in a destination, shopping has become a major leisure activity for many tourists and one of the most significant expenditure categories on vacations and trips (Littrell et al., 1994; Snepenger et al., 2003; Yuksel and Yuksel, 2007). In many circumstances, the shopping opportunity functions as the attraction itself (Reisinger and Turner, 2001). Shopping activity is particularly important to international tourists, who are likely to spend larger amounts of money on souvenirs and goods that may not be readily available or affordable in their home country (Dimanche, 2003; Jansen-Verbeke, 1991; Timothy and Butler, 1995).

Shopping in tourism settings can result from various motives, including diversion, self-gratification, learning about local traditions and new trends, and sensory stimulation (Tauber, 1972). Tourists' shopping motivations, as stated by Jansen–Verbeke (1994), can be divided into 4 categories: (a) strengthening social ties, (b) taking advantage of the unique goods provided or bargaining prices offered, (c) purchasing goods and products that represent the identity of the destination, and (d) being motivated by a favorable exchange rate. Similarly, Geuens et al. (2004) suggest that tourists' shopping motivations can be divided into functional motivation, social motivation, and experiential motivation. In addition, the leisure and functional nature of tourism shopping can be explained by hedonic and utilitarian shopping values, which have been applied in tourism settings (Yuksel and Yuksel, 2007). When shopping during the trip, tourists perceive certain attributes to be important in affecting their buying intentions. Product quality (LeHew and Wesley, 2007), price (Keown et al., 1984), uniqueness (Littrell et al., 1994), and personal factors (Reisinger and Waryszak, 1994) are reported to be most valued by tourists. However, tourists of different nationalities may have diverse shopping preferences (Reisinger and Turner, 2001; Lehto et al., 2004) and hold varied values (Wong and Law, 2003).

14.2 OVERVIEW OF CHINESE OUTBOUND TOURISM SHOPPING

Chinese tourists have been recognized as heavy spenders in tourism expenditures (UNWTO, 2014). For example, the Australia Bureau of Tourism Research (2003) found that 81% of Chinese outbound tourists engaged in shopping when traveling in Australia, making it the most popular activity (Chow and Murphy, 2008). Chinese tourists to the United States also demonstrated the fastest growth rate (53% increase in 2010 compared with that in 2009) and the highest spending amount (US$7,200 on average in 2008, topping all international markets to the United States, with a grand average of US$4,000) (Office of Travel and Tourism Industries [OTTI], 2012).

Although shopping was found to be one of the most favorite tourism activities by outbound Chinese tourists, very limited studies investigated outbound Chinese tourists' shopping behavior. Hong Kong is one of the few first outbound destinations for Mainland Chinese tourists (Qu and Lam, 1997; Zhang and Heung, 2002). A few studies shed light on Mainland Chinese tourists' shopping experience in Hong Kong. For example, Mainland Chinese tourists are found to have high expectation on quality of goods, service quality, variety of goods, and price of goods when shopping in Hong Kong (Wong and Lau, 2001). Qu and Li (1997) indicated that Mainland Chinese tourists in Hong Kong were satisfied with the shopping facility, variety of choices, convenience to the shop, and staff service but were not satisfied with both the product quality and the price offered. The most significant complaints among Mainland Chinese visitors to Hong Kong are found to be the mandatory shopping trips provided by tour guides (Zhang and Chow, 2004; Zhang et al., 2009). Furthermore, Choi et al. (2008) examined a wide range of aspects of shopping, including locations, brand preference, tendency to purchase new brand, decision-making style, product attributes, shopping environment, and sales service, of individual visitors from Chinese Mainland to Hong Kong.

Chinese outbound tourists' shopping behavior has drawn increasing attention in both academia and industry. The majority of prior studies examined the Chinese shopping behavior in terms of shopping motives, preference, satisfaction, and experience (Choi et al., 2008; Heung and Cheng, 2000; Wong and Law, 2003; Lehto et al., 2004; Lin and Lin, 2006; Xu and McGehee, 2012). It is suggested that Chinese tourists prefer purchasing electronics and famous brand-name items for their extended network of friends, family, and even acquaintances when traveling overseas (Guo et al., 2007). Chow and Murphy (2011) examined the predictive power of psychographic and demographic variables on Chinese outbound tourists to Australia and found that "Shopping and dining" is significantly related to a combination of psychographic and demographic contributors. More recently, Xu and McGehee (2012) conducted in-depth interviews with Chinese tourists who had participated in group tours of East Coast cities of the United States during the spring of 2008. The qualitative research revealed Chinese tourists' product choices, shopping motivation, shopping experience, and problems during the shopping trip. In addition to the academic research, the Western retail industry has started making efforts to attract more Chinese tourists, including hiring Mandarin-speaking employees, providing Chinese labels/instructions in stores, releasing Chinese-version Web sites, accepting Chinese credit and debit cards as payment, celebrating Chinese festivals and holidays in stores, and organizing stronger exposure/promotion campaigns in China (Kumar, 2014; Clark, 2014). However, due to the limited research, much is still unknown about this important, fast-growing market.

14.3 EMPIRICAL RESEARCH OF THIS STUDY

The major empirical results of this particular study come from a larger project examining Chinese outbound tourists' shopping behavior. Survey research was employed with the target population defined as adult Mainland Chinese citizens (18 years old and above) who had undertaken an outbound leisure trip outside Asia in the previous 3 years. To make the sampling more relevant to overseas tourism shopping, the potential respondents must have had previous overseas tourism shopping experience by spending >6,000 Chinese Yuan (about US$1,000) on shopping activities undertaken by the tourist. Data collection was conducted in Beijing, Shanghai, and Guangzhou, the 3 major gateway cities that are the primary outbound-tourist-generating areas generally categorized as "Tier-I" cities of China (Canadian Tourism Commission [CTC], 2007). A professional marketing research company was contracted in China to conduct the fieldwork in May 2013. Systematic intercept sampling method with face-to-face interview was utilized in major business districts and premium commercial shopping areas in each city to approach professional or middle-to-upper class residents. Two screening questions were asked at the beginning to only include those who had undertaken at least 1 overseas travel with shopping expenses >6,000 Yuan in the past 3 years. As a result, 300 completed, usable responses were collected and utilized for the data analysis.

The survey recorded respondents' most recent overseas tourism shopping characteristics, perception of shopping attributes, attitudes, perceived shopping barriers, and future intentions for tourism shopping. Demographic and socioeconomic characteristics of the respondents – age, gender, education, income and occupation, and so on– were also collected. Gender and age group were carefully monitored to generate a balanced data set.

In addition to the quantitative survey research, empirical findings from another qualitative study with similar research purpose were used to provide more detailed understanding of Chinese tourists' shopping behavior and experience in the United States. A total of 20 in-depth, face-to face interviews were conducted in May 2012; each interview lasted around 40–45 minutes. The interview sites covered 3 first-tier cities, Beijing, Shanghai, and Guangzhou, as well as 1 midsized city (Zhoushan) in Southeast China. Interview participants were adult Mainland Chinese citizens who had undertaken at least 1 leisure trip of 4 or more nights in the United States in the past 3 years and spent a minimum of $1,500 exclusively on shopping during the trip. The recruitment process ensured the balance of the demographic characteristics, such as age, gender, purpose of travel, and shopping spending amount.

14.4 RESULTS AND DISCUSSION

14.4.1 DEMOGRAPHIC CHARACTERISTICS

Of the 300 respondents in the quantitative study, the residency cities were evenly allocated among Beijing, Shanghai, and Guangzhou, each of which had an equal number of respondents, namely 100. The sample had a balanced gender profile, with 51% females and 49% males. Most respondents were married or had a partner

(77%), and 21% were single. Most of the respondents were within the age range of 26–55 years (77%), among which the 26- to 35-year age group represented 28% of the sample, 46–55 years old accounted for 25%, and 36–45 years old accounted for 24%. About 66% of the respondents had received a college degree (35%), had an associate degree, or had attended some college (31%). The remaining respondents had had high school/vocational/technical school education or below (27%) or graduate work/master's/doctoral degree (7%). The respondents had varying amounts of monthly household income. The largest group earned between 10,001 and 20,000 RMB (about US$1,501–3,100) per month (41%), followed by the group with monthly income of 20,001–30,000 RMB (about US$3,101–4,700) (19%) and 7,000–10,000 RMB (about US$1,100–1,500) (18%). The majority of the respondents were employed full time or part time (82%), and 11% were retired. The remaining respondents were students, people who were temporarily unemployed or looking for work, and housewives (househusbands)/homemakers (5%). Table 14.1 summarizes respondents' demographic information. The demographic characteristics of the sample are generally consistent with those in previous studies on Chinese outbound tourists (Sparks and Pan, 2009; Li et al., 2013).

TABLE 14.1 Demographic Characteristics of the Respondents (N = 300)

Variable	Frequency	Percentage
Gender		
Male	146	49
Female	154	51
Residency city		
Beijing	100	33
Shanghai	100	33
Guangzhou	100	33
Marital status		
Single	63	21
Married/Partnered	232	77
Separated/Divorced/Widowed	5	2
Age (years)		
18–25	29	10
26–35	83	28
36–45	72	24
46–55	74	25

56–65	21	7
≥66	21	7
Education level		
High school/vocational/technical school or below	81	27
Associate degree or some college	93	31
College degree	106	35
Graduate work/Master's/Doctoral degree	20	7
Monthly Household Income (RMB)		
4,001–7,000	6	2
7,001–10,000	53	18
10,001–20,000	123	41
20,001–30,000	56	19
30,001–40,000	30	10
40,001–50,000	17	6
≥50,001	15	5
Employment status		
Employed full-time/part-time	246	82
Housewife (househusband)/homemaker	2	1
Temporarily unemployed/looking for work	2	1
Retired	34	11
Student	10	3
Others	6	2

The qualitative study also had a generally equal gender distribution of participants, with 11 males and 9 females. Most of them were middle-aged, married, and college-educated people. About half of them were first-time visitors to the United States, and the remaining had visited the United States 2–3 times. Most had a monthly household income between US$5,000 and US$8,000 and their tourist shopping expenditures varied between US$3,000 and US$7,500.

14.4.2 TRAVEL- AND SHOPPING-RELATED CHARACTERISTICS

In the quantitative study, most of the respondents had previous outbound tourism experience to Europe (36%), followed by North America (30%), Australia/New

Zealand (30%), and Africa (21%). When asked about how many times did they spend >6,000 RMB on shopping during an overseas trip, the majority (65%) answered 1 time and 30% answered 2–3 times (30%). Only 3 respondents (1%) indicated 6 times or above. The largest group spent between 10,001 and 20,000 RMB (38%) on shopping during their most recent overseas trip, with shopping expenditures >6,000 RMB. The second-largest group spent between 6,000 and 10,000 RMB (21%) on shopping, followed by the groups spending 20,001–30,000 RMB (17%), 30,001–40,000 (10%) RMB, ≥50,000 RMB (8%), and 40,001 –50,000 RMB (7%). As for the length of stay in the overseas destination, most of the respondents spent 1–2 weeks (68%) and 3–4 weeks (26%). Only 5% of the respondents spent less than a week for an overseas trip, and only 3 respondents (1%) indicated that the length of stay was more than 1 month. The majority of the respondents took the packaged tours (56%) and another 44% traveled as fully independent tourists (FIT) (44%). Respondents indicated that they purchased a wide range of products: apparel, shoes, and handbags were found to be the top choices (79%), followed by jewelry and accessories (65%), souvenirs (64%), cosmetics and beauty care (58%), electronics (42%), and health care products (25%). Most of the respondents spent 6–10 hours (41%) or 11–20 hours (30%) on shopping during an overseas trip. More than half of the respondents were accompanied by family members (54%) and friends or relatives (46%) while shopping; about 22% of the respondents indicated that the tour group or tour guide was their shopping companions, and 16% shopped with coworkers. When asked about the allocation of shopping expenditure, approximately 77% of the shopping money was spent on buying gifts for family, friends, and relatives, whereas 23% were spent on buying on others' behalf. Table 14.2 summarizes respondents' travel- and shopping-related information. The survey findings were validated by the qualitative study, in which interview participants reported a large variety of branded merchandise purchased during their trip, including clothes, handbags, shoes, cosmetics and fragrance, health care products, jewelry and accessories, electronics, souvenirs, toys, and daily-use products.

TABLE 14.2 Travel- and Shopping-Related Characteristics of the Respondents (N = 300)

Variable	Frequency	Percentage
Overseas destination visited		
Europe	108	36
North America	91	30
Australia/New Zealand	91	30
Africa	62	21
Number of times spent 6,000+ on shopping (RMB)		
1 time	196	65

2 to 3 times	91	30
4 to 5 times	10	3
≥6 times	3	1
Amount spent on shopping during such a trip (RMB)		
6,000–10,000	62	21
10,001–20,000	113	38
20,001–30,000	51	17
30,001–40,000	31	10
40,001–50,000	20	7
≥50,001	23	8
Length of stay in the destination		
Less than a week	15	5
1–2 weeks	202	68
3–4 weeks	80	26
More than a month	3	1
Type of the tour		
Packaged tour	169	56
Full Independent Tour (FIT)	131	44
Primary products bought		
Apparel/shoes/handbags	237	79
jewelry/accessories	194	65
Souvenirs	192	64
Cosmetics/beauty care	174	58
Electronics	125	42
Health care products	74	25
Hours spent on shopping during the trip		
5 hours or less	36	12
6–10 hours	124	41
11–20 hours	91	30
21–30 hours	30	10
≥31 hours	19	6

Shopping companion		
Family members	162	54
Friends/Relatives	138	46
Travel group/Tour guide	67	22
Coworkers	48	16
Shopping expenditure allocation		*Mean*
Gifts for family, friends and relatives, coworkers, etc		77
Purchase on others' behalf (with payment)		23

14.4.3 PERCEIVED IMPORTANCE OF SHOPPING ATTRIBUTES

In the quantitative study, the importance of 14 shopping attributes were rated in a 5-point Likert scale ranging from 1 = very unimportant to 5 = very important. The 14 items were derived from the tourist shopping literature (i.e., Jansen-Verbeke 1991; Keown et al., 1984; Wong and Law 2003) and the Chinese tourist shopping literature (Heung and Cheng, 2000; Wong and Law, 2003; Lin and Lin, 2006; Xu and McGehee, 2012). They are "product trustworthiness," "genuine branded goods," "good value for the money," "high product quality," "attractive product price," "fashion and novelty," "wide variety of product," "hospitable service," "product uniqueness," "access to world-known brand," "good store environment," "unavailable in my own country," "commemoration of the trip," and "bringing back gifts for others." All 14 items received high ratings with scores >3.90, meaning that all of the shopping attributes are considered to be important to respondents for their overseas tourism shopping intentions. However, certain attributes were rated comparatively higher. For example, "product trustworthiness" topped the list, followed by "genuine branded goods." This indicated that Chinese outbound tourists value genuineness of products very much. They may believe that the chance of buying spurious or counterfeit branded goods in a developed Western country is very small; thus, they are more likely to shop directly in overseas merchandise stores. The third important shopping attribute was "good value for the money," similar as the fifth attribute, "attractive product price," meaning that Chinese outbound tourists may shop overseas in order to take advantage of the price difference. "High product quality" was ranked number 4. Detail information is provided in Table 14.3. Similarly, in the qualitative study, interview participants reported that they shopped mainly for world-known brands with competitive pricing and high quality and trustworthy products compared with China, access to unique and trendy US brands that are unavailable in China, and good service experience in the US shopping. Some common notions included "globally known brands," "better quality," "much lower price," "better design of product," and "more variety," which were consistent with the survey findings.

TABLE 14.3 Ranked Importance of Shopping Attributes (N = 300)

Shopping Attributes	Rank	Mean	Std. Deviation
Product trustworthiness	1	4.43	0.611
Genuine branded goods	2	4.41	0.640
Good value for the money	3	4.41	0.732
High product quality	4	4.38	0.603
Attractive product price	5	4.34	0.762
Fashion and novelty	6	4.30	0.653
Wide variety of products	7	4.28	0.629
Hospitable service	8	4.21	0.637
Product uniqueness	9	4.21	0.739
Access to world-known brand	10	4.16	0.774
Good store environment	11	4.12	0.650
Unavailable in my own country	12	4.02	0.788
Commemoration of the trip	13	3.99	0.824
Bringing back gifts for others	14	3.97	0.784

14.4.4 PERCEIVED SHOPPING BARRIERS

In the quantitative study, 6 shopping barriers were asked in the question about the problems with shopping during their most recent overseas trip (5-point Likert scale ranging from 1 = strongly disagree to 5 = strongly agree). The 6 shopping barriers are "shopping cost (money spent on shopping)," "language problem," "limited shopping time," "mandatory shopping stops," "limited payment methods," and "inconvenience in transportation." Overall, all listed barriers were considered to be "agreed" or "strongly agreed" with mean scores >3, except the last item "inconvenience in transportation." Shopping cost was considered to be the biggest shopping barrier with a mean score of 4.06, followed by language barriers (mean = 3.68) and limited time to shop (mean = 3.34). As the cost issue largely depends on tourists' own budget, destination retailers and travel agents should take the second and third barriers into consideration, that is, language barriers and limited time for shopping. Detailed ranking of the shopping barriers is listed in Table 14.4. Again, the perceived barriers were validated by the qualitative study, in which interview participants reported multiple barriers when shopping in the United States, including language barriers, unclear sales promotion information in store, limited payment methods available to Chinese travelers, sales tax, inconsistent apparel size measure, and inconvenient public transportation to reach shopping malls.

TABLE 14.4 Ranked Shopping Barriers (N = 300)

Shopping Barriers	Rank	Mean	Std. Deviation
Shopping could cost me a lot	1	4.06	0.775
I have language/communication barriers	2	3.68	0.894
There is limited time to shop	3	3.34	0.871
I have to go for the mandatory shopping arranged by travel agents	4	3.25	1.010
There are limited payment methods for shopping	5	3.10	0.923
There is inconvenience in transportation for shopping	6	2.95	0.883

14.4.5 FUTURE SHOPPING INTENTIONS

In the quantitative study, respondents were asked to rate their future overseas shopping intentions by 4 similar statements on a 5-point Likert scale ranging from 1 = strongly disagree to 5 = strongly agree. The 4 shopping intentions are "I intend to shop during my future overseas trip," "I plan to shop during my future overseas trip," "I desire to shop during my future overseas trip," and "I probably will shop again during my future overseas trip." The results showed high intentions with all the mean scores >4.0, indicating that Chinese outbound tourists had strong willingness to shop again during their future outbound travels. Among the 4 statements, "I plan to shop during my future overseas trip" received highest mean score, meaning that respondents very much intended to carry on future overseas shopping that they had planned in mind already. They were certain at some level that they would conduct overseas shopping again. Detailed information is provided in Table 14.5. Similarly, in the qualitative study, interview participants showed the same strong interest and desire to shop again in the United States. Despite the barriers and constraints, they reported that their shopping experience was very pleasant and comfortable, especially with the low price and high quality of goods and service.

TABLE 14.5 Future Overseas Shopping Intentions (N = 300)

Shopping Intentions	Mean	Std. Deviation
I intend to shop during my future overseas trip	4.09	0.562
I plan to shop during my future overseas trip	4.15	0.638
I desire to shop during my future overseas trip	4.01	0.761
I probably will shop again during my future overseas trip	4.13	0.756

14.5 CONCLUSION

The results of the 2 empirical studies revealed that respondents (Chinese outbound tourists) belong to the typical middle or upper class in China: generally middle-aged, married, with good education (had college degree or had attended some college) and professional jobs. They had prior overseas travel experience and spent substantial amount of money on shopping during the outbound trip (>6,000 RMB or US$1,000). They bought a wide variety of products during their outbound trip, including apparels, shoes, handbags, jewelry, accessories, souvenirs, cosmetics, beauty care, electronics, and health care products. Respondents spent 6–30 hours on shopping activities depending on their time availability. In addition to shopping for themselves, most of the respondents spent money for gifts for family, friends, and relatives or purchased on others' behalf based on received request.

Product features such as trustworthiness, genuine brands, high quality, and good price (value for money) were the most important shopping attributes valued by the respondents. The top 5 important shopping attributes render the implication for destination retailers that they may focus more on physical aspects of products such as decent craftsmanship, durable quality, genuine brand name, and comparatively lower price when targeting the Chinese outbound tourist market. Meanwhile, respondents reported language barriers and limited shopping time as their external shopping barriers. Therefore, to better serve the Chinese outbound tourist shoppers, Mandarin-speaking shopping assistants should be hired and product labels and product information should include Mandarin for Chinese shoppers' reference. In addition, destination marketers and travel agencies should work out more flexible and individualized travel itinerary for either package tours or independent travelers for their particular shopping needs.

With the growing population of the Chinese middle class, Chinese outbound tourists will continue to increase rapidly, and so is their purchase capability during the overseas trip. Chinese tourists' interest in global luxury brands will remain strong or even become stronger; with increasing disposable income of Chinese citizens, they will continue to be heavy spenders in tourism expenditures, particularly in overseas shopping. To the Chinese tourists, the benefit of genuine luxury products, high product quality, and price difference between overseas shopping malls and Mainland Chinese stores will continue to be the most important shopping attributes to motivate them to purchase during the outbound trips.

The current middle- and upper-class overseas travelers comprise the major force of Chinese outbound tourism shopping. Among them, middle-aged and young adults (25–45 years old), who comprise the post-70s and post-80s generations, serve as the primary market, whereas the "post-90s" will be a fast-growing purchase force due to the one-child policy and more open access to Western brands and consumerism. In addition, as Chinese tourists become more experienced in overseas travel, they would be less dependent on packaged tour or travel agencies; thus, the proportion of FIT will continue to increase fast. Shopping will continue to be the most popular activity for Chinese outbound tourists and accounts for the largest proportion of their tourism expenditure. With the market growth of the young generation, Chinese outbound tourists will be more sophisticated shoppers with better language skills and education. They would become more confident and require more individualized and personalized service in their overseas shopping experience. The limitation associated with packaged tour, such as limited shopping time and mandatory stops arranged by tour guides, may be considered to be less influential as the number of independent tourists increases. They would become more aware of their shopping needs, become more prepared with a shopping plan before the trip, and would require individualized experience. In the planning phase, word-of-mouth and social networks would be the most important channels for gathering product information and shopping tips or strategies. Recommendations from family, friends, relatives, and coworkers, as well as travel Web sites, online blogs, and social networks (e.g., Weibo or WeChat) would play more important roles in purchasers' planning and decision making. At the same time, with the increased number of independent tourists in outbound travel, family, friends, and relatives will remain the largest group of shopping companions and therefore have substantial influence on the shopping decision on site. In addition, Chinese outbound tourists will become more diverse in terms of demographic characteristics and socioeconomic status, as increasing number of outbound visitors are generated from the second- and third-tier cities and demonstrate varied shopping needs including both luxury brands and more practical, better value-for-money products (China Tourism Academy [CTA], 2014).

Therefore, destination marketers and retailers need to better understand the behavior and future trends of the Chinese tourist shoppers so as to provide the best shopping experience to these visitors. To better promote the destination shopping to Chinese tourists, marketers and retailers need to focus on genuine brands, high product quality, and attractive price (value for money) of the products and offer more personalized service such as Mandarin-speaking shopping assistants, product labels and store promotion notice with the Chinese translation, flexible store hours, and even transportation service to and from major shopping malls for independent tourists who may not have their own cars during the trip. To better promote word-of-mouth among potential consumers, especially through online platforms, destination marketers and retailers should invest on creating targeted brand awareness programs by advertising on blogs and forums, as using social media to connect and cultivate Chinese tourist shoppers has become more essential in today's marketing environment. It is important to understand that Chinese tourists do not necessarily have homogeneous shopping needs and have demands based on their age, gender, education, income, past travel experience, and even residential cities (e.g., first-, second-, or third-tier cities). More individualized and personalized shopping experience should be provided to meet the varying needs of Chinese tourist shoppers.

REFERENCES

Australian Bureau of Tourism Research. *International Visitors in Australia: Annual Results of the International Visitor Survey 1999-2002;* 2003, Australian Tourism Forecasting Council.

Canadian Tourism Commission (CTC). *Consumer Research in China.* Canadian Tourism Organization: Ottawa, Canada. http://www.corporate.canada.travel/docs/research_and_statistics/market_knowledge/AsiaPacific/Final_Consumer_V2_2007.pdf. (accessed Oct 3, 2013).

China National Tourism Administration (CNTA). *Chinese Citizens' ADS Destinations have Reached 140.* http://www.cnta.gov.cn/html/2011-4/2011-4-13-16-30-11231.html (accessed May 12, 2014).

China National Tourism Administration (CNTA). *2012 China Tourism Industry Statistics Report.* http://www.cnta.gov.cn/html/2013-9/2013-9-12-%7B@hur%7D-39-08306.html (accessed May 12, 2014).

Choi, T.; Liu, S.; Pang, K.; Chow, P. Shopping behaviors of individual tourists from the Chinese Mainland to Hong Kong. *Tour. Manage.* 2008, 29, 811-820.

Chow, I. and Murphy, P. Travel activity preferences of Chinese outbound tourists for overseas destinations. *J. Hospital Leisure Market.* 2008, 16, 61-80.

Chow, I. and Murphy, P. Predicting intended and actual travel behaviors: an examination of Chinese outbound tourists to Australia. *J. Travel Tour. Market.* 2011, 28, 318-330.

Clark, N. Catering to the Chinese shopper's grand tour. *The New York Times.* http://www.nytimes.com/2014/02/04/business/international/catering-to-the-chinese-shoppers-grand-tour.html?_r=0. (accessed Sep 25, 2014).

Dimanche, F. The Louisiana tax free shopping program for international visitors: a case study. *J. Travel Res.* 2003, 41, 311-314.

Geuens, M.; Vantomme, D.; Brengman, M. Developing a typology of airport shoppers. *Tour. Manage.* 2004, 25, 615-622.

Guo, Y.; Seongseop, K.; Timothy, D. J. Development characteristics and implications of Mainland Chinese outbound tourism. *Asia Pacific J. Tour. Res.* 2007, 12, 313-332.

Heung, V. and Cheng, E. Assessing tourists' satisfaction with shopping in the Hong Kong special administrative region of China. *J. Travel Res.* 2000, 38, 396-404.

Jansen-Verbeke, M. Leisure shopping: a magic concept for the tourism industry? *Tour. Manage.* 1991, 12, 9-14.

Jansen-Verbeke, M. The synergy between shopping and tourism: the Japanese experience. In *Global Tourism: The Next Decade*; Theobald, W. F., Ed.; Butterworth-Heinemann: Oxford, 1994.

Keown, C.; Jacobs, L.; Worthley, R. American tourists' perception of retail stores in 12 selected countries. *J. Travel Res.* 1984, 22, 26-30.

Kumar, K. Mall of America looks to China for more shoppers. *StarTribune.* http://www.startribune.com/local/west/279637882.html. (accessed Oct 20, 2014).

LeHew, M. and Wesley, S. C. Tourist shoppers' satisfaction with regional shopping mall experiences. *Int. J. Cul. Tour. Hospital. Res.* 2007, 1, 82-96.

Lehto, X. Y.; Cai, L. A.; O'Leary, J. T.; Huan, T. Tourist shopping preferences and expenditure behaviours: the case of the Taiwanese outbound market. *J. Vacation Market.* 2004, 10, 320-332.

Li, X., Meng, F., Uysal, M. & Mihalik, B. (2013). Understanding the Chinese long-haul outbound travel market: An overlapped segmentation approach. *Journal of Business Research.* 66(6). 786-793.

Lin, Y. H. and Lin, K. Q. R. Assessing Mainland Chinese visitors' satisfaction with shopping in Taiwan. *Asia Pacific J. Tour. Res.* 2006, 11, 247-268.

Littrell, M.A.; Baizerman, S.; Kean, R.; Gahring, S.; Niemeyer, S.; Reilly, R.; Stout, J.A. Souvenirs and tourism styles. 1994, *J. Travel Res.*, 33, 3-11.

Office of Travel and Tourism Industries (OTTI). *Top 10 international Markets: 2011 Visitation and Spending.* 2011. http://www.tinet.ita.doc.gov/pdf/2011-Top-10-Markets.pdf (accessed Oct 9, 2012).

Qu, H. and Lam, S. A travel demand model for Mainland Chinese tourists to Hong Kong. *Tour. Manage.* 1997, 18, 593-597.

Qu, H. and Li, I. The characteristics and satisfaction of Mainland Chinese visitors to Hong Kong. *J. Travel Res.* 1997, 35, 37-41.

Reisinger, Y. and Turner, L. W. Shopping satisfaction for domestic tourists. *J. Retail. Consum. Serv.* 2001, 8, 15-27.

Reisinger, Y. and Waryszak, R. Tourists' perceptions of service in shops: Japanese tourists in Australia. *Int. J. Retail Distrib. Manage.*, 1994, 22, 20-28.

Snepenger, J. D.; Murphy, L.; O'Connell, R.; Gregg, E. Tourists and residents use of a shopping space. *Ann. Tour. Res.* 2003, 30, 567–580.

Sparks, B. and Pan, G. W. Chinese outbound tourists: understanding their attitudes, constraints and use of information sources. *Tour. Manage.* 2009, 30, 483-494.

Tauber, E. M. Why do people shop? *J. Market.* 1972, 36, 46-49.

Timothy, D. J. and Butler, R. W. Cross-border shopping: a North American perspective. *Ann. Tour. Res.* 1995, 22, 16-34.

Wong, S. and Lau, E. Understanding the behavior of Hong Kong Chinese tourists on group tour packages. *J. Travel Res.* 2001, 40, 57-67.

Wong, J. and Law, R. Difference in shopping satisfaction levels: a study of tourists in Hong Kong. *Tour. Manage.*, 2003, 24, 401-410.

World Tourism Organization (UNWTO). *UNWTO Tourism Highlights (2013 Edition)*. http://dtxtq4w60xqpw.cloudfront.net/sites/all/files/pdf/unwto_highlights13_en_lr_0.pdf (accessed May 12, 2014).

Xu, Y. and McGehee, N. G. Shopping behavior of Chinese tourists visiting the United States: letting the shoppers do the talking. *Tour. Manage.* 2012, 33, 427-430.

Yüksel, A. and Yüksel, F. Shopping risk perceptions: effects on tourists' emotions, satisfaction and expressed loyalty intentions. *Tour. Manage.* 2007, 28(3), 703-713.

Zhang, H. Q. and Chow, I. Application of importance-performance model in tour guides' performance: evidence from mainland Chinese outbound visitors in Hong Kong. *Tour. Manage.* 2004, 25, 81-91.

Zhang, H. Q. and Heung, V. The emergence of the mainland Chinese outbound travel market and its implications for tourism marketing. *J. Vacation Market.* 2002, 8, 7-12.

Zhang, H. Q.; Heung, V.; Yan, Y. Q. Play or not to play – an analysis of the mechanism of the zero-commission Chinese outbound tours through a game theory approach. *Tour. Manage.* 2009, 30, 366-371.

CHAPTER 15

CHINESE OUTBOUND TOURISTS' LUXURY CONSUMPTION

WAN YANG

CONTENTS

15.1 OVERVIEW OF CHINESE OUTBOUND TOURISTS' LUXURY CONSUMPTION

As China's outbound tourism continues to grow, Chinese travelers are now the top source of tourism cash in the world according, to a recent report by the United Nations World Tourism Organization (UNWTO). Mafengwo.cn (2014), a leading travel social media Website in China, reports that Hong Kong, Macau, Korea, Thailand, and the United States were the most popular outbound destinations, and the average spending per outbound tourist in 2013 was about US$1,350. There were more female tourists (63.31%) than male tourists (36.69%), and about 68% of Chinese outbound tourists were between the ages 25 and 44. They spent most of their money on shopping (52.6%) and transportation (26.8%).

In fact, among other tourist activities, Chinese travelers prefer shopping overseas, especially luxury shopping. According to the World Luxury Association, Chinese outbound tourists spent US$7.2 billion on luxury products overseas in January 2012, and Chinese visitors purchased 62% of luxury products sold on the European continent in 2011 (World Travel Online, 2012). Bain & Company also suggested that the number of Chinese luxury tourists was steadily growing and about 50% total sales of luxury products in many European countries (e.g., London, Paris, Milan) could be attributed to Chinese outbound tourists (Shukla, 2014).

Compared to sightseeing and other tourist activities, Chinese tourists might be more interested in luxury purchases. Why is that? Market reports have revealed several reasons such as price discrepancy, product selection, and product authenticity. For example, due to the high-level luxury tax, the average price of a luxury product in China is more than 50% higher than that in the United States. Such a huge price gap is the main driver of Chinese outbound tourists' luxury consumption. In addition, some Chinese customers prefer shopping for luxury products overseas because of the greater brand availability, better product selections, and the availability of limited editions.

As suggested by the General Manager of the Spring International Travel Agency in Shanghai, most Chinese outbound tourists consider luxury shopping as the main purpose of their trips overseas, and such a trend is set to continue (Rhodes, 2014). Meanwhile, wealthy Chinese consumers are shifting their luxury buying overseas. Larson (2014) reported that the overall luxury spending in Mainland China increased only 2% in 2013, but he argued that the number did not reflect all luxury consumptions by Chinese consumers as oversea luxury shopping was growing fast. A survey conducted by Chinese news site iFeng (2013) revealed that 43% of Chinese luxury consumers shopped for luxury products during their outbound trips. The most popular overseas luxury shopping destinations include Hong Kong, Europe, the United States, and Japan; Luxury watches, handbags, and jewelry are the top product categories purchased by Chinese overseas shoppers (Albatross, 2012).

Taken together, as the overseas luxury shopping continues to grow, it is important to examine Chinese outbound tourists' luxury consumption – Who are they? Where do they spend the money? What do they say about the trip? In the following sections, a framework called "Luxury 4Ps" will be introduced to obtain a deeper understanding of Chinese luxury tourists, and then several empirical studies based on the Luxury 4Ps framework will be discussed.

15.2 CONSUMER'S NEED FOR STATUS AND LUXURY 4PS FRAMEWORK

15.2.1 CONSUMER'S NEED FOR STATUS

Veblen's theory of The Leisure Class is considered the first to shed light on status seeking and luxury consumption (1899). In the chapter "Pecuniary Emulation," Veblen explains the most important premise of his book: People strive for social status through comparing and competing with each other for material resources (1899). In other words, people live to show off their wealth. In this pecuniary competition system, people try to distance themselves from people in a lower class, while mimicking the behaviors of people in a higher class. In addition, people tend to put their wealth on display to demonstrate how wasteful they can afford to be, thereby elevating their social status. This behavior is termed "conspicuous consumption."

Later on, Sociologist Georg Simmel proposed a similar theory called the upper class theory of fashion (1904). He states that there are 2 conflicting forces that drive fashion change. First, lower classes adopt the status symbols of the classes above them as they attempt to climb the ladder of social status. Second, the upper classes abandon fashions that are adopted by lower classes as they attempt to distinguish themselves from the lower classes. The theories proposed by Veblen and Simmel are jointly termed "trickle-down theory."

Recent studies on luxury consumption tend to use the term "need for status" instead of "pecuniary emulation." According to Eastman et al. (1999), status consumption can exist in all communities, independent of social class membership. It is defined as "the motivational process by which individuals strive to improve their social standing through the conspicuous consumption of consumer products that confer and symbolize status both for the individual and surrounding significant others" (Eastman et al., 1999, p. 42).

15.2.2 LUXURY 4PS FRAMEWORK

Although status seeking is a major motivation of luxury consumption, not all consumers purchase luxury products to signal social status. In their innovative Luxury 4Ps framework, Han et al. (2010) categorized consumers into 4 groups based on their wealth and need for status: Patricians, Parvenus, Poseurs, and Proletarians (see

Fig. 15.1). Patricians possess significant wealth and tend to purchase inconspicuously branded products. They are low in need for status and tend to use subtle signals to associate with their in-groups. Similar to Patricians, Parvenus also possess significant wealth, but they actively seek social status through luxury consumption. They tend to broadcast their luxury consumption by using conspicuous signals, such as prominent brand logos and labels, and their first concern is to dissociate themselves from less-affluent consumers. Poseurs, like Parvenus, seek status and prefer conspicuous products. However, they do not possess enough wealth to afford authentic luxury goods and tend to purchase counterfeits. Proletarians are less affluent and low in need for status. This group does not have a strong motivation to purchase either conspicuous or inconspicuous luxury products.

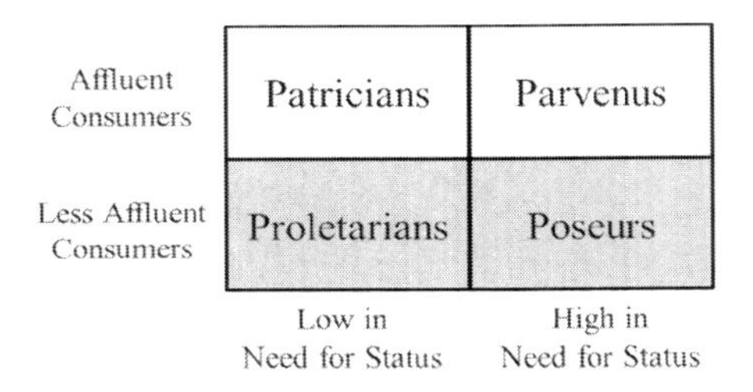

FIGURE 15.1 Luxury 4Ps Typology. Adapted from Han et al., 2010, p, 17.

In summary, both Patricians and Parvenus are wealthy consumers, but the latter seek status through consumption, whereas the former do not. Poseurs and Proletarians are less-affluent consumers; the former value social status but purchase counterfeits because they cannot afford authentic luxury products, while the latter are not interested in signaling social status and therefore are not motivated to purchase luxury products.

In the Luxury 4Ps framework of Han et al., consumers' need for status was measured as an individual-level trait (five 7-point Likert questions), and some marketers may find it difficult to gather such information. In fact, the need for status can also be captured through sociodemographic and geographic information. As suggested by the literature on geographic market segmentation, consumer needs or consumption patterns vary geographically (Beane and Ennis, 1987). Individuals living in the same country, state, or neighborhood will share similar attitudes, needs, and characteristics (Goyat, 2011). Therefore, marketers may also use city tiers or zip codes to portray the Luxury 4Ps framework.

15.2.3 LUXURY 4PS AND CHINESE LUXURY CONSUMERS

According to Gu (2013), a reporter at *The Wall Street Journal*, only 3% of the Chinese have a passport. In other words, Chinese tourists who can afford to travel overseas are only the top of the relatively better-off Chinese tourists. Since most Chinese outbound tourists are relatively wealthy, the following discussion in this chapter will mainly focus on Patricians and Parvenus.

The word Patrician originally referred to a group of ruling-class families in ancient Rome who had high social status and privileged nobility. The word Parvenu is originated from the French, and it is also called "nouveau riche" or "new money." The term "Parvenu" usually refers to people who have humble backgrounds, have rapidly gained wealth, and have quickly climbed up the social ladder (Augustine, 1994; Bourdieu, 2000; Walkerdine, 2003). For example, Holt (1998) portrayed Parvenus as those who have relatively lower cultural capital (e.g., family upbringing, education, and occupation) and a lack of connoisseurship and taste.

Compared with most Western countries, China has newly advanced and growing economic development. There is ample evidence to show that the majority of Chinese luxury consumers are similar to "Parvenus" and tend to purchase conspicuous luxury products to signal social status (Yang and Mattila, 2014). While there are a number of Chinese outbound tourists acting similar to Patricians, most Chinese outbound tourists tend to purchase luxury products as status symbols and not necessarily because of taste or aesthetics (Shukla, 2012a; 2014).

So how do Patricians and Parvenus behave differently? In the next section, 3 empirical studies will reveal some interesting behavior patterns of Parvenus, and implications will be provided to luxury hospitality practitioners who would like to target Chinese outbound tourists.

15.3 CHINESE OUTBOUND TOURISTS' LUXURY CONSUMPTION

15.3.1 PRODUCT PREFERENCES: LUXURY HANDBAG OR FINE DINING RESTAURANT?

According to a recent report, while Chinese outbound tourists splash out a lot of money on luxury goods such as handbags and watches, they tend to save on food and hotels (Gu, 2013). The article suggests that Chinese outbound tourists usually travel in groups, prefer to stay in economic hotels, and tend not to spend too much on food. Is this true across all Chinese outbound tourists? Do they indeed prefer luxury handbags to luxury hospitality experiences? And why is that?

Yang and Mattila (2013) conducted an empirical study and compared product preferences between Patricians and Parvenus. According to the extant literature, consumers who are high in need for status prefer highly visible products over less visible ones. In other words, when making consumption decisions, Parvenus (high in need for status) will prefer choices that are highly visible in public. For example, Chao and Schor (1998) reveal that consumers are more likely to purchase luxury brand lipsticks (more visible) than luxury brand facial cleansers (less visible) to signal social status. Consumers feel that they can pull out their luxury brand lipstick anytime they want a retouch to make their luxury possessions highly visible in public, but facial cleansers are relatively less visible in public.

Similarly, Fisman (2008) examines African-Americans' spending behaviors and explains why this group spends more on expensive sneakers than on education. According to Fisman, for people who engage in status consumption, wealthy signals need to be easily observed by audiences that they are trying to impress. Usually, the audiences include close companions as well as strangers on the street. Both audiences can easily see expensive sneakers, but the latter would have a difficult time inferring how much people spend on education.

In a recent national survey on visibility of consumer expenditures, Heffetz (2012) reveals that in general, "physical objects—durable and nondurable goods—have the highest visibility average (scores), while the less tangible service-related expenditures have the lowest visibility averages (scores)" (pp. 21). Therefore, as compared to luxury goods such as handbags and watches, the intangible nature of luxury hospitality services makes them more difficult to display and less likely to attract attention from public.

As illustrated in the Luxury 4Ps framework, Parvenus are high in need for status, and their consumption decisions are mainly motivated by audience's reactions (Eastman et al., 1999; Carter and Gilovich, 2012). Therefore, compared to luxury goods, the less-conspicuous nature of luxury hospitality services makes them less attractive in terms of status-enhancing attributes. In other words, Parvenus will prefer purchasing luxury goods in order to impress others.

Indeed, Yang and Mattila (2013) surveyed 228 American affluent consumers who frequently purchased luxury products and then grouped them into Patricians and Parvenus based on their scores of need for status measurement. All participants were instructed to imagine that they received a US$2,000 bonus and were considering one of the 2 luxury consumption options: (a) A luxury good such as a Louis Vuitton handbag or an Omega watch; or (b) A weekend getaway in a luxury hotel such as The Ritz-Carlton. The study results reveal that compared with Patricians, Parvenus are more likely to purchase luxury goods than luxury hospitality services. As most Chinese outbound tourists act similar to Parvenus and tend to purchase conspicuous luxury products to signal social status, it is reasonable to claim that the majority of Chinese outbound tourists tend to spend their money on luxury goods instead of luxury hotels and restaurants.

So what can luxury hospitality practitioners do to attract Chinese outbound tourists? As suggested by Yang and Mattila (2013), Parvenus rely heavily on tangible evidence of their luxury consumption. Therefore, for luxury hotels and fine dining restaurants that aim at Chinese outbound tourists, the key to success is to incorporate some forms of tangible evidence into their service offerings. For example, luxury hotels can offer complimentary products such as creative photos and videos, bumper stickers, coffee mugs that showcase the luxury experience to others either during the service consumption or as follow-up "thank you for your business" gifts. Therefore, Chinese outbound tourists will have certain physical evidence to display their luxury hospitality consumptions after they return home.

In addition, luxury hospitality practitioners may think outside the box to create innovative strategies and partner with luxury retailers. For example, besides stocking the minibar with regular drinks and snacks, Quin, a luxury art-themed hotel in Manhattan, provides a "provisions cabinet" that sells selected luxury products such as high-end jewelry, handbags, and some exclusive designer products. The hotel general manager claims that the provisions cabinet will give hotel guests an opportunity to get a sense of the latest fashion trends and shop for exclusive designer products with a decent discount (usually 15%–20% off retail price) (Vora, 2013). If used smartly, such a strategy can effectively attract Chinese outbound tourists as staying in the hotel can provide them easy access to shop for the latest luxury products and exclusive designs/limited editions.

15.3.2 LUXURY BRANDS GO MASS MARKET: DOES IT MATTER?

As China's gross domestic product (GDP) continues to grow, Chinese consumers are getting more familiar with luxury brands and are more likely to purchase a luxury product, either authentic or counterfeit. More Chinese customers can now afford luxury products due to the notion of mass luxury or through the abundance of counterfeit products. Luxury brands have implemented market expansions during the past decade, and less-affluent consumers can now purchase real luxury brand products through this mass luxury strategy (Kapferer and Bastien, 2009). Consequently, many customers have abandoned or exhibited negative attitude toward their favorite luxury brands due to the loss of prestige or exclusivity of the brand. For example, Gu (2013) suggests that Chinese outbound tourists do not want to purchase classic Louis Vuitton handbags that their maid can also buy (either real or fake) in China. They tend to purchase luxury products that are not available in China or the ones that are not easily copied by less-affluent consumers. Therefore, we know that in the context of luxury goods consumption, Chinese outbound tourists tend to seek status identities and tend to abandon their original choice if the brand or product becomes too popular or starts attracting less-affluent consumers in China. What about mass luxury in hospitality services? Will Chinese outbound tourists also react strongly when luxury hotels/restaurants become available to the mass market?

As extant studies have mainly focused on tangible possessions and failed to capture the differences between luxury hospitality services and luxury goods, Yang and

Mattila (2014) conducted a study of luxury goods against luxury hospitality services and examined how Parvenus react to less-affluent consumers' mimicking behaviors. In fact, a growing number of luxury hotels and restaurants in the United States have started to implement discounted pricing such as special packages and flash sales (Piccoli and Dev, 2012). These types of price promotions may inevitably induce bookings from the less affluent, thus changing the composition of the customer mix.

Compared to Patricians, Parvenus have a strong motive to dissociate themselves from those who cannot afford luxury products. Therefore, they prefer choices that can effectively communicate their desired identities (Han et al., 2010). If less-affluent consumers start purchasing the same luxury product, Parvenus will compare the consumption objects and perceive the luxury image of that object to be contaminated. Consequently, they tend to have negative attitudes toward the luxury brand or abandon their original choices in order to avoid the costs of misidentification (Berger and Heath, 2008).

However, Yang and Mattila (2014) further argue that hospitality services are more difficult to compare and more resistant to invidious comparison than tangible goods. Due to the less conspicuous nature of hospitality services, consumers will have a difficult time comparing their hospitality purchases to someone else's (Carter and Gilovich, 2010). Since hospitality services are more resistant to social comparisons, Parvenus are less likely to react negatively to mass luxury in the hospitality industry. They may discover that less-affluent consumers start mimicking their choices by staying at the same hotel with a discount price, but the abundance of intangible attributes (e.g., employee's attitude, interactions with other customers, availability of special services, etc.) makes direct comparisons between the 2 experiences very difficult. In addition, consumers in general tend to reconstruct their life experiences favorably in their memories (Mitchell et al., 1997; Dunn et al., 2011). In other words, hospitality consumption is a unique life experience to each consumer, and people tend not to make direct comparisons on personal life experiences. Indeed, Yang and Mattila (2014) reveal that although Parvenus tend to distinguish themselves from a lower class through consumption, they are less likely to abandon their luxury hospitality choices than their luxury goods choices when faced with mimicking behaviors by less-affluent consumers.

Whenever a luxury brand extends to a less-affluent market, there is a risk of contaminating the brand's luxury image and losing its original customers. Fortunately, the findings of the study of Yang and Mattila (2014) suggest that luxury hospitality services are less vulnerable than luxury goods. In other words, Parvenus are less likely to change their attitudes toward their favorite luxury brands when the consumption object is a hospitality service rather than a material possession. Therefore, luxury hospitality companies that target Chinese outbound tourists may find it easier to implement price promotions and expand to less-affluent markets than their luxury goods counterparts.

However, Yang and Mattila (2014) further suggest that such a strategy needs to be carefully implemented because consumers tend to evaluate hospitality services as a multidimensional experience that is composed of both tangible products and the intangible service environment. Besides the tangible benefits (e.g., hotel room, restaurant food, etc.), customer-to-customer interaction is also an important component in the hospitality experience (Baker et al., 1994; Tombs and McColl-Kennedy, 2003; Wall and Berry, 2007). For example, consumers may use the appearance of other customers as an important dimension to evaluate luxury hotel experience, and they expect their fellow consumers to be dressed properly and come from a similar social class (Walls et al., 2011). In other words, the presence of less-affluent consumers may interfere with the luxury consumption environment. Therefore, when entering the less-affluent market or using a discriminatory pricing strategy, luxury hospitality practitioners that target Chinese outbound tourists should be cautious and effectively use rate fences that can separate different groups of consumers.

15.3.3 WORD-OF-MOUTH COMMUNICATIONS

Word-of-mouth (WOM) is an effective and powerful marketing tool, and it has strong impact on consumer preferences and behaviors. Will Chinese outbound tourists actively spread the word on their overseas luxury consumption experiences to their family and friends? If yes, will they talk about their luxury goods purchases or luxury hospitality purchases?

According to the Luxury 4Ps framework, Parvenus are high in need for status and tend to seek wealthy identities through luxury consumption. Their consumption satisfaction derives from audiences' reactions to the consumption object, rather than from the positive attributes of the object itself (Eastman et al. 1999; Veblen, 1899). Therefore, Parvenus tend to conspicuously display their consumption objects with the intention to achieve stronger audiences' reactions (Veblen, 1899). Similarly, Parvenus also tend to actively tell their friends and family members about their luxury purchases because WOM is an effective tool to reach more audiences and trigger strong audience reactions. Using 228 American affluent consumers, Yang and Mattila (2013) reveal that Parvenus indeed exhibit stronger intentions to spread positive WOM about their luxury consumption than Patricians. In other words, the majority of Chinese outbound tourists who act similar to Parvenus are expected to actively talk about their overseas luxury consumption experiences. In addition, unlike Patricians who are more likely to talk about luxury hospitality purchases than luxury goods purchase, Parvenus indicated an equally strong intention to broadcast their luxury consumptions regardless of the purchase type (goods or services). Therefore, the majority of Chinese outbound tourists who are high in need for status tend to actively spread positive WOM on both overseas luxury hospitality purchases and luxury good purchases.

Since Chinese outbound tourists tend to talk about their overseas experiences, marketers of luxury hotels and fine dining restaurants that target Chinese outbound tourists can take advantages of the power of WOM. They may provide more marketing campaigns and venues for Chinese outbound tourists to actively engage in WOM communications. For example, popular social media Web sites in China such as Weibo and WeChat could be effective platforms for customers to share their consumption experiences and help promote the brands. Luxury hospitality companies should make sure that they have presence on those social media Web sites and actively interact with Chinese outbound tourists (Yang and Mattila, 2013).

15.4 A FINAL NOTE

In an emerging country with a large number of newly wealthy citizens, the majority of Chinese outbound tourists are expected to act similar to Parvenus at this stage. However, as China's GDP continues to grow and Chinese luxury consumers continue to mature, the desire for conspicuous products will cool off in the future. In addition, Shukla (2014) argues that the Luxury 4Ps framework is not a watertight concept, and he believes that a consumer can be either a Patrician or a Parvenus depending on the specific situation. In other words, although the majority of Chinese outbound tourists act similar to Parvenus at this point, they may exhibit different consumption behaviors due to changes in social and political environments.

15.4.1 ANTICORRUPTION CAMPAIGNS

Since the election of the new leadership in China in 2013, President Xi Jinping has launched a series of anticorruption campaigns to eliminate bureaucracy, formalism, and lavish spending. These campaigns have had a dampening effect on conspicuous luxury consumption in China overall. Market reports show that sales of luxury goods in China's first-tier cities have suffered, ranging from various degrees of slowdown to outright decline. However, the impact of anticorruption campaigns on Chinese outbound tourists' luxury consumption is not clear. Frizell (2014) argues that the campaigns do not entirely discourage luxury consumption and Chinese consumers are now spending their money overseas. He said thus:

The crackdown hasn't entirely discouraged spending, and the Chinese are figuring out new ways to spend. Chinese consumers are now purchasing more luxury goods abroad, traveling to New York and Paris to buy handbags and pens, buying more than 60% of their luxury goods outside of the country. (In 2012, they were already far and away the world's largest international tourism spenders, dishing out a total of US$102 billion.) And they're buying fewer flashy goods at home.

Therefore, it is possible that Chinese outbound tourists will spend more money on luxury products in the near future; however, this author suspects that they will act more similar to Patricians rather than Parvenus, as lavish spending and conspicuous displays are not appropriate during the anticorruption campaigns.

15.4.2 YOUNGER OUTBOUND TOURISTS

As shown in the Annual Report of China Outbound Tourism Development, a younger group of outbound tourists with ages between 25 and 34 is growing fast, and it now accounts for >50% of the total China outbound tourists (China Tourism Academy, 2014). This group of tourists in general is well educated, with college education or higher, and they prefer more authentic and in-depth travel experiences. This new wave of Chinese outbound tourists is less concerned about status and prestige, and they tend to spend more on luxury travel products such as lodging, dining, and entertainment activities (China Tourism Academy, 2014). For example, as revealed by a recent report by Hotel.com (2014), although the majority of outbound tourists preferred midscale and economic hotels and only 17% of outbound tourists chose 5-star luxury hotels on an international trip, the growth rate of the luxury hotel segment was the highest. In other words, younger outbound travelers are expected to spend more on luxury travel products in the next few years. Besides shopping for luxury products such as clothes and jewelry, they are willing to spend more on authentic and exotic activities such as skydiving, surfing, concerts, and opera tickets. Therefore, hospitality practitioners may consider developing niche travel products such as fashion tours in Italy, luxury car driving in Germany, wine tasting in France, or skiing in Canada to attract the new wave of Chinese outbound tourists.

REFERENCES

Albatross. 7 reasons why Chinese luxury consumers prefer to buy abroad. 2012. http://www.albatrossasia.com/7-reasons-why-chinese-luxury-consumers-prefer-to-buy-abroad/ (accessed May 20, 2014).

Augustine, D. L. *Patricians and Parvenus: Wealth and High Society in Wilhelmine Germany*; Berg Publishers: Oxford, 1994.

Baker, J.; Grewal, D.; Parasuraman, A. The influence of store environment on quality inferences and store image. *J. Acad. Mark. Sci.* 1994, 22, 328-339.

Beane, T.P. and Ennis, D.M. Market segmentation: a review. *Eur. J. Mark.* 1987, 21, 20-42.

Berger, J. and Heath, C. Who drives divergence? Identity signaling, outgroup dissimilarity, and the abandonment of cultural tastes, *J. Pers. Soc. Psychol.* 2008, 95(3), 593-607.

Bourdieu, P. *Pascalian Meditations*; Stanford University Press: Redwood city, 2000.

Carter, T. J. and Gilovich, T. The relative relativity of material and experiential purchases. *J. Pers. Soc. Psychol.* 2010, 98, 146.

Carter, T. J.; Gilovich, T. I am what I do, not what I have: the differential centrality of experiential and material purchases to the self. *J. Pers. Soc. Psychol.* 2012, 102, 1304.

Chao, A. and Schor, J. B. Empirical tests of status consumption: evidence from women's cosmetics. *J. Econ. Psychol.* 1998, 19, 107-131.

China Tourism Academy. *Annual Report of China Outbound Tourism Development 2014*. China: Tourism Education Press, 2014.

Dunn, E. W.; Gilbert, D. T.; Wilson, T. D. If money doesn't make you happy, then you probably aren't spending it right. *J. Consum. Psychol.* 2011, 21, 115-125.

Eastman, J. K.; Goldsmith, R. E.; Flynn, L. R. Status consumption in consumer behavior: scale development and validation. *J. Mark. Theory Pract.* 1999, 7, 41-52.

Fisman, R. Cos and effect: bill Cosby may be right about African-Americans spending a lot on expensive sneakers—but he's wrong about why. 2008. http://www.slate.com/articles/business/the_dismal_science/2008/01/cos_and_effect.html (accessed Jun 20, 2014)

Frizell, S. Despite slowdown, the cult of luxury grows in China. http://business.time.com/2014/02/13/despite-slowdown-the-cult-of-luxury-grows-in-china/ (accessed May 27 2014)

Goyat, S. The basis of market segmentation: a critical review of literature. *Eur. J. Bus. Manag.* 2011, 3, 45-54.

Gu, W. What are Chinese tourists buying? 2013. http://www.chinaspeakersbureau.info/2013/04/what-are-chinese-tourists-buying-wei-gu/ (accessed May 2, 2014)

Han, Y. J.; Nunes, J. C.; Drèze, X. Signaling status with luxury goods: the role of brand prominence. *J. Mark.* 2010, 74, 15-30.

Heffetz, O. Who sees what? Demographics and the visibility of consumer expenditures. *J. Econ. Psychol.* 2012, 33, 801-818.

Holt, D. B. Does cultural capital structure American consumption? *J. Consum. Res.* 1998, 25, 1-25.

Hotel.com. Chinese international travel monitor. 2014. https://press.hotels.com/content/themes/CITM/assets/pdf/CITM_UK_PDF_2014.pdf (accessed Sep 4, 2014)

IFeng. Luxury Purchase Power. 2013. http://fashion.ifeng.com/zh/special/2013nzch/ (accessed Sep 4, 2014)

Kapferer, J.-N. and Bastien, V. The Specificity of Luxury Management: Turning Marketing Upside down. *J. Brand Manag.* 2009, 16, 311-322.

Larson, C. More Chinese luxury shoppers prefer to buy overseas. 2014. http://www.businessweek.com/articles/2014-01-14/more-chinese-luxury-shoppers-prefer-to-buy-overseas (accessed Sept 4, 2014)

Mafengwo. China outbound tourism report. 2014. http://www.mafengwo.cn/gonglve/zt-574.html (accessed Sep 4, 2014)

Mitchell, T. R.; Thompson, L.; Peterson, E.; Cronk, R. Temporal Adjustments in the Evaluation of Events: The "rosy View." *J. Exp. Soc. Psychol.* 1997, 33, 421-448.

Piccoli, G. and Dev, C. Emerging marketing channels in hospitality: a global study of internet-enabled flash sales and private sales. *Cornell Hospital. Rep.* 2012, 12, 6-19.

Rhodes, F. The rise of the Chinese luxury consumers. https://www.credit-suisse.com/ch/en/news-and-expertise/news/economy/sectors-and-companies.article.html/article/pwp/news-and-expertise/2013/09/en/the-rise-of-the-chinese-luxury-consumer.html (accessed May 20, 2014)

Rova, S. New Minibar offerings: handbags and jewelry. http://intransit.blogs.nytimes.com/2013/11/20/new-minibar-offerings-handbags-and-jewlery/ (accessed Dec 12, 2013)

Shukla, P. The influence of value perceptions on luxury purchase intentions in developed and emerging markets. *Int. Mark. Rev.* 2012a, 29, 574-596.

Shukla, P. What is your Chinese new year luxury strategy? http://www.pauravshukla.com/what-is-your-chinese-new-year-luxury-strategy (accessed May 20, 2014)

Simmel, G. Fashion, Donald N. L. Ed., *Individuality and Social Forms*. Chicago: University of Chicago Press, 1904, 294-323.

Tombs, A. and McColl-Kennedy, J. R. Social-servicescape conceptual model. *Mark. Theory.* 2003, 3, 447-475.

Veblen, T. *The Theory of the Leisure Class*. New York: Penguin, 1899.

Vova, S. New Minibar offerings: handbags and jewelry. http://intransit.blogs.nytimes.com/2013/11/20/new-minibar-offerings-handbags-and-jewlery/ (accessed Dec 12, 2013)

Walkerdine, V. Reclassifying upward mobility: femininity and the neo-liberal subject. *Gend. Educ.* 2003, 15, 237-248.

Wall, E. A. and Berry, L. L. The Combined effects of the physical environment and employee behavior on customer perception of restaurant service quality. *Cornell Hotel Restaur. Adm. Q.* 2007, 48, 59-69.

Walls, A.; Okumus, F.; Wang, Y.; Kwun, D. J.-W. Understanding the consumer experience: an exploratory study of luxury hotels. *J. Hosp. Mark. Manag.* 2011, 20, 166-197.

World Travel Online. Chinese tourists splurge US$7.2b overseas for luxuries during the Spring Festival 2012. http://lvyou168.cn/tools/doPrint_en.aspx?main_id=201223105723672 (accessed May 20, 2014)

Yang, W. and Mattila, A. S. The Impact of Status Seeking on Consumers' Word of Mouth and Product Preference: A Comparison Between Luxury Hospitality Services and Luxury Goods. *J. Hosp. Tour. Res.* 2013. DOI: 10.1177/1096348013515920. Published Online: December 18, 2013. http://jht.sagepub.com/content/early/2013/12/17/1096348013515920.abstract (accessed Jun 20, 2014)

Yang, W. and Mattila, A. S. Do affluent customers care when luxury brands go mass?: the role of product type and status seeking on luxury brand attitude. *Int. J. Contemp. Hosp. Manag.* 2014, 26, 526-543.

CHAPTER 16

APPLICATION OF SOCIAL MEDIA AMONG CHINESE OUTBOUND TOURISTS: PLATFORMS AND BEHAVIORS

HAN SHEN and XING LIU

CONTENTS

16.1 INTRODUCTION

According to the data from CNNIC, China's Internet penetration increased from 42.1% in 2012 to 45.8% at the end of 2013. By the end of 2013, China had 618 million Internet users, among whom 500 million users surf the Internet via mobile phones. In China, 90% of network users have at least 1 social media account, and the number of social media users amounts to 550 million.

Due to the rapid development of China's social media and the continuous update and growing demands from users, China's social media platforms have demonstrated clear product and market differentiation. Table 16.1 summarizes the major social media in China.

TABLE 16.1 The Major Social Media in China

Platform Category	Platforms in China	The Corresponding International Platforms
Instant communication	QQ, Fetion, Alitalk	Skype
Music and Video	Tudou, PPLive, tv.sohu.com, , iQIYI, Youku	YouTube
Blog	Sohu blog, Sina blog	Blog
Microblog	Sina Weibo, Tencent Weibo	Twitter
Social network	Renren, QQ space	Facebook
Forum	Liba.com, Post Bar	BBS
Mobile social networking	WeChat	WhatsApp
Social life	Dianping, Douban	Yelp
E-Commerce	Taobao, JD, Tmall	Amazon
Picture social networking	TuHai	Instagram
Enterprise social networking	UU	Yammer
Business social networking	Belink	LinkedIn
Dating social networking	Baihe.com, Jiayuan.com	match.com
Tourism social networking	Mafengwo, Qyer	TripAdvisor

Source: Adapted from the Statistical Report on Internet Development in China, published by China Internet Network Information Center, CNNIC, 2014.

Different types of social media platforms have different marketing values. First, in terms of social media platforms like instant messenger and video and/or music, they had users of 530 million and 430 million, respectively, by the end of 2013, among whom Internet users accounted for 86.2% and 69.3%, respectively (CNNIC, 2014). With such a large number of users, it is very suitable to carry out brand promotion and establish brand popularity so as to quickly create brand recognition among widespread users in a short time.

Second, as to platforms featuring high interactivity and rich information like Blogs, Weibo, social networks, and forums, they are easy to spread in-depth information and increase interconnection so as to create groups of fans with close connection for certain brands.

Third, the development trend of more and more platforms featuring mobile social, socialized life, and e-commerce is to provide service and transactions; for example, dianping.com is developing to offer group purchase service and transactions so as to provide more attractive value-added transactions in addition to providing business information and user comments.

Fourth, in terms of target market segmentation, China's social media can be segmented into such platforms concerning travel, dating, business, enterprises, pictures, and questions and answers. And the corresponding groups of users have developed into vertical consumers with similar interest, and this provides a precise user location for brand promotion.

16.2 CHINESE SOCIAL MEDIA IN TOURISM

16.2.1 SINA WEIBO TRAVEL

Sina Weibo is a platform introduced by Sina.com, offering microblog service similar to Twitter. Users can release news or upload pictures via web pages, WAP pages, mobile clients, text messages, and MMS. Users can write what they see, hear, or think of in a sentence or send a picture and share it with friends whenever and wherever possible via computers or mobile phones for discussion together; they can also see the news released by their friends immediately.

Though Sina Weibo experienced a serious running off of users due to external impacts in 2012, the tourism part remains an important social media platform for promoting tourism. Sina Weibo data show that by the end of 2012, active users of microblog travel came to 62 million persons, of whom 45.82 million are fans of travel agencies, 21.61 million are users with travel labels, and 21.7 million are travel enthusiasts. Among these users, those born after the 1970s account for 11%, those born after 1980s account for 36%, and those born after the 1990s account for 50%, with females making up 67% and males 33%. And these people usually are highly educated with 54.35% of them holding a bachelor degree or above and 29.45% holding a junior college diploma. Average topics released concerning tourism surpassed 5 million, and during the national day holidays, the daily average travel topics released amounted to 14 million. Users on weekdays log in and release as many as 18.35 million blog articles via mobiles and at weekends release 17.69 million blog articles in relation to travel (Sina, 2013).

At different stages of tourism, Weibo plays different roles, influencing ways of information search, sharing, and evaluation of tourists (Figs. 16.1–16.3).

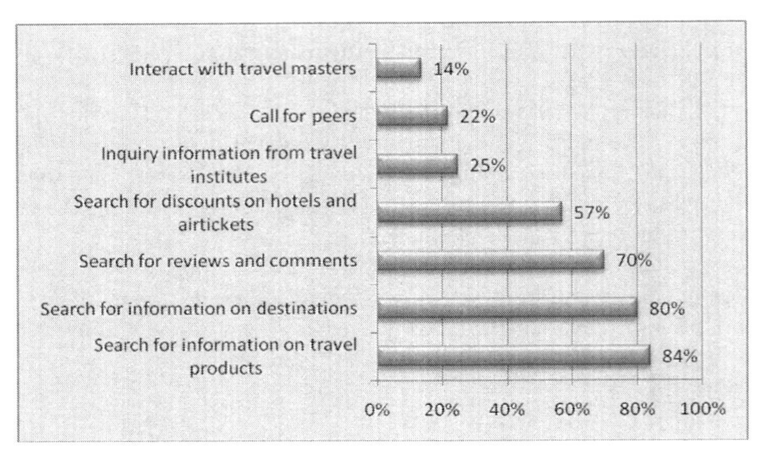

FIGURE 16.1 Weibo User's Behavior before Traveling.
(From Sina Tourism, Sina Weibo Data Center, 2013.)

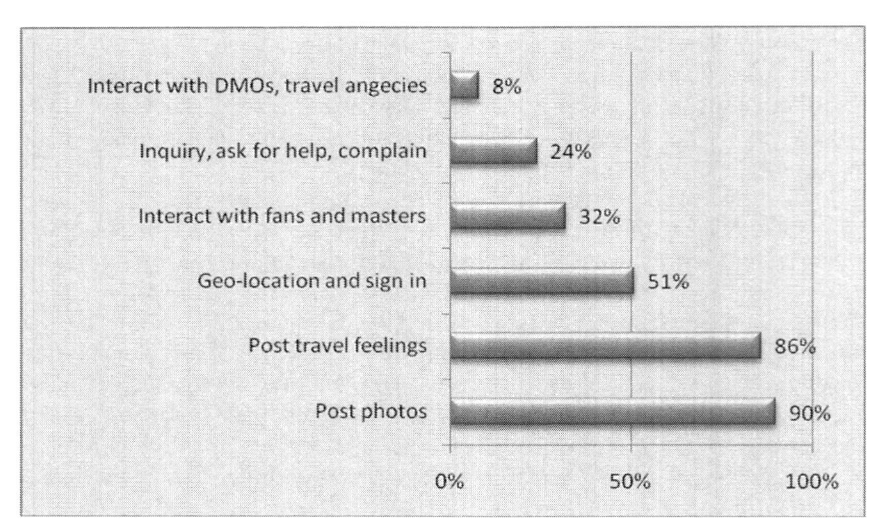

FIGURE 16.2 Weibo User's Behavior while Traveling.
(From Sina Tourism, Sina Weibo Data Center, 2013.)

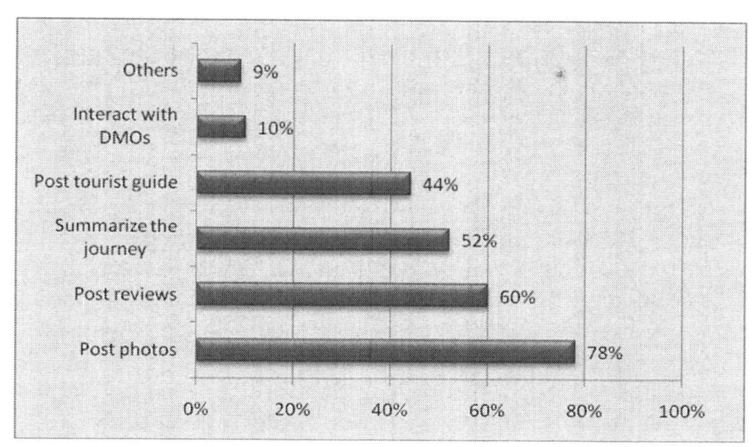

FIGURE 16.3 Weibo User's Behavior after Traveling.
(From Sina Tourism, Sina Weibo Data Center, 2013.)

The travel channel of Sina Weibo is closely connected with industries, and such online travel agencies (OTAs) as Ctrip and eLong and the like have all opened an account number for Sina Weibo, creating tourism products based on the Weibo system. In addition, some agencies in some important tourism destinations both at home and abroad have also opened their official account for Sina Weibo, thus providing rich information for tourists, promoting all kinds of marketing information, enhancing brand names of sightseeing places, and attracting more tourists.

So, the information in Weibo travel channel comprises not only information released by users themselves but also all kinds of commercial information and official news from government departments, thus greatly enriching the quality and quantity of Weibo travel information. Now Weibo has become an important source of travel information for Chinese tourists.

16.2.2 TOURIST COMMUNITY WEBSITE

Tourist community websites are important platforms for travel social media, and the 2 most well-known websites in Chinese travel communities are mafengwo.com and qyer.com, which have developed into travel strategy websites since PC times. Qyer.com, which was created in 2004, is China's biggest travel community for outbound travel, providing users with value-added products related to outbound travel, like such core products as itinerary assistant, forum, questions and answers, destinations, tips for travel with a shoestring via community-based platforms and UGC, and an easy communication mode. These websites also help users make a travel plan and offer outbound tourists effective travel tips and information. Mafengwo.com is a travel platform for mutual cooperation and sharing, which provides travel lovers wonderful itinerary strategies for more than 80,000 travel destinations worldwide and offers all kinds of travel information including introductions to destinations, exquisite pictures, travel notes, transport, delicious food, and shopping. It provides tourists with a communication platform, opportunities for seeking persons to travel together, and exchange of their photos.

The core products of travel community websites are travel strategies, and on the basis of sharing strategies, the websites offer users online reservation services for travel products and make a profit by doing so. Advertising revenue and commission revenue constitute the major source of income. The advertising revenue is mainly the income acquired by promoting related enterprises' brands and products, and the commission revenue is mainly the commission acquired by providing hotel and air tickets booking platforms for OTAs.

When we look at the decision-making step for users' travel plan, we can see that the users are usually influenced by their friends and relatives or all kinds of media and travel information when they choose their destinations, make a travel plan again, and book air tickets and hotels. Yang (2012) found that the information searching procedure for tourists was basically in agreement with the information searching model by Kuhlthau (2004) and Ellis (1987). On the basis of this, a usual procedure for searching online travel information was put forward: determining the information source, searching, browsing, assessing, retrieving, and follow-up searching. The information searching procedure for Chinese tourist is as follows: search engines (baidu, google), – travel community travel information websites (qyer, mafengwo), and travel booking websites (Ctrip, eLong, qunar, booking).

The in-depth information from travel community websites can markedly influence users' decisions when users make decisions about travel. In recent years, travel community websites have also adopted such methods as travel group buying and special offers to increase users' purchasing activities. In 2014, Alibaba group made strategic investment in qyer.com and its business model changed into " community+ search+e-commerce," Alibaba's alipay with its escrow advantage was inlaid into the procedure of sharing and searching community information, thus enabling users to have more convenience when purchasing. Moreover, based on having confidence in the prospect of online travel development, Alibaba group has upgraded its Travel Business Department's Taobao Travel to an independent brand "Qua." "Qua" will use wireless network, service, creativity, and platform as its 4 strategies, and, based on the advantage of mass flow of Alipay and Taobao, promote online travel industry to upgrade from sales to service by paying attention to travel demands of consumers.

16.2.3 WECHAT

WeChat is a free application, similar to WhatsApp, launched by Tencent in 2011 to provide instant message service for smart terminals. WeChat supports cross-communication operators and cross-operating system platforms to send free voice short messages, videos, pictures, and words rapidly through websites, and at the same time, information through sharing streaming media and social plug-ins based on positions can be used. WeChat provides a public platform, friend circle, news pushing, and other functions, and users can add friends and pay attention to the public platform through such functions as "shake it off," "search number," "people nearby," "scan 2-dimensional code," and meanwhile WeChat shares the contents among friends and shares the wonderful contents that users have seen among WeChat friend circles.

WeChat, as a newcomer in mobile social media, has experienced a rapid development. By July 1, 2014, the number of registered users of WeChat platform has exceeded 600,000,000, the number of monthly active users is 355,000,000, the number of opened public accounts is more than 2,000,000, the daily average number of registered public accounts is 8,000, and the number of certified public accounts is more than 50,000. Every day hundreds of millions of pieces of information are exchanged in the public accounts (CNNIC, 2014).

For WeChat, its major functions like "voice chat," "friend circle," and "public account" cover all the mainstream functions of social networks. WeChat realizes information exchange via its instant messenger and supports all multimedia functions like voice, images, video, address, and name card information, and it can realize the transfer of HD pictures and video shooting with 1 click, thus greatly changing the habits and modes of interpersonal communications. The social networks of acquaintances of "friend circle" and the information sharing circle and the new channel for making friends through such functions like "shake it off." "sweep it over," and "nearby" have made WeChat as a social network more interesting.

Consumers focus on sharing experiences with WeChat, and people are inclined to acquire travel information from Weibo and travel community websites and share their travel experiences with WeChat. WeChat friend circle is moderately strong network with "I" as the center spreading outward in all directions, whose relative closed structure makes users more frank to share experiences and voice opinions. WeChat's payment function has provided a good opportunity for the development of OTAs. In addition, a large number of tourism consulting firms create travel user databases, carry out direct, effective, and accurate point-to-point contact and point-to-whole travel information transmission activities, and build aggregation platforms for travel media resources.

16.3 SOCIAL MEDIA APPLICATION BEHAVIORS OF CHINESE TOURISTS

16.3.1 CHARACTERISTICS OF CHINESE SOCIAL MEDIA USERS

The number of social media users in China exceeded 550,000,000 at the end of 2013 (CNNIC, 2014), and in such aspects as social needs, social privacy, and social media use, an increasing differentiation trend has appeared.

First, in terms of the characteristics of individual users, users born after 1970 are the majority, accounting for 93.4%; they are well educated, 43.3% of whom have received university education and above; single users make up 48.6% of them and married ones 47.8% (CNNIC,2014).

Second, in terms of preferences of users of different ages, mature users tend to favor Weibo and younger users prefer to use WeChat. The proportion of Weibo users above 36 is much higher than that of WeChat users, the proportion of WeChat users younger than 35 is higher than that of Weibo users, and the proportion of WeChat users under 18 and those between 19 and 24 (born in the1990s) is far higher than that of Weibo users (Dratio, 2013).

Third, in the aspect of social needs, Chinese consumers mainly through social media achieve social communication, entertainment, news information, and other functional services, like learning about product information, catering information, and employment information. In the aspect of information focus, female users tend to focus on daily life and like browsing contents concerning recreation, travel, life, and education, while male users pay more attention to contents in relation to news, finance, sports, science and technology, automobiles, and games (Dratio, 2013).

Fourth, in terms of privacy of social contact, with the increasing popularity of instant messenger and social circles from WeChat, the trend of using real names to share personal-related information has become increasingly obvious, showing that it has a relative real social network basis.

Fifth, from the regional perspective of social media use, social media use in such first-tier cities as Beijing and Shanghai is the most active, while in fourth-tier cities like Zhongshan and Xuzhou, the proportion of social media use is markedly lower than that of other cities (CNNIC, 2014).

Sixth, in terms of the use preference of social media, the use coverage rate of instant message social media like WeChat and QQ among all netizens is 61.7%, the social network coverage rate is 61.7%, and the Weibo coverage rate is 43.6%. Each of the platforms has its focus, and they are complementary to each other. Social media websites and instant messengers are mainly for communication, exchange, and interactivity, while Weibo is mainly for information transmission, so 33.7% of the users use social websites, Weibo, and instant messenger at the same time to meet their different needs and users have a high degree of overlap (Dratio, 2013).

16.3.2 SOCIAL MEDIA APPLICATION BEHAVIORS OF CHINESE TOURISTS

Tourism consumption behavior can be divided into 3 stages: tourism decision-making before purchasing, tourism consumption during purchasing, and summary feedback after purchasing. The stage "tourism decision-making before purchasing," is the process in which tourism consumers choose appropriate ways of travel and goals in relation to destinations, including the generation of tourism motivation, query and search of tourism information, and analysis of advantages and disadvantages. The stage "tourism consumption during purchasing," is a process in which tourists implement their travel plans and decisions or a process where tourists experience tourism. This process lasts from tourists' departure from where they live to their destinations when they have completed their experience of travel.

The stage "feedback summary" is a process in which tourism consumers go back where they live and they sum up whether they had a good time while traveling, whether travel service was satisfactory, and whether tourism needs have been fully met, and this feedback process can directly affect whether the tourism consumers will have another tourism consumption.

The use of online travel resources runs through the whole process of Chinese tourists' travel. Before the trip, they make decisions and book itinerary online; in the course of or after the trip, they share photos and experiences with their family members and friends. Social media play different roles in the consumption process of tourism consumers and their specific roles are given in Table 16.2.

TABLE 16.2 Chinese Tourist's Behavior on Social Media

Stage	The Use of Platform	User's Behavior
Before traveling	Tourist's social networking platform (e.g., Qyer)	Collect information and purchase tourism products
	Tourist's review websites (e.g., Lvping.com)	Look up tourist's reviews
	Tourist Groupon websites (e.g., lvmama.com)	Groupon of travel products
	Online travel agency (e.g., Qunar.com)	Book air ticket and hotel
	Wiki Platform (e.g., Baidu Encyclopedia)	Ask questions
	Social networking platform (e.g., WeChat)	Share and exchange tourism information

While traveling	Tourist's social networking platform (e.g., Qyer)	Query travel information and broadcast travel journey in real-time
	Microblog (e.g., Weibo.com)	Share pictures and real-time journey
	WeChat	Share pictures and real-time journey, express their feelings
	Tourist's mobile APP (e.g., Ctrip APP)	Search travel product information
After traveling	Tourist's social networking platform	Share travel experiences, provide travel notes and tips
	Tourist's review websites (e.g., Lvping.com)	Comment on attractions and hotels
	Microblog (e.g., Weibo.com)	Share pictures and tourist's experience
	WeChat	Share pictures and real-time journey, express their feelings

Source: Adapted from Hao (2012).

16.3.3 SOCIAL MEDIA APPLICATION HABITS OF CHINESE TOURISTS BEFORE TRAVELING

According to a survey by hotels.com, nearly half of the tourists think that hotel booking websites, travel websites, and online comment websites are their most dependable sources of information when they make a decision about travel. More than one-third (36%) of the respondents book hotel rooms online, and nearly one-fifth (17%) of the respondents book rooms via mobile applications (Hotels.com, 2014). Tourists' perceived credibility about social media platforms is essential to the construction of dependence relations (Sparks and Browning, 2011), and Chinese tourists think that real name comments can offer more help than anonymous comments and that the real name system in a virtual community contributes to the establishment of trust relations. Therefore, WeChat based on acquaintances relationships and dependence on social media websites has a great impact on tourists' behavioral intention (Huang, 2014).

Social media influence consumers' information searching process. On the one hand, the potential consumers are stimulated by social environmental factors to have more network tourism consumption demands; on the other hand, as there are many types of sources of information provided by social media, including Weibo,

WeChat, Wiki, SNS, and online travel communities, information searching routes for tourism consumers have changed and the effective amount of information has become larger. According to a survey conducted by Hao in 2012, in the motivation stage, over half of the Chinese tourists take social media as important tools and reference sources of searching and collection of travel information; Chinese tourists pay attention to 5 aspects of social media information: price information, Internet word-of-mouth, travel strategy, catering characteristics, and traffic information.

16.3.4 SOCIAL MEDIA APPLICATION BEHAVIORS OF CHINESE TOURISTS WHILE TRAVELING

To share one's tourism experiences while traveling cannot be realized through traditional modes of travel. In the course of travel experience by tourism consumers, WeChat, Weibo, and tourism mobile clients all have a certain impact on the travel process of consumers. During a trip, travelers can show others what they meet in the course of travel through words, images, and videos; some travelers can even use mobile media to ask for help from others when encountering difficulty while traveling.

The report of Text 100 Public Relations (2013) shows that among 4,600 respondents in 13 countries including China, the Chinese respondents knew best how to make use of social websites to share their travel experience. If destinations provide free WiFi, Chinese tourists will use social websites to share what they see and hear with their friends immediately and are more likely to use smart phones to seek the direction and information of scenic spots. When making a purchase in scenic spots and stores, Chinese tourists hope to have free access to WiFi and social media to have voice chat and image communication with their relatives and friends and then make a decision whether to make a purchase or not.

16.3.5 SOCIAL MEDIA APPLICATION BEHAVIORS OF CHINESE TOURISTS AFTER TRAVELING

After the completion of a trip, many Chinese tourists are keen on displaying how they plan their itinerary before the trip on social media and show others all the information they search online about the destination, and in this way, sharing information helps other tourists save time and energy. To share tourism photos is an important part of tourism-related content, and tourist like to take photos when traveling to display themselves.

In addition, Chinese tourists are generally inclined to make a narration, especially telling the acquaintance circle in WeChat about their true feelings for the trip and making comments at the comment platforms. Such electronic word-of-mouth provides very detailed information for others to make decisions about tourism and also greatly influences their decisions.

16.4 TREND OF CHINESE TOURISTS

Replacing text messages with images and videos is the major development trend in the future. Information sharing in WeChat, Weibo, and travel community websites tend to become more refined and more visible. The trust tendency of Chinese tourists for acquaintance relations will be projected into social media, so more real name systems and social media with strong acquaintance relations will also be strengthened (Hao, 2012).

With the increase of mobile smart phone users in China year by year, social media sharing based on mobile technology will become an important trend. Many OATs, scenic spots, and management organizations in destinations have launched mobile platforms to meet the need of consumers. Surfing the Internet at any time and maintaining interaction anywhere are important features of Chinese tourists, and the real-time information searching and sharing will be realized with the popularization of 3G and 4G technologies. Tourists' behavior and decisions will become more casual and information searched at random from social media will have an impact on tourists' itinerary, which will create new demands for the contents to be provided by tourism suppliers and marketers.

ACKNOWLEDGMENT

This study is supported by China Scholarship Council and China National Social Sciences Foundation (Project No. 14BGL202).

REFERENCES

China Internet Network Information Center (CNNIC). Statistical Report on Internet Development in China. 2014. https://www.cnnic.net.cn/hlwfzyj/hlwxzbg/hlwtjbg/201403/t20140305_46240.htm.

Dratio. Research on Social Media Users Clustering and Factors. 2013. http://www.wrating.com/2013/0703/192267.html.

Ellis, D. *The Derivation of a Behavioural Model for Information Retrieval System Design*. University of Sheffield: Sheffield, 1987.

Hao, J. *A Study on the Impact of the Act of Tourists by Social Media*. Capital University of Economic and Business: China, 2012.

Hotels.com. Chinese International Travel Monitor 2014. 2014. http://press.hotels.com/citmcn/.

Huang, Y-H. Research on tourists social network behavior from the perspective of postmodernism. *Tour. Tribune*, 2014, 08: 9-11.

Kuhlthau, C. C. *Seeking Meaning: A Process Approach to Library and Information Services*, 2nd ed.; Libraries Unlimited: Westport, CT, 2004; p 82.

Sina Tourism, Sina Weibo Data Center. Micro-blog tourism white paper of 2012. 2013. http://data.weibo.com/report/reportDetail?id=160.

Sparks, B. A. and Browning, V. The impact of online reviews on hotel booking intentions and perception of trust. *Tour. Manage.* 2011, 32 (6): 1310-1323.

Text 100 Public Relations. The Media Use Behavior Research Report of Global Tourism Consumers 2012. 2013. http://www.travelweekly-china.com/the-2012china-hawaii-tourism-promotion-will-be-held-successfully/21854.

Yang, M. Online travel information search: demand, behavior and mechanism. Shanxi Normal University, 2012.

CHAPTER 17

YOUR STORIES ONLINE, THE HEURISTICS OF MY NEXT JOURNEY

XIN YANG, DAN WANG, and BRIAN KING

CONTENTS

17.1 INTRODUCTION

The remarkable economic growth of China over recent decades has brought transformations in many aspects of society such as socioeconomic structures, personal welfare, lifestyles, Internet popularity, and values ("What China Wants," 2014). Three decades of almost continuous double-digit growth have benefited Chinese households, with a 26,955 yuan average per capita disposable income for urban households in 2013 (National Bureau of Statistics of China, 2014). Economic growth has also impacted communications. The total Internet penetration rate in China reached 46.9%, and by mid-2014, there were 632 million Chinese Internet users (China Internet Network Information Center [CNNIC], 2014). Changing Chinese society has also been reflected through the interests of Chinese tourists who are setting out to experience the world outside China in diversified ways (Wu and Pearce, 2014a).

China Tourism Academy (CTA) and Ctrip (an influential online travel agency) (2013) identified 2 main trends in the development of Chinese outbound tourism: the increasing number of independent outbound tourists and a high dependence on the Internet for travel planning. While group tours have continued to dominate Chinese outbound tourism in recent years, the number of outbound independent tourists has been increasing (Pearce et al., 2013; Xiang, 2013), and their proportion of Chinese outbound trips reached 70% in 2012 (CTA & Ctrip, 2013). Another critical issue for Chinese outbound tourism is that independent travelers are more likely to be tech-savvy and empowered by social media (Arlt, 2013; Wu and Pearce, 2014b). Chinese independent tourists are heavy users of social media channels such as travel blogs, travel social network sites (SNS), virtual communities, microblogs, and products' review websites for experience sharing and travel planning (CTA, 2014; Kristensen, 2013; Zhang et al., 2013). It has been recognized that understanding social media channels in China is a key to knowing the contemporary Chinese market, particularly in the case of independent Chinese outbound tourists (Wu and Pearce, 2014a,b).

This chapter explores the role of social media in travel planning by Chinese outbound independent tourists. In particular, the researchers are interested in the impact of social media on traveler information search strategies. They examine the impact of social media on the 3 dimensions of travel information search strategy: (a) sources – the combinations of different information sources, (b) degree – the time spending for information search on each type of information sources, and (c) spectrum – the extent of browsing or directed search in travel information search. The chapter describes the integration of social media in the information search process of Chinese outbound independent tourists and offers insights into the penetration of social media channels on the travel planning process.

17.2 TRAVELER INFORMATION SEARCH STRATEGIES

Information search can be defined as "the motivated activation of knowledge stored in memory or acquisition of information from the environment" (Engel et al., 1995, p. 494). It is a function that helps travelers to reduce risks or uncertainties, thus improving their travel experience (Jang, 2005). Previous studies have explored tourist information search strategies from 3 perspectives: (a) the sources that tourists use such as internal and external search, mixed and single sources; (b) the degree to which tourists devote resources to searching (e.g., amount of time); and (c) the spectrum across which tourists browse or search directly until they satisfy their information needs.

A combination of internal and external information sources is necessary when selecting destinations and making on-site decisions such as attractions, location activities, and lodging (Fodness and Murray, 1997). Internal information search is defined as retrieving information stored in one's long-term memory, which has been actively acquired from previous experience, past information searches, or passive exposure to repeated marketing stimuli (Fodness and Murray, 1998; Gursoy and McCleary, 2004; Jang, 2005; Money and Crotts, 2003). External search refers to collecting information from the marketplace. An external search is unnecessary when the internal search provides sufficient information about a trip decision (Beatty and Smith, 1987). Beatty and Smith (1987) categorized external information search into 4 types: (a) personal (e.g., friends, relatives, and colleagues), (b) marketer-dominated (e.g., advertisements and promotions), (c) neutral (e.g., third-party such as travel agents and travel guides), and (d) experiential sources through connections with retailers.

Fodness and Murry (1997) identified another dimension of information search strategy: degree (the number of information sources and the amount of time devoted to search activities). In their examination of Florida inbound leisure tourists, they identified 3 levels of degree: active (using a wide variety of sources, most likely auto clubs, brochures, highway, welcome centers, local tourism offices, state travel guides, and travel agents), passive (using less external search, heavy reliance on friends or relatives, and highway welcome centers), and possessive (reliance on either friends or relatives or personal experience). Hyde (2008) further developed Fodness and Murry's (1997) study and identified 7 levels that describe the different combinations of information sources.

With the prevalence of the Internet, Rowley (2000) identified another dimension of information search strategy in e-shopping: browsing or directed search. "Browsing" refers to "instances when consumers are not sure, or if, how their shopping requirements can be met . . . they have a less precise view of the product information that may be available, needed or used, thus seek out information in more of an exploratory fashion" (Detlor et al., 2003, p. 72). In contrast, directed search refers to situations in which consumers know what they are seeking and usually possess some information about the relevant products and as a result actively seek out product information to make a purchase decision (Chiou and Ting, 2011; Detlor et al., 2003).

According to Rowley (2000), searchers can be placed along a spectrum ranging from a direct or purposeful searching to aimless or general browsing. Sometimes a search that begins as general browsing may turn out to have a focused outcome. Also, when the first search leads to an unsatisfying outcome, the searchers are likely to broaden or narrow their search strategy. In the context of travel, browsing occurs when (a) tourists do not know where to look for relevant information, so they may use search engines to explore, and (b) tourists go to websites—such as SNS—to explore the destination without the intention of purchasing anything. In contrast, "directed search" refers to when (a) tourists use search engines to find specific product such as admission tickets and car rentals, and (b) tourists go to certain online domains to compare prices and quality with the intention to buy.

17.3 METHODOLOGY

The overall goal of this chapter is to explore the role of social media in the travel planning of Chinese outbound independent tourists. Qualitative research methods are used. The informants were selected based on the following 2 criteria: (a) Chinese tourists who have independently planned (instead of package tour) at least 1 outbound trip during the previous year, and (b) Chinese tourists who used social media channels for travel planning. Snowball sampling was used to obtain qualified informants until no new information emerged. In total, 21 semi-structured in-depth interviews were conducted across different age groups. All interviews were tape-recorded and conducted in Chinese. Each interview lasted between 30 and 60 minutes. The questions consisted of 2 main sections: general travel information and information search strategy (travel planning process, sources, degree, and spectrum; Table 17.1).

TABLE 17.1 Interview Protocol

Section 1: General Travel Information
How many trips did you take during the last 12 months?
What destinations have you visited during the last 12 months?
Can you please describe your last independent trip? What was the duration? Who did you travel with? What was your itinerary?
Section 2: Information Search Strategy
General Travel Planning Process
How did you plan your last independent trips?
What were your information search steps?
Sources

What travel information sources did you use?

How did you use information sources?

Did you make use of information sources in other languages? If yes, how did you feel about these information sources? If no, why?

Were you aware of the official destination websites of the relevant destination marketing organizations or local governments? If yes, how did you use it? If no, Why?

Degree

How much time did you spend on planning your last trip?

How long in advance did you start your information search process?

How detailed was the trip planning?

Spectrum

After you decided your destination, do you prefer general browsing, direct searching, or both?

How much time and effort do you devote to the 2 kinds of search?

Nvivo10 was used for data analysis, which followed the procedures of the data analysis spiral (Creswell, 2007). The analysis spiral consists of the stages of data managing, reading memos, describing/classifying/interpreting, and representing.

17.4 RESULTS

Informants first described their general travel information search pattern (i.e., the information required for different travel components such as flight and accommodation, and the sequence of information search for the different components). According to informants, the information that was sought during the travel planning process included 3 types of content: flight information, accommodation information, and destination information. Not all informants included all 3 parts throughout their travel planning process. Most informants (13) planned their trips in this order: flight, accommodation, and destination itinerary design. Some informants started the process by briefly browsing information on the destination and then searching for flights and accommodation, followed by an additional in-depth destination information search. A few informants mentioned that they preferred to seek out accommodations prior to flight information. Some informants conducted their travel planning by first searching for flights and destination information but not searching for accommodation.

Overall and based on the 3 types of contents that were sought by informants, 4 types of general sequential travel planning process emerged: (a) flight information→accommodation information→ destination information; (b) brief destination information→flight information →accommodation information→detailed destination information; (c) destination information→accommodation information→flight information; and (d) flight information →destination information. Most informants (13) used the first planning pattern, 3 informants used the second pattern, and 2 informants used the 2 remaining types. The following section illustrates the 3 dimensions of travel information search strategy (i.e., sources, degree, and spectrum) for different components of trips.

17.4.1 SOURCES

Table 17.2 illustrates the number of informants that used different kinds of sources. All 21 informants used travel SNS, and 19 informants used online travel agencies (OTAs) for reviews and ratings.

TABLE 17.2 The Use of Information Sources by Chinese Outbound Independent Tourists

Information Sources	No. of Informants Used
Social media channels	
Travel social networking sites (Travel SNS)	21
Online Travel Agencies (OTAs) with reviews and ratings	19
Microblogs	7
Mobile social applications	4
General social networking sites (General SNS)	4
Blogs	2
Bulletin board system (BBS)/Forum	2
Other online sources	
Search engines	21
OTAs for airlines	19
Airline official websites	12
Destination Marketing Organization (DMO) websites	10
Hotel official websites	5
Offline sources	
Friends	10
Travel books	4
Travel agents	2
Travel exhibitions	1

Informants said that they used travel SNS to search the personal itineraries of other travelers to obtain destination information such as attractions and activities for their own itineraries. Informant 5 said,

These 2 websites [travel SNS] have many articles posted by independent travelers like me. The information is abundant and organized and includes information about the culture at the destination and famous attractions. They are very professional and provide a full summary of the destination including travel tips for you to download when you don't have much time to read all kinds of itineraries.

Informants stated that the similarity of social media contributors and themselves in terms of background is a more important reason for them to use SNS. Informant 4 commented thus:

On these websites [travel SNS], the background, culture, and taste of independent travelers are similar to me, so there are many situations in which their suggestions are applicable to me. For example, if I wanted to go to some suburban beaches in Japan, I knew I couldn't use my passport to rent a car. So I would go to those websites to find out how other travelers with the same problem were able to travel to those suburban beaches. The information on travel SNS is relevant to my situation.

In this investigation, the OTAs with reviews and ratings refer to OTAs with hotel booking functions according to the informants, since flight OTAs do not provide reviews and ratings. The informants used this kind of social media channels to compare prices, review ratings, and reviews and make a reservation. However, some informants reported the concerns of using OTAs for hotel reservation. Informant 12 said,

I used the map to decide which district I would be staying at and then compared the prices and read others' comments and ratings. But I didn't make reservations there. I still prefer to use the hotels' official websites to make the reservation because I was worried that there might be a chance that no room would be available when I arrive there, so I booked the rooms through the hotels' official website just in case. Sometimes, I don't trust these online travel agencies.

Some travelers appear to be more concerned about the reliability of the transaction, especially the risk of not having a reservation when arriving at the destination.

Another social media outlet used by most of informants is microblogs. The informants used microblogs to obtain 3 types of information: (a) general destination information (3 informants), (b) updated destination information (3 informants), and (c) discounts or promotion information (2 informants). Informant 16 said,

We mostly seek destination information on microblogs to get other travelers' itineraries such as where to stay and visit.

Informant 11 used microblogs to obtain updated destination information. She commented:

There are some emergent situations at the destination that can influence my itineraries. Microblogs are a good source to get updated and instant information to supplement the itineraries on those travel SNS.

And informant 18 explained her usage of microblogs for both general destination information and updated destination information:

I did use microblogs when planning my trip to Thailand. I heard there is a 'Loy Krathong Festival' during my travel period, so I searched this keyword in microblogs to see what kind of things others did for the festival and pictures of the destination. Also, any updated information that is shared by other travelers who are currently at the destination can also be helpful, especially during the time of their political instability.

Some informants followed travel accounts on microblogs to keep up with promotion information. Informant 12 said,

There are some accounts I follow on microblogs. Mostly, they update information on flight discounts or package discounts, which links to certain websites or OTAs. If some deals are really good, I will definitely consider them.

Informants used general SNS such as Kaixin.com (A Chinese equivalent of Facebook) to obtain supplementary destination information and promotional messages. Informant 17 mentioned using general SNS as a supplementary source:

I would review my friends' pictures on Kaixin if I know they recently have been to the destination, but nothing else is useful except for the images because they usually don't talk about their itineraries.

Informant 3 stated the details of her usage of general SNS:

I followed some accounts that provide updated travel promotions such as flight discounts on Kaixin. If I see great deals, I sometimes purchase the ticket even before I search for destination information.

In addition, handheld mobile devices such as smartphones and tablets provide access to travel information. Four informants mentioned having adopted them to receive promotional messages from airlines and OTAs via the mobile application WeChat. Informant 15 mentioned,

My friends on WeChat [a popular mobile social application] sometimes snapshot the discount or promotional information from the Internet and share it through WeChat. When I see their sharing, I will go to the original sites to check out.

Another informant (3) also said,

I followed some airline accounts on WeChat, so I receive information passively because they send the flight promotions to me. My friends on WeChat sometimes share discounts and promotional information as well.

Besides social media outlets, travelers also used other online resources and tools for their travel planning. All 21 informants used search engines such as Google (www.google.com) or Baidu (www.baidu.com) to find their preferred sites. They used search engines briefly at the destination selection stage and mostly at the itinerary design stage after purchasing flights. Search engines appear to be used in 4 kinds of scenarios. First, the majority of informants used search engines to search for other tourists' itineraries as a reference when planning their own travel. Second, they only used search engines to find the official websites of specific destinations. Informant 10 said,

I used Google to search for the visa application process because Cambodia doesn't have a clear official website, and I don't know which website is the real official one.

Third, they only used search engines to search for local transportation information. Finally, they used search engines to search for all types of content when they did not know where to find the information.

I found Kayak.com [an air travel OTA], Tufeng.com [a BBS platform], Tripadvisor.com [an OTA with reviews and ratings] all through Google. In other words, I normally found out all kinds of information like flight, hotels, others' itineraries, and local travel packages through Google.

The informants also used air travel OTAs or car rental OTAs, the official websites of airlines and hotels, and the websites of destination marketing organizations (DMOs).

17.4.2 DEGREE

The degree of information search refers to (a) time spent on searching for different types of information content during the pre-trip information search process and (b) the level of details that the travel plans contain. The length of time spent on the search process varies from 1–2 nights to 1 month long. The majority of informants used percentages to indicate their degree of information search. Six informants devoted over 40%–80% of their overall search effort to seek destination information to plan their own itineraries. Informant 4 described why destination information took such a long time:

I normally type in the destination name into Google, and based on the search result, I sometimes go to the DMO websites, but they usually don't have as much details as others' itineraries do. I just use it as a reference. Usually I read others' itineraries, mostly on Lvmama.com and Mafengwo.com [both are travel SNS]. I would look for the destination pictures they shared, read at least 10 to 20 itineraries to get a sense, and then if I'm particularly interested in something, I will Google it. Planning the itinerary is a very time-consuming part.

Four other informants also spent over 40% of their overall search effort on accommodation decision-making. Informant 5 mentioned,

Hotel booking decisions are the most time-consuming because I normally go to at least 3 OTAs to check their reviews and ratings. Besides Tripadvisor.com, I also go to Booking.com, Hotels.com, and other similar sites. I feel that the more comments I read the more comprehensive information I get. So it normally takes me a lot of time to compare.

In terms of how long in advance the pre-trip information search started, the answers also varied significantly. Five informants started their information search 3–6 months in advance because "*the earlier you book the flight ticket, the cheaper it is. As the departure time gets closer, the flight price goes up day by day*" (Informant

8). Five informants started their information search between 2 weeks and 1 month in advance. For instance, Informant 3 said, "*Hotel prices don't change too much in Vancouver, so I didn't book the rooms until around 2 weeks before the trip.*"

How detailed the travelers conduct their information search is another aspect to investigating travelers' information search degree. Four informants said that their information search would not stop until they filled up the daily time blocks on their itineraries. Informant 19 mentioned,

I normally search on RenRen.com [a general SNS] and Mafengwo.com [a travel SNS] to plan my trip. These sources have enough information that I can fully fill my travel itinerary. So I don't need to go to any other sources.

On the other hand, 8 informants were satisfied as long as they could plan general attractions and the number of days at each city. They did not create detailed schedules of when to visit which attractions. Informant 11 said,

After I booked my flight and hotels, I briefly sought out others' itineraries to see how many ways I can travel around this destination so I could have an idea which attractions I should visit. I didn't search too much because I think more information will unfold after I arrive at the destination.

One informant only planned out the number of days at each city without previously planning out activities at the destination. She (Informant 10) said,

I knew that I wanted to spend 2 days in this city, but I was flexible to visit any attractions or doing nothing based on my will. I didn't plan where I would be staying each night; it all depended on whether I would find travel companions on Douban.com [a general SNS] or maybe meet anyone along the way.

17.4.3 SPECTRUM

Table 17.3 identifies 6 types of spectra. Search engines and social media channels are identified separately because they are used by all informants. The other online domains are categorized as "other online domains."

TABLE 17.3 Spectrum of Information Search Strategy

		Online Sources		
		Search Engines	**Social Media Outlets**	**Other Online Domains**
Spectrum	Browsing	Browsing initiated through search engines	Browsing initiated through social media channels	Browsing initiated through other online domains
	Directed Search	Specific websites search initiated through search engines	Specific travel product search initiated through social media channels	Specific travel product search initiated through other online domains

Six informants used browsing strategy to initiate search through both search engines and social media channels. Informant 15 said,

I went to Baidu.com [a search engine] and typed in "Thailand vacation" just to browse the relevant content and what other travelers shared. Then I went to some more organized travel SNS to see what kind of information is available. Through browsing different things, I can get a more comprehensive idea of the destination.

Similarly, another informant (17) said,

Because the culture at the destination is very different from mine, I Googled basic information of this country and obtained information such as cultural taboos. Then I went to Mafengwo.com [a travel SNS] to see others' itineraries, what attractions they went to, what the attractions looks like.

In terms of directed search, all informants conducted directed search because they intended to make a purchase or need to make decisions on travel activities. Thus, all 3 kinds of sources have been used for directed search. Most informants started their directed search on social media channels. OTAs with reviews and ratings were the most popular information sources that can influence accommodation purchase decisions. Informant 13 said,

I normally go to at least 3 OTAs to check their reviews and ratings. Besides Tripadvisor.com, I also go to Booking.com, Hotels.com, and other similar sites . . . when I was deciding where to stay and which attractions to go to, the most important factor is others' comments.

Interestingly, microblogs were also used as a directed search tool. Informant 16 said,

I didn't search on Baidu [a search engine] for destination information because nowadays either Weibo.com [a microblog platform] or Mafengwo.com [a travel SNS] is the place to find others' sharing and itineraries; there is no need to use search engine. We basically just go on Weibo.com to search certain restaurants or hotels; some even provide details and pictures to show other travelers which corner or floor the place is located.

In summary, 18 informants used both browsing and directed searches. The other 3 informants only used directed searches. Informants seemed to browse on search engines in conjunction with social media outlets. Browsing plays a major role in seeking out destination information, whereas participants used directed searches for accommodation and flight information. Informants mentioned social media outlets—travel SNS in particular—most often among the websites they use to browse. As noted above, social media platforms (such as OTAs with reviews and ratings and microblogs) were critical to making decisions on renting accommodations and selecting attractions or activities for their trip.

17.5 CONCLUSIONS

In this chapter, the researchers have shown that social media channels are exerting a comprehensive and in-depth influence on the travel planning process of independent

Chinese outbound tourists. This group is most likely to use social media channels as their information sources for comments of specific travel products such as hotels, itineraries, deals of air tickets, travel tips, and as the gateway websites to search for relevant information. Table 17.4 summarizes the use of social media channels for planning major travel activities and the roles of these channels in the 3 dimensions of travel information search strategy. In general, Chinese outbound independent tourists make heavy use of social media channels to plan their outbound travel activities such as the acquisition of flight deals, decision-making on accommodations, and the arrangement of destination activities. This is reflected in the 3 dimensions of travel information search strategy. Most informants took social media channels as their major travel planning sources such as travel and general SNS, OTAs with reviews and ratings, microblogs, and blogs. They spent at least 40% of overall planning time on these channels. Most informants reported using social media channels as the gate (e.g., the websites to initiate search activities) to initiate information search activities and access to other relevant information sources.

TABLE 17.4 Impacts of Social Media on the Information Search Strategies of Chinese Outbound Independent Tourists

	Travel Planning Components		
Three Dimensions of Information Search	**Flight Information (No. of informants)**	**Accommodation Information (No. of informants)**	**Destination Information (No. of informants)**
Resources	Social media channels: • Mobile social applications (4) • General SNS (2) • Travel SNS (3) • Microblogs (1) Other sources: • OTAs for air tickets (19) • Airline companies' official websites (12) • Travel agents (2) • Search engines (1)	Social media channels: • OTAs with reviews and ratings (19) • Travel SNS (5) Other sources: • Hotels' official websites (5) • Travel books (1) • Travel agents (1)	Social media channels: • Travel SNS (21) • Microblogs (7) • General SNS (3) • Blogs (2) • BBS (2) Other sources: • Search engines (19) • DMO websites (10) • Friends (10) • Travel books (3) • Travel exhibition (1)
Degree	40%–50% time allocation (2)	40%–50% time allocation (4)	40%–80% time allocation (6)

Spectrum (the combination of browsing and direct search)	• Browsing initiated through social media channels. (9) • Browsing flight information initiated through search engines. (1) • Specific flight search initiated through other online domains. (18)	• Browsing initiated through social media channels. (5) • Specific accommodation search initiated through social media channels for decision-making. (19) • Specific accommodation search initiated through other online domains for reservation. (18)	• Browsing initiated through social media channels. (11) • Specific search of destination activities and itineraries initiated through social media channels for decision-making. (19) • Specific search for destination information initiated through search engines. (9) • Browsing destination information initiated through search engines. (10)

In this chapter, the researchers have indicated that social media channels play different roles when Chinese outbound independent tourists plan alternative travel activities. For example, social media channels weigh less heavily in the planning of air tickets, because flight products are price-driven and tourists require less information about the experience. For flight products, the informants mainly used social media channels such as mobile social application WeChat to keep updated on deals. The OTAs with hotel product reviews and ratings are dominant in the case of accommodation products for the decision-making process of Chinese tourists. Informant 13 explained, "*The hotels are usually most expensive part in the destination. I want to make sure a good experience. I am not familiar with the hotel brands there, so I only can judge based on the reviews.*" The social media channels are extremely important for arranging destination activities. Travel SNS are particularly influential because Chinese outbound independent tourists rely on these websites for itineraries, and they prefer to get heuristics from other people's experience and story to make their own itineraries. Most informants shared the opinion that travel SNS is critical for planning independent outbound trips, because they can identify the background of post-contributors (e.g., age group, lifestyle, professions, personality, etc.). Because of this, they felt greater relevance to planning their own trips, based on the stories and suggestions of these contributors. With the support of smartphones and other mobile devices such as tablets, the informants reported that they use travel SNS and microblogs to plan activities after their arrival at destinations. Informant 7 shared that her busy schedule left insufficient time to plan details prior to travelling. For this reason, she usually examined several possible itineraries on the plane, which she printed from some travel SNS such as mafengwo.com. She then searched for the specific attractions in the hotel through the travel SNS to find all relevant web posts and made decisions accordingly. Another informant 11 said, "*I used lv-mama.com to keep updated on the local activities after I arrived at Bangkok. I usually has very flexible schedule there because I just want to relax. So I check out the local activities when I want to do something.*"

In this chapter, the researchers have described the information search strategy of Chinese outbound independent tourists when planning their outbound trips and have identified the role of social media channels in the travel planning process. There are several potential implications for tourism businesses and for DMOs. First, due to culture differences when making travel decisions, Chinese independent outbound tourists heavily rely on the opinions of other tourists who have previously been to the destinations and have consumed products. They adopt this approach in preference to relying on information provided by official destination marketing websites or outbound tourism businesses. On this basis, it is suggested that DMOs and tourism businesses should engage proactively in online reputation management. It is important to understand China's major social media channels and monitor the relevant discussions in Chinese, considering that the popular social media channels in China are quite different from the ones (e.g., Facebook, Twitter, etc.) in other parts of the world (Yang & Wang, 2015). Second, effective cooperation with popular social media channels such as travel SNS and OTAs in China can leverage the marketing efforts. Informant 5 suggested that

It would be great if destinations or companies' websites could include (Chinese) travelers' comments. In addition, I hope they can provide rankings, ratings of attractions by using points or stars such as 5 stars, 4 stars, and they can explain why certain places are more worth visiting than others from a Chinese travelers' perspective instead of just listing out attractions and providing brief introductions. If they'd done so, we would consider using their official website more often when searching travel information.

Most informants indicated that the relevance of suggestions is important because they prefer to trust people with similar culture backgrounds. The destinations and business, which intend to attract more Chinese outbound independent tourists, may consider their presence and appearance on major Chinese social media channels through paid advertisement or the integration of the content of these channels on the websites of DMOs or tourism businesses.

REFERENCES

Arlt, W. G. The second wave of Chinese outbound tourism. *Tourism Plann. Dev.* 2013, 10(2), 126–133.

Beatty, S. E.; Smith, S. M. External search effort: an investigation across several product categories. *J. Consum. Res.* 1987, 14(1), 83–95.

Chiou, J. S.; Ting, C. C. Will you spend more money and time on internet shopping when the product and situation are right? *Comput. Human Behav.* 2011, 27(1), 203–208.

China Internet Network Information Center (CNNIC). *34th Statistical Report on Internet Development in China*, July 23, 2014. http://www1.cnnic.cn

China Tourism Academy (CTA). *Annual Report of China Outbound Tourism Development 2013*, 2013.

CTA, & Ctrip. *Report on China Independent Tourism Development (2012-2013)*; CTA (China Tourism Academy): Beijing, 2013.

Creswell, J. W. *Qualitative Inquiry and Research Design: Choosing Among Five Perspectives*; Sage Publications: CA, 2007.

Detlor, B.; Sproule, S.; Gupta, C. Pre-purchase online information seeking: search versus browse. *J. Electron. Commer. Res.* 2003, 4(2), 72–84.

Engel, J. F.; Blackwell, R. D.; Miniard, P. W. *Consumer Behavior*, 8th ed.; Dryder: New York, NY, 1995.

Fodness, D.; Murray, B. Tourist information search. *Ann. Tourism Res.* 1997, 24(3), 503–523.

Fodness, D.; Murray, B. A typology of tourist information search strategies. *J. Travel Res.* 1998, 37(2), 108–119.

Gursoy, D.; McCleary, K. W. An integrative model of tourists' information search behaviour. *Ann. Tourism Res.* 2004, 31(2), 353–373.

Hyde, K. F. Information processing and touring planning theory. *Ann. Tourism Res.* 2008, 35(3), 712–731.

Jang, S. The past, present, and future research of online information search. *J. Travel Tourism Mark.* 2005, 17(2–3), 41–47.

Money, R. B.; Crotts, J. C. The effect of uncertainty avoidance on information search, planning, and purchases of international travel vacations. *Tourism Manage.* 2003, 24(2), 191–202.

National Bureau of Statistics of China. *China's Economy Showed Good Momentum of Steady Growth in the Year of 2013*, January 20, 2014. http://www.stats.gov.cn

Rowley, J. Product search in e-shopping: A review and research propositions. *J. Consum. Mark.* 2000, 17(1), 20–35.

What China Wants. *The Economist*, August 23, 2014. http://www.economist.com

Wu, M. Y.; Pearce, P. L. Chinese recreational vehicle users in Australia: a netnographic study of tourist motivation. *Tourism Manage.* 2014a, 43, 22–35.

Wu, M. Y.; Pearce, P. L. Tourism blogging motivations why do Chinese tourists create little "lonely planets"? *J. Travel Res.* 2014b. DOI: 0047287514553057.

Yang, X.; Wang, D. The exploration of social media marketing strategies of destination marketing organizations in China. *J. China Tourism Res.* 2015, 11(2), 166–185.

Zhang, G.; Li, J.; Bi, L.; Pang, L. An analysis of information exchange characteristics about tourism synchronous virtual community. *Tourism Trib.* 2013, 28(2), 119–126.

CHAPTER 18

THE ROAD LESS TRAVELED: REGIONAL DISPERSAL OF CHINESE TOURISTS IN AUSTRALIA[1]

BYRON W. KEATING and MARGARET DEERY

[1]This study is based on an invited presentation at the 2013 Australian Regional Tourism Network Conference in Margaret River, Western Australia. Special thanks to Therese Phillips from Tourism and Events Queensland for contributions to the presentation and directions of this study.

18.1 INTRODUCTION

The growth in outbound tourism from China has excited the imagination of destinations around the world. Outbound tourists from China exceeded 97 million in 2013, with an annual expenditure of more than $130 billion (UNWTO, 2014). This distinguishes Chinese travelers as the world's highest travel spenders. Industry sources also predict that the number of outbound Chinese tourists will exceed 200 million by 2020, with more than US$300 billion in overseas spending (UNWTO, 2013).

It should come as no surprise that this unprecedented growth has created enormous competition among destinations keen to share in the economic spoils (Keating and Kriz, 2008). However, even when destinations manage to attract Chinese tourists, the economic advantage is often concentrated quite narrowly within gateway cities and around major landmark attractions of host countries. For instance, a recent study of Chinese tourists in Australia revealed that less than 6% of overnight stays occurred outside of the major metropolitan centers (Koo et al., 2012). This is significant as the economic benefit associated with international tourism is highly correlated with visitor nights (Song et al, 2012).

International destinations, such as Australia, have started to recognize that increasing dispersal beyond the gateways has important implications for their regional economies. In particular, Koo et al. (2012) highlight that greater regional dispersal can contribute to greater employment opportunities, broadened economic activity, and generate higher standards of living. Within our study, "regional dispersal" is defined as the proportion of international visitors who travel beyond the gateway cities to outlying regional destinations, where the term "gateway city" refers to a metropolitan center that is serviced by direct international flights. For example, international travel into the State of Queensland in Australia is serviced by 3 gateways cities – Brisbane, Gold Coast, and Cairns. These gateway cities account for around 22% of all international visitor arrivals into Australia. Including the regions incorporating these gateway cities, Queensland has 13 regions that are marketed and managed as discrete destinations.

Despite the benefits of dispersal and the growing importance of China as a source market, there is a scarcity of research on the drivers and motivations for regional dispersal by Chinese travelers. With the notable exception of Koo et al. (2012), who adopt a market perspective to study the dispersal characteristics of 7 source markets including China, we could find no other study that has addressed the issue of regional dispersal of Chinese travelers and no study that explores the issue of regional dispersal of Chinese travelers from a destination perspective.

To better understand the nature of regional dispersal of Chinese travelers, we propose to conduct a case study of regional destinations within the State of Queensland in Australia. To motivate this case study, we identified 2 research questions:

RQ1: Which regions are most efficient at attracting Chinese visitors?

RQ2: What can inefficient regions do to improve their attractiveness?

Beyond the reasons already mentioned, addressing the first research question also has important practical implications for destinations. A better understanding of regional dispersal can (a) increase the carrying capacity of a country, (b) provide a greater range of products that can contribute to product diversification, and (c) open new markets and generate new revenue streams to support investment in infrastructure that can be of benefit to visitors and locals alike. Regional dispersal also helps to internationalize outlying regions, as they confront and respond to the needs of international visitors from different countries. Resolving the second research question will assist less-efficient regions to improve their prospects as they position for a share of the highly competitive Chinese tourism market.

The next sections of this study will provide an overview of the regional dispersal research, before moving on to briefly discuss the data envelopment method and present the findings from our analysis using this approach. We will conclude with a consideration of the implications and future research directions.

18.1.1 REGIONAL DISPERSAL – AN OVERVIEW

A considerable volume of tourism dispersal literature has amassed since the early 1980s. In summarizing this literature, Koo et al. (2012) suggest that prior tourism dispersal studies can be classified as either descriptive or causal. The descriptive studies focus on the dependent variable, which in this case is the number of visitors and associated visitor nights. In contrast, causal studies focus on the factors that cause or drive visitation. Within this research we make a unique contribution, in that our study can be considered a hybrid approach. While we focus on the identification of an index, which is typical of descriptive studies, through inclusion of multiple inputs and outputs, we also provide some insight into the causal factors driving dispersal.

The work of Koo et al. (2012) provides a nice starting point for understanding the various inputs and outputs influencing dispersal by different source markets, including China. Yet a review of the variables used in their analysis suggests that they failed to distinguish between traveler and destination-specific factors, or what Um and Crompton (1990) refer to as push and pull factors. As our study focuses on regional destinations, we will concentrate exclusively on destination-specific "pull" factors. This resulted in the identification of 3 input categories for inclusion within our modeling – appeal, access, and affordability.

The appeal factor draws on Plog's allocentric motivation where tourist behavior is closely linked to the variety and concentration of activities offered within a destination (Lue et al., 1993). Appeal has also been shown to influence spatial dispersal (Weaver, 2006). The access factor reflects the time constraints of tourists. At the destination level, a good proxy for access is the distance of the regional destination to the gateway (Landau et al., 1981). The affordability factor relates to the relative cost of visiting 1 destination vis-a-vis another. Affordability has been shown to be a major influence on dispersal patterns and transport options, particularly for travelers from less economically developed countries (Koo et al., 2012).

18.2 METHOD

Within our study, we utilize a non-parametric, linear programming technique called data envelopment analysis (DEA) to understand how our 3 input factors impact on visitation-related outputs. While a comprehensive discussion of the DEA approach is outside of the scope of this study[2], we believe that this approach is particularly well suited to the study of dispersal as it overcomes one of the inherent limitations of the existing dispersal ratio approach through the inclusion of multiple inputs to identify those regions that are most efficient at attracting visitors and generating visitor nights.

An acknowledged challenge of the existing dispersal ratio approach is choosing the focus for the evaluation. Do you evaluate the proportion of time spent in the region vis-à-vis the gateway city? Do you consider the distance from the gateway? Or do you consider the amount spent in the region as a proportion of the overall travel budget? Our approach allows for the inclusion of multiple inputs to capture all of these considerations in order to identify a more elaborate dispersal index.

Despite the aforementioned advantages and a rich history chronicling the use of DEA in cognate disciplines such as hospitality (e.g., Wang et al., 2001; Barros, 2006) and its application to the study of travel agencies (e.g., Koksal and Aksu, 2007; Barros and Matias, 2006) and transport issues (e.g., Sarkis and Talluri, 2004; Charles and Paul, 2001), our study represents the first time that DEA has been applied to the study regional dispersal of international tourists.

DEA provides a mathematical procedure for comparing different decision-making units (DMUs) based on their efficiency in transforming inputs into outputs (Charnes et al., 1981). The technique assumes that DMUs (in our case, regional destinations) have the same classes of inputs and outputs. DEA is particularly relevant for the study of dispersal, as it represents an extension of traditional ratio analysis such as is common within the study of tourist dispersal. The goal of this study is to identify regions that are more (or less) efficient in their capacity to translate destination-specific inputs (appeal, access, and affordability) into outputs (visitors and visitor nights). Accordingly, we can define an efficient region as one that is capable of producing higher outputs with the same inputs (output-orientated), or one that is able to produce the same outputs with less inputs (input-orientated).

The efficiency score of a DMU can vary between 0 and 1, where a score approaching 1 indicates greater efficiency. Benchmark DMUs located on the efficiency frontier are identified by a score of exactly 1. It is noteworthy that efficiency scores

[2]We would encourage interested readers to consult Cook and Seiford's (2009) discussion of DEA developments over the past 30 years.

are independent of the units in which outputs and inputs are measured, thus providing greater flexibility in the choice of inputs and outputs for inclusion in a study.

Based on an assumption of common purpose across the different regional destinations, we can denote a given DMU (DMU_j, where $j = 1, \ldots, 9$), where the focus of this study is on 9 regional destinations beyond the gateway cities (see Fig. 18.1). For the purpose of this study, we excluded Brisbane and the Gold Coast as the 2 main gateway cities. However, we were reluctant to remove the region containing Cairns due to its size and diversity. As such, we combined the balance of this region (TNQ) with Townsville to create a large northern region. In addition to being located adjacent to one another, justification for combining these 2 regions is based on their comparatively similar tourism products and natural attractions. We also combined the regions of Capricorn and Gladstone. While the justification for this combination was due to data sparseness, the 2 regions, once again, co-located and shared similar natural and tourism features.

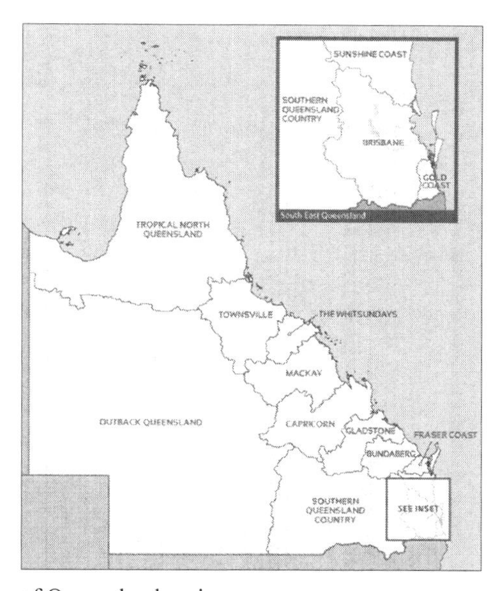

FIGURE 18.1 Map of Queensland regions.
(From Tourism & Events Queensland, 2014. Accessed 11 July 2014 http://teq.queensland.com/Destinations/Tropical-North-Queensland.aspx.)

Each DMU can be characterized by a vector of inputs where $X_j = (x_{1j}, x_{2j}, x_{3j})$ and a vector of outputs $Y_j = (y_{1j}, y_{2j})$, where linear programming is used to solve the Charnes, Cooper, and Rhodes objective function. This function seeks to maximize the efficiency score for each DMU given a set of constraints associated with different DEA models. According to established procedures of DEA-based efficiency estimation, 3 different DEA models can be obtained based on different restricting conditions – constant returns to scale (CRS), variable returns to scale, and non-increasing returns to scale (Coelli, 1996).

18.3 FINDINGS

To facilitate the DEA analysis, we first needed to identify suitable input and output data. For the appeal input, we relied on data on the number of tourism-related businesses in the respective regions. These data were considered to offer a reasonable proxy for the variety of tourism attractions available within a region. Data on the number of tourism businesses were obtained from Australian Government's Tourism Satellite Account (ABS, 2013). These data are collected based on international conventions relating to the identification of tourism characteristic industries. These are businesses that would either cease to exist in their present form or be significantly affected if tourism were to cease. While core lists of tourism characteristic industries are promoted under international tourism satellite conventions (OECD, 2000), the Australian enumeration also requires that at least 25% of the business output is consumed by visitors.

To measure access, we obtained Google map data for the travel time in minutes between the most populous suburb in each respective region and the nearest gateway city. The population data were obtained from the latest available Australian Census data (ABS, 2012). For the affordability dimension, we obtained data on the average hotel tariff for the respective region on data available from TripAdvisor (2013). The output data on visitor numbers and visitor nights were obtained from the Australian International Visitor Survey for Chinese travelers to each region during 2013 (TRA, 2013). A summary of this input and output data is presented in Table 18.1.

From the data presented in Table 18.1, we can see that the region with the highest appeal as measured by the greatest number of tourism businesses was the Sunshine Coast followed by the Darling Downs region. The Whitsundays was observed to have the highest number of visitors and visitor nights, as well as the lowest affordability as measured by average hotel tariff. The Outback region was observed to have the most significant barrier to access as measured by the greatest travel time, with the shortest travel time recorded by the Sunshine Coast. These data were subject to input-oriented, constant-return to scale DEA analysis using the DEA Frontier add-in for Microsoft Excel (Cook and Zhu, 2008).

TABLE 18.1 Input and Output Data

DMU Name	Inputs			Outputs	
	Tourism Businesses	Tariff ($AU)	Travel Time (mins)	Visitors	Visitor Nights
Bundaberg	1,046	118.91	246	764	764
Capricorn-Gladstone	2,090	152.90	432	826	16,897
Darling Downs	2,902	124.50	212	1,500	8,110
Fraser Coast	1,459	127.53	204	1,042	39,936
Mackay	1,333	155.24	508	428	15,030
TNQ-Townsville	2,215	132.48	250	1,046	33,894
Outback	658	117.85	776	530	967
Sunshine Coast	5,790	170.18	80	589	4,642
Whitsundays	634	262.26	440	3246	61,920

18.3.1 WHICH REGIONS ARE MOST EFFICIENT AT ATTRACTING CHINESE VISITORS?

To address the first research question, we examined the results of the DEA modeling (see Table 18.2). This analysis identified 2 efficient DMUs (benchmark regions) – Fraser Coast and Whitsundays. The other regions had efficiency scores of less than 1, meaning that they were located at various distances from the efficiency frontier. These other DMUs can be interpreted as inefficient as they could achieve the same outputs with a reduction of inputs. For instance, the region of Bundaberg had an efficiency score of 52%, meaning that it could reduce the linear combination of inputs by 48% and still achieve the same level of outputs (visitors and visitor nights). The region with the lowest observed efficiency was Mackay (34%) followed by the Outback region (36%).

TABLE 18.2 Efficiency Data (CRS Model)

DMU Name	Efficiency Score	Sum of Lambdas	RTS	Optimal Lambdas	Benchmarks
Bundaberg	0.519	0.235	Increasing	0.235	Whitsundays
Capricorn-Gladstone	0.453	0.293	Increasing	0.057	Fraser Coast
Darling Downs	0.973	0.462	Increasing	0.462	Whitsundays
Fraser Coast	1.000	1.000	Constant	1.000	Fraser Coast
Mackay	0.335	0.342	Increasing	0.279	Fraser Coast

TNQ-Townsville	0.865	0.794	Increasing	0.695	Fraser Coast
Outback	0.363	0.163	Increasing	0.163	Whitsundays
Sunshine Coast	0.998	0.181	Increasing	0.181	Whitsundays
Whitsundays	1.000	1.000	Constant	1.000	Whitsundays

One useful aspect of this analysis is that it provides insight into the benchmark regions that the less-efficient regions can learn from. For example, the data on the optimal lambda levels for Darling Downs suggest that they would be best placed to benchmark their performance against that of the Whitsundays region as they were the closest DMU that was on the efficiency frontier. In this way, DEA provides direction for future empirical research that could be undertaken to understand and unpack the areas of operations that have contributed to the performance of the regions on the efficiency frontier (such as Whitsundays).

18.3.2 WHAT CAN INEFFICIENT REGIONS DO TO IMPROVE THEIR ATTRACTIVENESS?

When you combine the information on the benchmark regions (Table 18.2) with the data in Tables 18.3 and 18.4, we obtain insights into the specific areas where the inefficient regions would need to reduce inputs to achieve the same efficiency as the benchmark regions. The analysis also provides an indication of the reduction in outputs that would be achieved from the current input levels.

TABLE 18.3 Input and Output Slacks

	Input Slacks			Output Slacks	
DMU Name	**Tourism Businesses**	**Tariff ($AU)**	**Travel Time (mins)**	**Visitors**	**Visitor Nights**
Bundaberg	393.766	0.000	24.140	0.000	13,809.900
Capricorn-Gladstone	713.199	0.000	79.984	0.000	0.000
Darling Downs	2,531.923	0.000	3.040	0.000	20,503.678
Fraser Coast	0.000	0.000	0.000	0.000	0.000
Mackay	0.000	0.000	85.774	66.492	0.000
TNQ-Townsville	839.770	0.000	30.921	0.000	0.000
Outback	135.569	0.000	210.120	0.000	9,143.166
Sunshine Coast	5,663.364	122.251	0.000	0.000	6,593.638
Whitsundays	0.000	0.000	0.000	0.000	0.000

TABLE 18.4 Input and Output Efficiency Targets

	Efficient Input Target			Efficient Output Target	
DMU Name	**Tourism Businesses**	**Tariff ($AU)**	**Travel Time (mins)**	**Visitors**	**Visitor Nights**
Bundaberg	149.222	61.727	103.561	764.000	14,573.900
Capricorn-Gladstone	232.709	69.201	115.533	826.000	16,897.000
Darling Downs	292.976	121.192	203.327	1,500.000	28,613.678
Fraser Coast	1,459.000	127.530	204.000	1,042.000	39,936.000
Mackay	446.906	52.046	84.540	494.492	15,030.000
TNQ-Townsville	1,076.841	114.633	185.401	1,046.000	33,894.000
Outback	103.518	42.821	71.842	530.000	10,110.166
Sunshine Coast	115.042	47.588	79.840	589.000	11,235.638
Whitsundays	634.000	262.260	440.000	3,246.000	61,920.000

Table 18.3 provides information on the input and output slacks. These are the areas of shortage or excess in the inputs and outputs. When combined with the original input and output levels (Table 18.2), we are able to identify a set of performance targets that would be needed to move the region to the efficiency frontier (Table 18.4).

For example, the results presented in Table 18.4 suggest that there is a surplus availability of tourism businesses in the Sunshine Coast for the current visitation levels. This means that the destination would only require 115 tourism businesses in order to reach the efficiency frontier. But this information may not be useful as widespread structural change may create unintentional impacts in other areas of the region's economy. It may be more desirable to reduce the travel time or the hotel tariff. In the case of the Sunshine Coast, it would need to reduce the average hotel tariff by $122–$47 to achieve the same result. Alternatively, the current level of inputs would need to generate an additional 6,593 visitor nights for the current level of visitation in order to become efficient.

18.4 CONCLUSION

Maximizing regional dispersal of international visitors from China has emerged as an important issue within the Australian tourism industry in recent years. Beyond addressing issues of economic inequality, regional dispersal also has the potential to broaden the relative attractiveness of a national destination through diversification of product offerings. Against this backdrop, our study provides a case study of regional dispersal of Chinese travelers within the State of Queensland in Australia.

Using DEA, we identified 2 benchmark regions – Whitsundays and Fraser Coast – which were observed to be most efficient among Queensland's 9 non-gateway regions at attracting Chinese visitors given 3 destination-specific inputs related to appeal, access, and affordability. Using these 2 benchmark regions, we were then able to identify different levels of inputs needed to bring the non-efficient regions into line with an optimal benchmark region.

This research makes a number of important theoretical and practical contributions. Our study presents the first evaluation of how destination-specific characteristics impact on the relative efficiency of different regions in dispersing Chinese visitors within a destination. This is an important contribution as prior research has mostly ignored regional differences within destination markets, even though the economic advantages that flow from increased dispersal of international tourists are well established (e.g., Koo et al., 2012).

The findings presented here also challenge traditional conventions that support investment in tourism infrastructure as a key driver of destination appeal. From a structural perspective, the results suggest that tourism marketing has a critical role to play in creating appeal, reinforcing that dispersal efficiency is not deterministic. That is, simply increasing the number of attractions or reducing the cost and travel time is not sufficient in itself to guarantee more visitors from China. Another possible explanation relates to the quality and/or relevance of customer service. All things being equal, the more efficient regions provide a package of services that is relatively more attractive to Chinese tourists. Future research should delve deeper with the efficient regions to better understand what it is that they are doing which is having this impact on visitation.

Finally, the findings of this study also provide a valuable methodological contribution to the measurement of dispersal, promoting a new method and identifying a more sophisticated descriptive index (efficiency score). We acknowledge that the selection of input and output variables is an imprecise science and that the absence of an important variable could impact on the efficacy of the findings. Nonetheless, by drawing on the latest literature, we are confident that the method as presented has included some key variables that are important to dispersal decisions. We will leave it to future research to build on this study by exploring the inclusion of different variables and application to different destinations. To this end, we recommend that consideration be given to extending the current model through inclusion of a variable that captures destination-specific information on marketing and service-related activity that is of relevance to Chinese visitors.

There were also a number of assumptions regarding the input variables. For instance, we acknowledge a potential inter-relationship between the input categories. That said, correlation between input variables should not impact on the results as it is likely to affect all regions in the same way. Likewise, we have made 2 additional assumptions. The first is that Chinese tourists are likely to pay the same rates as non-Chinese tourists. In the event that a differential pricing strategy is employed, we would not expect that this practice would privilege one region over another. Future research could test these assumptions.

REFERENCES

ABS. *Census of Population and Housing (Cat. 2076.0)*; Australian Bureau of Statistics: Canberra, 2012.

ABS. *Australian National Accounts: Tourism Satellite Account 2012-13 (Cat. 5249.0)*; Australian Bureau of Statistics: Canberra, 2013.

Barros, C. Analyzing the rate of technical change in the Portuguese hotel industry. *Tourism Econ.* 2006, 12(3), 325–346.

Barros, C.; Matias, A. Assessing the efficiency of travel agencies with a stochastic cost frontier: a Portuguese case study. *Int. J. Tourism Res.* 2006, 8(5), 367–379.

Charles, K.; Paul, S. Competition, privatization and productive efficiency: evidence from the airline industry. *Econ. J.* 2001, 111(473), 591–619.

Charnes, A.; Cooper, W.; Rhodes, E. Evaluating program and managerial efficiency: an application of data envelopment analysis to program follow through. *Manage. Sci.* 1981, 27(6), 668–697.

Cook, W.; Zhu, J. *Data Envelopment Analysis: Modeling Operational Processes and Measuring Productivity*; CreateSpace: Charleston, SC, 2008.

Keating, B.; Kriz, A. Outbound tourism from China: literature review and research agenda. *J. Hosp. Tourism Manage.* 2008, 15(1), 32–41.

Koksal, C.; Aksu, A. Efficiency evaluation of A-group travel agencies with data envelopment analysis (DEA): a case study in the Antalya region, Turkey. *Tourism Manage.* 2007, 28(3), 830–834.

Koo, T.; Wu, C.; Dwyer, L. Dispersal of visitors within destinations: Descriptive measures and underlying drivers. *Tourism Manage.* 2012, 33(5), 1209–1219.

Landau, U.; Prashker, J.; Hirsh, M. The effect of temporal constraints on household travel behavior. *Environ. Plann. A* 1981, 13(4), 435–448.

Lue, C.; Crompton, J.; Fesenmaier, D. Conceptualization of multi-destination pleasure trips. *Ann. Tourism Res.* 1993, 20(2), 289–301.

Sarkis, J.; Talluri, S. Performance based clustering for benchmarking of US airports. *Trans. Res. Part A* 2004, 38(5), 329–346.

Song, H.; Dwyer, L.; Li, G.; Cao, Z. Tourism economics research: a review and assessment. *Ann. Tourism Res.* 2012, 39(3), 1653–1682.

TRA. *International Visitors to Australia*; Tourism Research Australia: Canberra, 2013. http://www.tra.gov.au (accessed July 10, 2014).

TripAdvisor. *Various Search Results Based on Specific Regions*; 2013. http://www.tripadvisor.com.au/ (accessed July 10, 2014).

Um, S.; Crompton, J. Attitude determinants in tourism destination choice. *Ann. Tourism Res.* 1990, 17(3), 432–448.

UNWTO. *China – The New Number One Tourism Source Market in the World*; United Nations World Tourism Organisation Tourism Barometer, April, 2013; Vol. 11. http://media.unwto.org/en/press-release/2013-04-04/china-new-number-one-tourism-source-market-world (accessed July 10, 2014).

UNWTO. *Tourism Highlights – 2014 Edition*; 2014. http://dtxtq4w60xqpw.cloudfront.net/sites/all/files/pdf/unwto_highlights14_en.pdf (accessed July 10, 2014).

Wang, Z.; Zhou, W.; Li, S. An analysis for market area of Chinese national park based on railway corridor. *Acta Geogr. Sin.* 2001, 56(2), 206–213.

Weaver, D. B. *Sustainable Tourism: Theory and practice*; Routledge: London, 2006.

CHAPTER 19

AN EMPIRICAL STUDY ON TRAVEL INTENTION TO JAPAN: A CASE STUDY OF MAINLAND CHINESE CITIZENS AFTER THE DIAOYU ISLAND POLITICAL CRISIS

YINGZHI GUO and YUN CHEN

CONTENTS

19.1 INTRODUCTION

In recent years, with the development of the economy, the sustained growth of income, the improving transportation system, and favorable policies, the outbound tourism in China greets its peak development. Japan, as the neighbor of China, becomes an important destination of Mainland Chinese citizens. In the 1990s, outbound tourism to Japan by Mainland Chinese tourists developed at a slow pace. In 1990, only 105,993 tourists traveled to Japan, while in 1999, it increased to 294,937, which was only 1.5 times of increase before 10 years. In 2000, China and Japan signed an ADS agreement. Since then, China and Japan witnessed a rapid tourism development. In 2013, tourists from China amounted to 1.314 million. Tourism, as a way of nongovernmental diplomacy, plays an important role in strengthening mutual understandings and friendships among nations, enhancing friendly cooperation and communications.

However, due to the sensitivity and vulnerability of tourism itself, it is highly susceptible to a variety of crisis, which not only affects the normal development of the destination, but also affects the confidence of tourists, thus inducing outbound tourists' risk perception, which leads to the great changes of structure in market shares (Dai, 2011). In addition to the general restrictive factors of outbound tourism, emergency crises such as natural disasters and the deterioration of political relations will also have a huge impact on tourists' travel decisions. China and Japan have the most complex and influential relation in the Asia–Pacific region. Thus, the relationship between China and Japan is fraught with uncertainty (Wang, 2011).

In 2011, from January to February, there was a rapid growth rate in the number of tourist arrivals, and more than 200,000 visitors were reported from the Mainland China. Due to the "3.11" East Japan earthquake, the number stagnated from March to May. In the same year, from June, the Japanese national tourism organization in collaboration with Japan airlines and tour operators launched a special campaign, and several effective marketing strategies were implemented such as discount airline tickets and discount routes. Thus, the number of visitors to Japan from China started to pick up, which was slightly increased by 1.2% compared with the year 2010 before the earthquake (Japan National Tourist Organization [JNTO], 2013). In November, 2012, the Diaoyu Island crisis erupted. As the situation became worse, there were increasing conflicts and anti-Japanese sentiment in China. A large number of people cancelled their tours to Japan with the percentage being up to 40% in November (Xinhua News Agency, 2012). The aim of this research was to study the perceived obstacle factors of Mainland Chinese citizens to travel to Japan after the crisis based on first-hand data.

19.2 LITERATURE REVIEW

19.2.1 TRAVEL DECISION INFLUENCING FACTORS

Decrop and Snelders (2005), Fesenmaier and JiannMin (2000), Jeng and Fesenmaier (2002), and Woodside and Dubelaar (2002) have all modeled the holiday and vacation decision-making process from multidimensional perspectives. Dellaert et al. (1998) considered the tourism decision-making process to be a complex phenomenon, because it consisted of many purchase decisions in a long time. Hyde (2008) studied a travel vacation decision-making structural model, confirming that the travel decision-making process included 3 distinct acts: information searching, vacation planning, and vacation reservation. Lim (2004) studied the influencing factors affecting the flow of foreign tourists between Australia and Korea, including income, relative prices, exchange rate, and seasonal change. She also studied the government's intervention factors affecting the development of the domestic economy and specifically, considered many bi-direction influencing factors. Bai et al. (2008) indicated on the basis of cognitive map that the income, price, and attitudes toward tourism and leisure time were the critical factors affecting the tourists' travel decision-making process; destination attractions and destination restrictions of space and time were important factors; outbound tourism policies and personal responsibility were less important factors; social development of the tourism destination, tourism facilities, and tourism products and services promotion were general factors. Many scholars have tried to construct travel decision-making models. Bian (2002) constructed the MICE tourist decision-making models to show the main factors affecting the process. Shen (2005) constructed tourist decision-making models in choosing tourism destinations and the tourism purchasing process to describe the decisive factors in tourism purchasing behaviors. Ma (2008) used models to analyze the factors affecting travel behaviors of tourists in China from America and Japan, including prices, time, facilities, transportation, hospitality, immigration and export procedures, and weather and climate of the destination. Jiao (2006) used his model to analyze preference and perceived risk of tourists.

19.2.2 TOURISM CRISIS INFLUENCE FACTORS

In order to have a better understanding of tourists under the crisis conditions, researchers brought in related theories of psychology, such as perception, experience, and image to tourism studies. They found that tourists' perceptions of a destination form the destination image. Whether tourists will travel to the destination or not is determined by the image rather than the reality (Gartner and Hunt, 1987; Dann, 1996; Baloglu, 1997). Sönmez and Graefe (1998) believed that the travelers' past travel experiences, types of risks, and the sense of security all exert an effect on the travelers' future travel behavior. Lepp and Gibson (2003) conducted a random survey among young people born in the United States. Variance analysis revealed that female tourists have higher risk perceptions about health and food, whereas experienced travelers have lower risk perceptions on terrorism. Wong and Yeh (2009) conducted an empirical study on 504 tourists on the basis of Structural Equation Modeling and revealed that tourists' risk perception had a positive impact on travel hesitation, while sufficient information can ease this impact. Bronner and de Robert (2012) studied the travel intention and behaviors of Dutch tourists and found that owing to the economic crisis, two-thirds of tourists have cut expenditure in summer vacations. Li (2008) set the "5.12" earthquake as an example to study the affecting factors of tourists' risk perceptions and found that the significant factors affecting the tourists' risk perceptions were gender, the understanding of earthquakes, the emergency of earthquakes, the severity of the consequences of the earthquake, and the impact of the surrounding people and government performance. Chai et al. (2011) used multiple regression models to analyze factors influencing risk perception of tourists and found 10 significant influencing factors. Zhang (2009) studied the main factors influencing risk perception of tourists in Tibet tourism and analyzed the effects of different factors on the risk perception of tourists.

19.2.3 CHAID DECISION TREE MODEL

Multidimensional decision tree model is a method of processing multidimensional data; the formation of each branch in decision tree is based on multidimensional space zoning rules, demonstrated as the tree structure chart. The commonly used models include Chi-squared Automatic Interaction Detector (CHAID), Classification and Regression Tree (CART), Commercial Version 4.5 (C4.5), Quick Unbiased Efficient Statistic Tree (QUEST), Iterative Dichotomiser Version 3 (ID3), etc., (Ture et al., 2009). CHAID model was first proposed by Kass (1980), with its former origin as AID algorithm model. Compared with CART and C4.5, CHAID could process categorical variables; its main characteristics are multi-branch and forward pruning. The advantages of CHAID are that it could produce multiple interaction branches and divide the branch from the point of view of statistical significance, thus optimizing the tree branching processes. When in the causal relationship investigation, it could facilitate achieving horizontal division of target variables.

Most of the decision tree models were applied in the market segmentation studies. Zhang and Yu (2014) used aggregate analysis and exhaustive CHAID analysis to understand people's decisions on travel behavior. Coussement et al. (2014) pointed out that decision trees are preferred over RFM analysis and logistic regression. Murphy and Comiskey (2013) used CHAID modeling to identify characteristics of clients experiencing statistically significant poor outcomes. do Valle et al. (2012) used CHAID to segment tourists according to their willingness to pay the environmental protection tax. Kim et al. (2011) designed a survey to assess the factors affecting Japanese tourists' shopping preference and intention to revisit Korea. Schultz and

Block (2011) studied how American visitors view in-store promotion activities and then used the decision tree to predict opinion about alternative promotions.

19.3 METHODOLOGY

19.3.1 QUESTIONNAIRE DESIGN

On the basis of literature review and in-depth interview with tourism operators, after 3 rounds of pretest and revision, the questionnaire was divided into 3 parts. The first part was designed to investigate Mainland Chinese citizens' travel intention to Japan and influencing factors. The 5-Point Likert scale was adopted in the first part of the questionnaire. Respondents were asked to tick between 1 and 5 (1 represents strongly disagree, 3 on behalf of the general level, and 5 represents strongly agree) to express their views. The second part is about residents' travel willingness characters including travel purpose to Japan, travel organization form, suitable travel time span, per capita budget, preferred travel time, most important factors for travel time consideration, channels for Japan tour information, times of visit to Japan, interested websites, microblogs, or BBS, and their attitude toward traveling to Japan. The third part of the questionnaire is about respondents' social demographic backgrounds, including 7 items: gender, marital status, occupation, education level, monthly income, family sizes, and age.

19.3.2 DATA COLLECTION

The aim of this study is to investigate the travel intention to Japan by Mainland Chinese citizens including both existing visitors who had been to Japan and potential visitors who had not. During the pretest phase, by interviewing and consulting staffs in travel agencies, it was found that most outbound tourists were urban citizens with economic capacity, as well as vacation time. Since the perception is implicit and changeable, survey investigation is necessary. The formal survey was conducted from January 10, 2013, to February 25, 2013, in 10 cities with a balance between the developed and underdeveloped regions; the Eastern and the Western cities; and big, medium, and small cities such as Shanghai, Fujian, Guizhou, Hubei, Jiangsu, and the Sichuan provinces. Street intercept surveys were carried out to invite citizens to do a self-administered questionnaire, as well as a brief interview by well-trained investigators. Through face-to-face interviews with a brief introduction of the study, qualified respondents were screened to fill in a structured questionnaire, with the wording such as "If you plan to visit Japan, what is your purpose?", and the respondents were asked to tick the options. A total of 800 questionnaires were distributed and 756 were collected. After deleting the incomplete and invalid questionnaires (the answers were wrongly filled or whole page was left blank), a total of 741 questionnaires were obtained with a response rate of 98%.

19.3.3 METHOD

In this study, the combination of qualitative and quantitative methods was adopted to analyze the different travel decision influencing factors of residents and their intention to travel to Japan. The following aspects were included: (a) using descriptive analysis and multiple response analysis to investigate the frequency and characteristics of Mainland Chinese citizens' travel intention to Japan; (b) using CHAID decision tree to simulate the intention disparity between different groups after the Diaoyu Island political crisis; and (c) proposing appropriate marketing strategies for targeted segmentation.

19.4 DATA ANALYSIS

19.4.1 DESCRIPTIVE AND MULTIPLE RESPONSE ANALYSIS OF RESIDENTS' TRAVEL DECISIONS TO JAPAN

Based on descriptive analysis, in terms of gender, the proportion of women respondents was slightly high, which was 51.3%. In terms of marriage, the proportion of the married was higher than the unmarried, which were 52.8% and 47.2%, respectively. In terms of age structure, young respondents were dominant, as the share of people from 20–29 and 30–39 were 36.8% and 27.6%, respectively. In terms of education, respondents with university/college degree were more than 50% and reached 57.8%. In terms of occupation, company staff enjoyed the highest proportion of 41.5%. As for monthly income, the number of respondents was evenly distributed. The respondents of monthly income from 3,001 to 5,000 Chinese RMB Yuan shared 20.1%, whose proportion was slightly larger than other income levels. In terms of family size, family of 3 members was the major composition (62.2%). Considering the age, occupation, income, and others, the distribution of the data was even, which covered a wide range of community and met the basic requirements for the sample surveys (Table 19.1).

As shown in Table 19.2, in terms of the purpose of travel to Japan, sightseeing accounted for the largest proportion of 79.8%. In terms of travel organization form, traveling with family members occupied the dominant position (41.6%), indicating that family travel is the main form of organization about traveling to Japan. As for channels of Japan tour information access, the travel agency and website (BBS/microblogging, etc.) constituted the largest proportion, accounting for 21.5% and 21.3%, respectively. In the aspects of selecting a suitable time span of the itinerary, 4–5 days and 5–6 days were the main choices, accounting for 24.3% and 22.9%, respectively. In choosing the travel time/season, the majority of respondents chose the national holidays and paid leave, accounting for 31.7% and 34.2%, respectively. The most important factors of travel time in consideration were "seasons of optimum climate" and "vacation time," which accounted for the major proportion,

34.3% and 34.0%, respectively. The per capita budgets were "5,001–6,000 RMB Yuan" and "9,001–10,000 RMB Yuan," which is 21.6% and 17.7%, respectively; this not only reflects the difference between monthly incomes, but also reflects the perceived price difference of Japanese routes. Travel companies, travel media, and tourism organizations are the main source of tour information platforms, which accounted for 28.2%, 25.6%, and 18.2%. Most of the respondents had not yet been to Japan (78.4%) and showed that their perception about traveling to Japan came from the introduction of media or friends, but with the increase in numbers of travel to Japan, there was an attitude change toward Japanese tourism resources and the public environment.

TABLE 19.1 Demographic Characteristics of Respondent Citizens' Travel Intention to Japan

	N	%		N	%
Gender			***Monthly Income***		
Male	361	48.7	1,000 or below	143	19.3
Female	380	51.3	1,001–3,000	135	18.2
Marital Status			3,001–5,000	149	20.1
Single	350	47.2	5,001–7,000	124	16.7
Married	391	52.8	7,001–9,000	79	10.8
Age			9,001 or above	111	14.9
19 or below	78	10.5	*Family Size*		
20–29	272	36.8	1	10	1.3
30–39	204	27.6	2	88	10.9
40–49	109	14.7	3	461	62.2
50–59	56	7.6	4	98	13.2
60 or above	21	2.8	5 or above	81	12.4
Educational Level			*Occupation*		
Primary school or below	10	1.3	Company employee	311	41.5
Junior middle school	37	4.9	Businessman	11	1.5
Junior high school/Technical secondary school/Technical school	167	22.3	Civil servant	29	3.9
University/Junior college/College	433	57.8	Farmer/Fisher	3	0.4

Master or above	102	13.7	Professional & technical	64	8.5
			Housewife	17	2.3
			Teacher	54	7.2
			Salesman/Service worker	37	4.9
			Private owner	18	2.4
			Student	161	21.5
			Soldier	2	0.3
			Others	42	5.6

TABLE 19.2 Frequency Analysis and Multiple Responses of Travel Intention Characteristics of Mainland Chinese Citizens

Travel Intention Characteristics	N	%	Travel Intention Characteristics	N	%
Travel purpose to Japan			*Travel organization Form*		
Sightseeing	591	79.8	Travel alone	74	10.0
Business	34	4.6	Travel with family	308	41.6
Visiting friends and relatives	16	2.2	Travel with relatives/ friends	141	19.0
Study tour	49	6.6	Travel with classmates/colleagues	79	10.7
Work tour	18	2.4	Travel with groups	109	14.7
Others	33	4.4	Others	30	4.0
Suitable travel time span (Multiple response)			*Per capita budget*		
2–3 days	43	4.9	3,000–4,000 RMB Yuan	83	11.6
3–4 days	67	7.6	4,001–5,000 RMB Yuan	131	17.7
4–5 days	215	24.3	5,001–6,000 RMB Yuan	160	21.6
5–6 days	203	22.9	6,001–7,000 RMB Yuan	67	9.0
6–7 days	187	21.1	7001–8000 RMB Yuan	66	8.9

More than 7 days	170	19.2	8,001–9,000 RMB Yuan	66	8.9
Preferred Travel Time (Multiple Response)			9,000–10,000 RMB Yuan	131	17.7
			Others	34	4.6
National holidays	315	31.7	*Most Important Factors for Travel Time Consideration*		
Monday to Friday	79	7.9			
Weekend	95	9.6	Season of optimum climate	254	34.3
Any time	165	16.6	Vacation time	252	34.0
Paid leave	340	34.2	The best time to travel for the destination	85	11.5
Channels for Japan Tour Information (Multiple Response)			When low-cost tours were promoted	78	10.5
			Others	72	9.7
Travel agency	348	21.5	*Times of Visit to Japan*		
Tour leader/ personnel in travel agency	100	6.2	0	581	78.4
Relatives/friends	219	13.5	1	102	13.8
Radio, TV, and movies	194	12.0	2	26	3.5
Newspapers, magazines, and books	213	13.2	3	8	1.1
Posters and travel brochures	164	10.1	More than 3 times	24	3.2
E-mail	34	2.2	*After the Diaoyu Island Crisis, what is your attitude toward traveling to Japan*		
Website(BBS/microblogging)	345	21.3			
Interested websites, microblog or BBS (multiple)			Travel as scheduled	144	19.4
			Postpone the trip	200	27.0
Tourism enterprises	369	28.2	Replace the destination	171	23.1
Tourism administration	186	14.2	Cancel the travel plan	190	25.6
Tourist celebrity	180	13.7	Others and missing	36	4.8
Travel media	336	25.6			
Tourism organization	239	18.2			

19.4.2 THE IMPACT OF DIAOYU ISLAND POLITICAL CRISIS ON THE MAINLAND CHINESE CITIZENS' TRAVEL DECISIONS TO JAPAN AND THE ANALYSIS OF CHAID

During the series of crises, the sensitivity of the Sino-Japanese political relations between the 2 countries is the biggest obstacle for tourism development. The perceived travel obstacle factors to Japan by the Mainland Chinese after the Diaoyu Island Crisis could be summarized as 6 aspects as shown in Table 19.3. As the option of "Other perceiving travel obstacles (please specify)" was rarely replied, only 6 main factors were calculated. Among which, "The political attitude of governments hinder Japan travel plans" (M = 3.85) ranked the first, and this was followed by "foreign media reports tend to hinder travel plans to Japan" (M = 3.76). "I am afraid travel to Japan would be unsafe" ranked the third (M =3.63). The government policies and media play a key decision-making influence in guiding the Mainland Chinese citizens to travel to Japan.

TABLE 19.3 Mean and Variance Analysis of Travel Intention Influence Factors of the Mainland Chinese Citizens to Japan

Influence Factors of Travel Intention to Japan	Mean	Standard Error	Standard Deviation	Variance
The political attitude of governments hinder Japan travel plans	3.85	0.041	1.108	1.227
The Chinese and Overseas media reports hinder Japan travel plans	3.76	0.043	1.168	1.363
The views of relatives and friends hinder Japan travel plans	3.40	0.044	1.185	1.404
I am afraid travel to Japan would be unsafe	3.63	0.043	1.158	1.341
I began to dislike Japan	3.60	0.044	1.195	1.428
I am afraid that Japanese would be hostile	3.41	0.044	1.189	1.414
Total mean	3.61			
Cronbach's alpha	0.876			

For analyzing the disparity between different groups, the common method ANOVA was used. Although this method can target difference between groups separately, the interaction between the variables could not be measured. Therefore, in this study, to measure the interaction between the variables, decision tree analysis and CHAID algorithm were used. Before using CHAID, dependent variables and key independent variables were first selected. CHAID analysis would select the independent factors with the strongest interaction with dependent variable as the first layer. If there is no significant difference between certain independent variables, the tree will merge into one branch automatically. The dividing point can be detected in the decision tree. The most relevant independent variable would be selected as predictors in each step, likewise until the convergence criteria. Since CHAID used the Bonferroni adjustment in internal statistical tests, the cross-validation of a segmentation model based on a set of data can be very predictable in a similar sample. In this study, the dependent variable is the attitude of Mainland Chinese citizens on traveling to Japan after the Diaoyu Island political crisis, which was classified into 4 categories, while the independent variables include demographic variables of the respondents. After preliminary analysis by ANOVA, 5 independent variables emerged, namely education level, monthly income, marital status, age, and times of visit to Japan. The maximum depth of the tree is 5 with a classification significance level of 0.05. Those designated as training samples were 80% and as verification samples were 20%. Deleting the subjective options "Others (please specify)" and missing values, 701 valid samples were left. The validation tree growth is consistent with the training samples, reflecting a good stability. When designating "postpone the trip" as a target response variable, the prediction accuracy of training samples was 67.1% and verification samples was 66.7%. Specific model parameters are shown in Table 19.4.

TABLE 19.4 Multidimensional Decision Model Parameters of Mainland Chinese Citizens Travel Intention to Japan

	Growth Method	CHAID
	Dependent variable	Your attitude about traveling to Japan after the Diaoyu island political crisis 1. Travel as scheduled, 2. Postpone the trip, 3. Replace the destination, 4.Cancel the travel plan

Designation	Independent variable	*a. Educational level* 1. Primary school or below, 2. Junior middle school, 3. Junior high school/Technical secondary school/Technical school, 4. University/Junior college/College, 5. Master or above *b. Monthly income (RMB Yuan)* 1. 1,000 or below 2. 1,001–3,000 3. 3,001–5,000 4. 5,001–7,000 5. 7,001–9,000 6. 9,001 or above *c. Marital status* 1.Single, 2.Married *d. Age* 1. 19 or below 2. 20–29 3. 30–39 4. 40–49 5. 50–59 6. 60 or above *e. Times of Visit to Japan* 1. 0 2. 1 3. 2 4. 3 5.more than 3 times
	Verification	Sample Split
	Maximum tree depth	5
	The minimum number of cases in the parent nodes	5
	The minimum number of cases of child nodes	2
Outcome	Independent variables include	Your age, your number of times travel to Japan, your monthly income, your marital status
	Node number	11
	Terminal nodes	7
	Depth	4

Considering the attitude about traveling to Japan after the Diaoyu Island political crisis, the majority would "cancel the travel plan" (29.0%), followed by "replace the destination," and "postpone the trip," while "travel as scheduled" only occupied

19.6%. It means the vast majority of the Mainland Chinese citizens hold a cautious and wait-to-see attitude on traveling to Japan.

First, the significant node in the first level was age (Adj. $P = 0.003$, chi-square $= 21.333$, and $df = 3$). In node 1, there was a total number of 101 people (35.3%) prone to cancel the travel plan, respectively in the subgroups of "19 years old and below," "30–39 years old," and "40–49 years old"; while in node 2, there was a total number of 80 people (30.7%) prone to "postpone the journey", respectively in the subgroups of "20–29 years old," "50–59 years old," and "60 years old and above." Thus, groups of "19 years old and below" and "30–49 years old" were more likely to cancel the travel plan and more sensitive to crisis, while groups of "20–29 years old" and "50 and above" were more likely to retain the plan but postpone the trip, which demonstrated that they have certain loyalty to Japan and are willing to wait for the right time to replan. So at the very beginning, it is necessary to focus on the groups with pull motivation. First, festival and event marketing strategies would be recommended by organizing large-scale tourism exhibitions to show Japan's cultural traditions, medical health care, shopping facilities, food, and other iconic elements to stimulate the travel intention. Second, in the medium and long terms, co-marketing would be useful for those groups aged "19 years and below" and "30–39," by combining the advantages of Japan and advantages of other destinations to renew tourism products and complement each other to meet different needs.

Second, the significant node in second level was times of visits to Japan (Adj. $P < 0.001$, chi-square $= 38.050$, $df = 6$). In node 3, 101 people who had never been to Japan were prone to cancel the traveling plan. While node 4 had 22 people who had been to Japan for 1 or 2 times and were prone to postpone the journey, and in node 5, 10 people (55.6%) who had been to Japan for 3 or more times were prone to travel as planned. According to the interview, the citizens who had been to Japan, especially those who had been to Japan for several times, held more objective and rational attitude to Sino-Japanese issue, which may be due to the information symmetry and communication with Japanese people, reflecting the important role of tourism as a way to promote civil diplomacy and enhance mutual understanding and communication between 2 countries. Therefore, after the tourism crisis, the repeat visitors are more likely to travel as scheduled, while others are more aggressive or full of anxiety and more uncertain about the journey. Though repeat visitors had a certain degree of customer loyalty to Japan and it is easier to restore their faith to go back to Japan, this is based on past experience, namely satisfaction and recognition with Japan tourism; thus, high standards would be maintained, and membership coupons or discounts would also be a nice choice. As for those who had never been to Japan, their understanding about Japan was mainly from media reports or friends, so it is vital to guide them to consider in a more positive direction, persuading them that the majority of Japanese people are friendly, and eliminate the fear and hostility due to the information gap.

Third, the significant node in third level was monthly income (Adj. $P < 0.001$, chi-square = 32.287, df = 6). Node 6 with monthly income "between 7,001 and 9,000" or "more than 9,000" had 19 people (28.4%) tending to postpone the journey. In node 7, 19 people (51.4%) whose monthly income was "3,001–5,000" or "less than 1,000" tended to replace the destination, and node 8 had 45 people (41%) with monthly income "1,000–3,000" or "5,001–7,000" apt to cancel the travel plan. Low and middle-income groups are more likely to replace the tourist destinations and cancel travel plan. When travel to Japan is unsafe or instable, those people may choose other destinations in Southeast Asia as alternatives, which means they are push-motivation-oriented. But for the people with high consumption ability, due to the unique scenery and high quality service in Japan compared with other Southeast Asian countries, the attraction is irreplaceable. So for those groups with high price sensitivity, it is suggested that the sales can be increased by use of price leverage, through design of low-cost tourist routes, while carrying out the flash sales, group purchase, combined with the use of anti-quarter sales during off season. For high-income groups, pay attention to provide high-end routes; as these groups had traveled to Japan several times, the conventional shopping or sightseeing itineraries would be difficult to meet their needs, but cultural tourism may be a good selling point.

Fourth, the significant node in fourth level was marital status (Adj. $P < 0.01$, chi-square = 13.534, df = 3). In node 9, 9 people (56.2%) who were single were prone to travel as planned. However, node 10 had 18 people who were married, and they were prone to cancel the travel plan. Compared with the married, the single were more prone to travel as planned, because they are more adventurous and have exploring spirits and also do not shoulder much responsibility and pressure, while the married will consider more comprehensive factors. The first suggestion is new media marketing. New media with detailed and comprehensive contents have become a necessary tool for the younger generations. Therefore, tour operators should pay attention to website and mobile app applications. Also the needs of all members of the family should be taken into consideration.

19.5 CONCLUSION

In this study, a questionnaire survey was conducted to 741 Chinese Mainland citizens after the Diaoyu Island Crisis from the tourism crisis perspective to explore their travel intention and characters. The major conclusions are as follows:

First, multidimensional characteristics of Mainland Chinese citizens' travel willingness to Japan were analyzed. In terms of travel purpose, sightseeing accounts for the largest proportion, while family travel is the main form for Japan tour organization. The main channels for information access are travel agencies and websites; 4–6 days itineraries are most suitable for Japan travel, while the majority of respondents plan to visit Japan during national statutory holidays and paid vacations; preferred travel budgets are "5,001–6,000" or "9,001–10,000" RMB Yuan; Tourism enterprises, tourism media, and tourism organizations are the main sources of communication channels.

Second, multidimensional responses of Mainland Chinese citizens' travel willingness to Japan were analyzed. Based on decision tree analysis, it was found that after the Diaoyu Island political crisis, most of the citizens would choose to "cancel the travel plan" (29.0%), followed by "replace tourist destination," and "postpone the travel." Only 19.6% would "travel as scheduled," indicating that the vast majority of citizens are cautious about Japanese travel. Four significant nodes were generated in tree simulation: age, number of times to Japan, monthly income, and marital status. In this study, detailed analysis about the causes has been carried out and targeted marketing measures for each group have been provided. To restore the tourism market, mixed measures should be implemented including short-term and long-term plans.

It should be noted that the vast majority of Mainland Chinese citizens hold a cautious and wait-to-see attitude on traveling to Japan after the Diaoyu Island political crisis. Regardless of the marketing strategies, only the 2 countries maintained a good political stability and peaceful environment that can provide guarantee for travel to Japan. People who had been to Japan 3 or more times tended to travel as planned. To some extent, tourism can enhance mutual understanding between citizens in 2 countries, thus contributing to clear misunderstandings and disagreements. In recent years, Japan had suffered from a series of crises such as financial crisis, an earthquake, a nuclear leak, and the Diaoyu Island political crisis. However, after the incidents, there is still potential market to tap; thus, companies need to make the appropriate response to crisis management and marketing through the introduction of specialty travel routes, bargain promotions, government assistance, and visa concessions to gradually restore tourism market.

Based on crisis decision-making and crisis marketing perspective, in this study, an exploratory multidimensional market segmentation of the Mainland Chinese citizens has been made to mine the heterogeneity and clustering features of the residents. As the study was carried out in a sensitive stage of the Diaoyu Island crisis, the outbound travel to Japan was severely inhibited; it is difficult to carry out a comparative study of the existing and potential tourist market perception gap as scheduled. This study tries to conduct sampling by objective and comprehensive criteria, including respondents of different regions, age, gender, occupation, levels of education, etc. Further research can be carried out to expand the cross-sectional data into longitude data and analysis from vertical and horizontal perspectives, and to compare time series and spatial distribution disparity to provide more timely and precise marketing advice.

ACKNOWLEDGMENT

This study was supported by Grand Key Project of China National Social Science Fund (No. 12&ZD024), the Third Period of 985 to Whole Promotion of Social Science Research and Humane Fund at Fudan University (No. 2012SHKXYB002), and China National Nature Science Funds (No. 71073029 & No. 71373054).

REFERENCES

Bai, K.; Sun, T.; Zheng, P., et al. Analysis of factors affecting the tourists' decision based on the cognitive map: an example of inbound tourism in Xi'an. *Resour. Sci.* 2008, 30(2), 313–319. (in Chinese).

Baloglu, S. The relationship between destination images and socio demographic and trip characteristics of international travelers. *J. Vacation Mark.* 1997, 3(3), 221–233.

Bian, X. Analysis of MICE tourism participants in the decision-making process and its influencing factors. *Tourism Trib.* 2002, 17(4), 59–62. (in Chinese).

Bronner, F.; de Hoog, R. Economizing strategies during an economic crisis. *Ann. Tourism Res.* 2012, 39(2), 1048–1069.

Chai, S.; Cao, Y.; Long, C. Evaluation study on the factors affecting the tourists risk perception based on analysis of multiple regression models. *J. China Ocean Univ. (Soc. Sci. Ed.)* 2011, 25(3), 55–62. (in Chinese).

Coussement, K.; Van den Bossche, F. A.; De Bock, K. W. Data accuracy's impact on segmentation performance: benchmarking RFM analysis, logistic regression, and decision trees. *J. Bus. Res.* 2014, 67(1), 2751–2758.

Dai, L. Impact analysis and countermeasure of the outbound tourism crisis event. *Tourism Trib.* 2011, 26(9), 8–9. (in Chinese).

Dann, G. M. Tourists' images of a destination-an alternative analysis. *J. Travel Tourism Mark.* 1996, 5(1–2), 41–55.

Decrop, A.; Snelders, D. A grounded typology of vacation decision-making. *Tourism Manage.* 2005, 26(2), 121–132.

Dellaert, B. G.; Ettema, D. F.; Lindh, C. Multi-faceted tourist travel decisions: a constraint-based conceptual framework to describe tourists' sequential choices of travel components. *Tourism Manage.* 1998, 19(4), 313–320.

do Valle, P. O.; Pintassilgo, P.; Matias, A.; André, F. Tourist attitudes towards an accommodation tax earmarked for environmental protection: a survey in the Algarve. *Tourism Manage.* 2012, 33(6), 1408–1416.

Fesenmaier, D. R.; JiannMin, J. Assessing structure in the pleasure trip planning process. *Tourism Anal.* 2000, 5(1), 13–27.

Gartner, W. C.; Hunt, J. D. An analysis of state image change over a twelve-year period (1971-1983). *J. Travel Res.* 1987, 26(2), 15–19.

Hyde, K. F. Information processing and touring planning theory. *Ann. Tourism Res.* 2008, 35(3), 712–731.

Japan National Tourist Organization. *Basic Data of Travel to Japan*; Japan National Tourist Organization (JNTO), 2013-2. http://www.jnto.go.jp/jpn/reference/tourism_data/basic_china.html,2013-2. (in Chinese).

Jeng, J.; Fesenmaier, D. R. Conceptualizing the travel decision-making hierarchy: a review of recent developments. *Tourism Anal.* 2002, 7(1), 15–32.

Jiao, Y. The analysis of decision-making process based on analysis of tourist preferences and perception of risks. *Tourism Trib.* 2006, 21(5), 42–47. (in Chinese).

Kass, G. V. An exploratory technique for investigating large quantities of categorical data. *Appl. Stat.* 1980, 29(2), 119–127.

Kim, S. S.; Timothy, D. J.; Hwang, J. Understanding Japanese tourists' shopping preferences using the decision tree analysis method. *Tourism Manage.* 2011, 32(3), 544–554.

Lepp, A.; Gibson, H. Tourist roles, perceived risk and international tourism. *Ann. Tourism Res.* 2003, 30(3), 606–624.

Li, F. Discrimination study on the factors influencing tourists risk perception based on Logit model: case study of Sichuan "5.12" earthquake. *Tourism Forum* 2008, 1(3), 341–346. (in Chinese).

Lim, C. The major determinants of Korean outbound travel to Australia. *Math. Comput. Simul.* 2004, 64(3), 477–485.

Ma, Y.; Zheng, P.; Bai, K., et al. A study on difference of inbound tourist decision: case study of Japanese and American tourists travel to China. *J. Arid Land Resour. Environ.* 2008, 22(1), 102–106. (in Chinese).

Murphy, E. L.; Comiskey, C. M. Using chi-squared automatic interaction detection (CHAID) modelling to identify groups of methadone treatment clients experiencing significantly poorer treatment outcomes. *J. Subst. Abuse Treat.* 2013, 45(4), 343–349.

Schultz, D. E.; Block, M. P. How US consumers view in-store promotions. *Journal of Business Research* 2011, 64(1), 51–54.

Shen, H. Tourist destination model analysis of selecting and purchasing decisions. *Tourism Trib.* 2005, 20(3), 43–47. (in Chinese).

Sönmez, S. F.; Graefe, A. R. Influence of terrorism risk on foreign tourism decisions. *Ann. Tourism Res.* 1998, 25(1), 112–144.

Ture, M.; Tokatli, F.; Kurt, I. Using Kaplan–Meier analysis together with decision tree methods (C&RT, CHAID, QUEST, C4. 5 and ID3) in determining recurrence-free survival of breast cancer patients. *Expert Syst. Appl.* 2009, 36(2), 2017–2026.

Wang, J. *The Impact of International Relations and Major Events on the Entry and Exit of Tourism*; Shaanxi Normal University: Xi'an, 2011. (in Chinese).

Wong, J. Y.; Yeh, C. Tourist hesitation in destination decision making. *Ann. Tourism Res.* 2009, 36(1), 6–23.

Woodside, A. G.; Dubelaar, C. A general theory of tourism consumption systems: a conceptual framework and an empirical exploration. *J. Travel Res.* 2002, 41(2), 120–132.

Xinhua News Agency. *Japan Tour Cancellation up to 40%*; Sina: Finance. http://finance.sina.com.cn/china/20120925/061913228059.shtml, 2012-09-05

Zhang, J. A study on risk perception of domestic tourists travel to Tibet. *J. Sichuan Norm. Univ. (Soc. Sci. Ed.)* 2009, 36(6), 111–118. (in Chinese).

Zhang, J., Yu, B., and Chikaraishi M. Interdependences between household residential and car ownership behavior: a life history analysis. *J. Trans. Geo*. 2014, 34(1), 165-174.

CHAPTER 20

CHINESE OUTBOUND STUDENT-TOURISTS – DEVELOPING A TASTE FOR INDEPENDENT TRAVEL

BRIAN KING and SARAH GARDINER

CONTENTS

20.1 INTRODUCTION

Young Chinese are traveling overseas to acquire educational qualifications in increasing numbers. Travel plays an important part in their international education experience, and many students travel independently around the destination where they are studying. Furthermore, most Chinese international students host friends and relatives from China over the course of their studies, often accompanying them on trips. This is largely an independent travel phenomenon. In this chapter, it is argued that independent travel by Chinese students and their visiting friends and family is a departure from the established pattern of group-based long-haul Chinese travel and represents a growth market for the tourism industry. Evidence of the transition from a group orientation to a more independent approach is potent in Australia, where there are strong ties between Chinese international education and inbound visitation. International education accounts for over half of Australia's total Chinese visitor nights and almost half of Chinese visitor expenditure, contributing US$1.8 billion to the Australian economy in 2011–2012 (Australian Government, 2013a). Much of the impetus for more independent travel among Chinese visitors to Australia is attributable to the international education sector. The contrasting travel styles of first time and repeat Chinese visitors to Australia exemplify this relationship.

Most Chinese first-time visitors to Australia travel as part of a tour group (51%) and use a travel package (53%). Repeat Chinese visitors were more likely to travel independently, with less than a third of repeat visitors to Australia opting for a tour group (28%) and packaged travel (28%) on their trip. Australia has also seen a change in the age profile of first-time versus repeat visitors from China. Repeat visitors from China tend to be younger (36% aged 15–29 years), spend more (US$16,459), and stay longer (over 4 months) than the first-time Chinese visitors (34% aged 45–59 years, spend US$3,911 per trip, and stay 17 nights) (Australian Government, 2012). The transition to independent travel among Chinese international travelers to Australia is thought to be stimulated by the emergent international student market. Therefore, prototypical behaviors by younger mobile Chinese of international students today may provide insights into the future of China outbound travel as Chinese travelers move from group-based travel to more independent travel options.

This study of 1,400 Chinese studying in Australia was undertaken in 2009 and the characteristics of these independent travelers, examining the role of travel in their international education experience, were studied. The profile and potential of the associated visiting friends and relatives are also highlighted. Following the completion of the survey, 6 focus groups were conducted with Chinese students studying in Australia to acquire a better understanding of their travel decision-making. Based on the findings, the researchers draw a distinction between independent student travelers and youth travelers and backpackers more generally. It is suggested that the travel behaviors of independent students may inform emerging conceptualizations of Chinese independent travel. The continuing growth of more individualized China outbound travel coincides with the global expansion of youth travel and international education. For many proponents, international education is no longer a "once in a lifetime" experience, but increasingly marks the start of a lifelong career of temporary migration (Hugo, 2004). International students generally and including those from China may be described as "new transnationals" whose lives, interests, and connections are based in multiple countries (Vertovec, 2004). The experience of international education has been described as having a transformational effect on young people. International education offers students an opportunity for self-discovery through escape from cultural and familial expectations (Brown, 2009). Such freedom has a long-lasting effect on the enhancement of self-efficacy – namely beliefs about one's capabilities (Milstein, 2005) and intercultural confidence. Thus, the nexus between travel and learning (Falk et al., 2012) and the expanding outlook of Chinese youth owing to international education may challenge the Chinese authorities in the event of potential conflict between nationalist impulses, self-interest, and a global cosmopolitan outlook.

The development of a more applicable typology of Chinese independent youth travelers could provide destination marketers and planners with a more solid basis for servicing this market. Product developers and destination managers who are seeking to retain competitiveness in the international youth tourism market will need to provide international students with a wider variety of travel products and more tailored bespoke experiences (Gardiner et al., 2013). Tourism operators and managers in destinations frequented by Chinese tourists will be increasingly confronted by the challenge of aligning infrastructural provision and business practices to the needs of Chinese independent youth travelers, including student tourists. This chapter reports on an investigation of the travel preferences and behaviors of international students studying in Australia. A national online survey was conducted in 2009 with international students enrolled at 12 universities. The support of the peak higher education body Universities Australia was secured and all member institutions (39) were invited to participate in the study. The various institutions subsequently disseminated an email and web link to their respective international students. A total of 5,991 responses were received of whom 1,414 (27.7%) were Chinese students. A statistical analysis of the resulting data provides a travel profile of this group. Key insights from the survey were subsequently investigated through 6 in-depth focus groups with Chinese students studying at universities in South East Queensland, Australia. In each session, 3–8 students participated, and sessions lasted approximately 45 minutes. The interviews were then transcribed and analyzed using key themes. The current and future travel behaviors of Chinese students are presented as the findings of this study.

20.2 THE TRANSFORMATIONAL POWER OF INTERNATIONAL EDUCATION AMONG CHINESE YOUTH

The development of the China outbound education phenomenon can best be understood in the context of changing relationships between China and the rest of the world. From the time of the Communist revolution in 1949, attitudes to the outside world, generally, and to the Western world, in particular, have evolved. Over recent times, China has moved from a closed approach (pre-1978) to an open approach (post 1978). There were shorter term distinctions with these 2 broad periods. During the "closed" period of the cultural revolution (from 1966 to about 1971), for example, the authorities viewed education as a bourgeois and counter-revolving activity. In accordance with the prevailing ideology, schools, and universities, the atmosphere closure was not conductive to the pursuit of scholarship inside or outside China.

Two major stages have characterized the subsequent period of openness. The first started in 1978 and concluded in1989, the date of the so-called "Tiananmen Square incident". Though outbound leisure travel flows had not yet gathered momentum in the late 1980s, many Chinese students were already studying at overseas institutions. Though brief in duration, a vigorous government crackdown on dissent followed the incident. The governments of Western nations and notably Australia offered extended residency to those who were enrolled as students. Many international students opted to remain permanently in their new country of residence. Following the 1989 crackdown, Chinese students have traveled to other countries in ever increasing numbers (amounting to 649,500 in 2011 according to UNESCO 2012). It is worth noting, however, that most students during the subsequent period have returned to China on completion of their studies. The rapid development of the Chinese economy has improved employment prospects for those graduating overseas (UNESCO, 2015). This pattern of return reinforces the view that Chinese international students have more in common with tourists than with permanent migrants.

During the closed period, officials viewed those pursuing studies overseas in the capitalist world with suspicion and mistrust. The numbers of Chinese traveling internationally were tiny, and overseas travel was largely confined to conducting diplomatic relations. The Chinese authorities have adopted a different view of international education and international travel following the open door policy. As various state controls over economic activity have been loosened, it has been increasingly acknowledged that acquiring knowledge and learning from overseas experience offer better opportunities for young Chinese to contribute to the development of nation. Over the past decades, an increasing number of Chinese students have been admitted to universities within the developed world and have been studying in English. Market growth has been propelled for by the reduction in Cold War tensions and increasing recognition within slower growth developed countries of the economic benefits associated with international students.

As the authorities were becoming more favorably disposed toward education, they also encouraged travel and tourism more generally. In 1999, Australia and New Zealand were the first "Western" countries to be granted Approved Destination Status (ADS). Since the time of the open door policy, the bulk of outbound tourism consistent with ADS-approved travel has been group-based travel. In contrast, the longer established flows of Chinese international students have been characterized by a more independent style of travel. Australia had been a major beneficiary of ADS visitors and received over 1 million Chinese ADS leisure group tourists between 1999 and 2013 (Australian Government, 2014). In 2013, Australia hosted 709,000 visitors from China, an increase of 14.5% relative to 2012, generating US$4.8 billion in expenditures. By 2020, total expenditure by the 860,000 Chinese visitors in Australia is expected to be worth between US$7.4 and US$9 billion per annum (Australian Government, 2011). For this reason, China is now viewed by Tourism Australia (Australia's destination marketing organization or DMO; Australian Government, 2011), as "Australia's most valuable inbound visitor market". (p. 1) (Fig. 20.1)

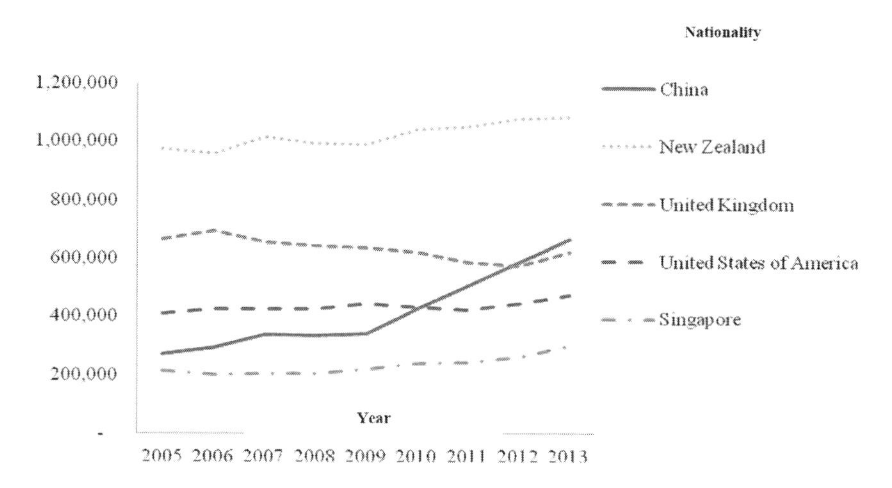

FIGURE 20.1 Top 5 Inbound Visitor Markets to Australia (2005–2013). (From Australian Government. International Visitors to Australia 2014. www.tra.gov.au/statistics/international-visitors-to-australia.html, accessed 10 June).

20.3 THE EMERGENCE OF CHINESE INTERNATIONAL STUDENTS

Paige has noted that international students reside temporarily "in a country other than their own citizenship or permanent residents in order to participate in international educational exchange as students" (Paige, 1990) (p162). Ward, Bochner, and Furnham (Ward et al., 2001) have described international students as one of the most intensively studied groups in the literature on cultural contact. Frequently investigated topics have included social capital, homesickness, satisfaction, contentment, and social connectivity. As "strangers in the strange land", it is commonplace for Chinese international students to form bonds with co-nationals and friendships, often extending into later life (Featherstone, 1995). This connection will have a potentially strong influence over future travel behaviors and word-of-mouth communication. Research conducted in Australia by Pyke, King, and Whitelaw (Pyke et al., 2014) demonstrated the power of Chinese alumni networks made up of graduates from various Australian universities who have returned to China following graduation.

There is an extensive literature on the experiences of international students emanating from different source countries, including China. Brown (2009) located international education in the context of the "sojourning" literature, which explores the extent to which travel experiences may be described as transformational (Milstein, 2005). There are parallels between the idea of education as a potentially transformational experience and the literature on backpacking and gap year studies. Citing Ward, Bochner, and Furnham, (Ward et al., 2001) and Brown (2009) defined an international sojourn as providing "temporary between-culture contact" (p184) that transforms individuals into "intercultural mediators who learn to grow beyond the psychological parameters of their own origin culture" (p184). Sojourners' experiences may be "mono-cultural, assimilationist, bicultural, marginalized or multicultural" (p185). Though they have the capacity to improve language capabilities (p189), such experiences may also exacerbate segregation because of pressures to conform that emanate from what has been described as "a faraway culture to which they were set to return" (p191).

Given China's rise as a global power, the attitude of Chinese authorities to both international education and tourism will play a role in shaping international relations to some extent. The acquisition of cross-cultural capabilities by international students who have relevance to their country of ultimate residence is important in the case of Chinese students who originate in a country of growing power and assertiveness. Cross-cultural capabilities also apply to the growing "gap year" market (Inkson and Myers, 2003). The international education experience involves adjustments into a new society and also at the time of homecoming. According to Brown (2009), the journey is not over until the returnees "negotiate the return to their old home world" (p. 518). She has suggested parallels between international students and long-stay tourist markets, in that both types of visitor to the new culture are motivated to make temporary adjustments and to learn culture-specific skills. Such parallels are relevant to the current volume and its exploration of China outbound tourism.

According to the Australia Education International, a government agency, Chinese students contributed 29.5% of all international enrolments in 2012. Almost two-thirds (63.5%) were enrolled in higher education programs with a further 17% undertaking English language programs (Australian Government, 2013a). Approximately 1.5 million Chinese students came to Australia between 2001 and 2011, more than half of all educational arrivals over the period. Though most students came with the exclusive purpose of pursuing their education, the role of the Chinese diaspora should not be discounted. Of about 88,000 Chinese who visited friends and relatives in Australia, a substantial proportion of them were visiting Chinese students studying at Australian educational institutions. China was Australia's fifth largest source of youth arrivals, and in 2012, education accounted for over half of Australia's Chinese arrivals in the 15–29 age group (the proportion for all markets is 18%; Tourism Australia, 2013). Despite the youthfulness of the market, China is not currently one of the countries granting working holiday visas. This visa permits young people to cover their Australian holiday expenses by accessing short-term employment. Australia currently maintains reciprocal working holiday visa arrangements with 19 countries. This program aims to promote travel to Australia, encourage cultural exchange, and build relationships between these countries and Australia (Australian Government, 2014). Working holiday visas are important for several youth tourism source markets including the United Kingdom, South Korea, Taiwan, Germany, and France (as shown in Table 20.1). The Australian Government granted 210,369 first Working Holiday Maker (WHM) visas and 38,862 second year WHM visas to youth travelers in 2012–2013. Demand is increasing, and the number of first year WHM visas granted in 2012–2013 rose by 14.2% and second WHM visas by 27.4% over the previous year. This pattern distinguishes Chinese international student from many Western counterparts. Demand for visas was particularly high among young Taiwanese, with a 57.6% increase in first year WHM visas granted and a 68.7% increase in second year WHM visas granted over this period (Australian Government, 2013b). Taiwanese are Putonghua-speaking people, and this indicates that the absence of taking working holiday among Chinese youth travelers to Australia should not be associated with cultural factors.

TABLE 20.1 Total Working Holiday (Subclass 417) Visa Applications Granted from Australia by Country of Citizenship

Citizenship Country	2011–2012 to 30/06/12	2012–2013 to 30/06/13	% Change from 2011–2012	2012–2013 as % of Total
United Kingdom	41 712	46 131	10.6%	18.5%
Taiwan	22 393	35 761	59.7%	14.3%
South Korea	32 591	35 220	8.1%	14.1%

Germany	22 499	26 184	16.4%	10.5%
France	20 086	24 788	23.4%	9.9%
Other	117 034	116 908	-0.1%	32.7%
Total	214 644	249 231	16.1%	100%

Source: Australian Government Working Holiday Maker visa program report; Canberra, 2013.

Various studies have investigated the extent to which Chinese students adapt to different study destination settings. Chirkov et al. (2007), for example, studied the students in Belgium and in China. Wang and Walker (2011) compared the effect of face concern between Chinese (n = 352) and Canadian (n = 295) university students when undertaking leisure travel activities. It is useful that various researchers have compared Chinese international students with those from other countries of origin. In their study of Taiwan, Jenkins and Galloway (2009) compared the adjustment problems encountered by students from China and other source markets. They found that Chinese respondents encountered more problems than non-Chinese respondents.

In seeking to understand the adjustments that students need to make during their studies, Smith and Khawaji (2011) examined the effectiveness of intervention programs. They found that forming friendships with local people help to combat loneliness. Research has shown that international students tend to maintain contact with those of their own ethnicity or with other international students (Williams and Johnson, 2011). The same authors also noted that it was "challenging and rare" to form friendships with American students (p. 41). Kim noted that different societies offer varying degrees of receptivity to strangers (Kim, 2001). In their study of the effects of intervention programs, Smith and Khawaji (2011) observed the adjustments that students made over the course of their studies. They found that friendships with locals contribute to combating loneliness. Zhang and Goodson drew a conclusion from the results of 64 studies in their review of psychosocial adjustment among international students in the United States (Zhang and Goodson, 2011). Research has shown that international students tend to maintain contact with other international students and particularly with those of their own ethnicity (Jenkins and Galloway, 2009). Symptomatic of a sense of solidarity Xiao's characterization of the development of so-called Chinese "red tourism" has been noted as a means of building national spirit and pride in Chinese heritage (Xiao, 2013).

In considering the context within which Chinese international students are operating, it is worth considering the scale of the permanent Chinese population in Australia. Inglis (1972) documented the history of the Chinese in Australia. Despite a history of diaspora dating back to the mid-nineteenth-century gold rushes, there were only 27,000 Chinese living in Australia in 1966, of whom one-third were born in Australia. By way of comparison, the Chinese accounted for 1.6% of Australia's total population (39,000) in 1881.

20.4 CHINESE OUTBOUND STUDENT TOURISM TO AUSTRALIA

Australia hosts over 410,000 international students annually and Chinese account for the largest share (about 29%; Australian Education International, 2013). There are approximately 120,000 Chinese studying in Australia on a student visa. After China, the next largest markets are India and Korea, which account for a more modest 8.8% and 4.9%, respectively, of the market. From these figures, it is evident that Chinese students are critical to Australia's international education sector. Since Chinese students live, work, and study in Australia over extended periods, they impact significantly on the economy. The study referred to previously found that international students typically spend US$42,000 per year when studying in Australia. These expenses are largely funded by parents. Most students reported that their parents paid their program fees (70.9%), living expenses (63.9%), and for travel and recreation (57.6%). However, income acquired from working (31.7%) and from personal savings (29.8%) contributed to travel and recreation-related spending.

The states of New South Wales, Victoria, and Queensland host 80% of all international students studying in Australia, with most students studying at universities in the state capitals, mostly Sydney, Melbourne, and Brisbane. Table 20.2 profiles international students studying in Australia by nationality, age, and gender. Most Chinese students are attending higher education (61.9%) or English language institutions (19.0%).The vocational (8.2%) and schools (6.0%) sectors represent only a small proportion of the market. Slightly more females (51.3%) than males (48.7%) were studying in Australia at the time of the study. Given the preponderance of university study, it is unsurprising that Chinese students are relatively youthful, typically aged in their 20s. In 2013, 70,553 Chinese students studying in Australia were aged 20–24 years and 26,125 were in the category 25–29 (Australian Education International, 2013). This youthful age profile indicates that Chinese students are an emerging part of the independent travel youth tourism market.

TABLE 20.2 Profile of Chinese and all International Students by Age, Gender, and Education Sector

	Students from China	**All International Students**
	% of Students (Number of Students)	**% of Students (Number of Students)**
Age (years)		
<18	4.2% (5 083)	3.4% (13 716)
18–19	9.3% (11 073)	6.8% (27 316)

20–24	62.7% (74 469)	46.3% (186 174)
25–29	20.5% (24 457)	27.4% (110 446)
30–34	2.3% (2 733)	10.5% (42 326)
≥35	1.0% (1 017)	5.6% (22 410)
Total	100% (118 832)	100% (402 388)
Gender		
Female	51.8%	47.6%
Male	48.2%	52.4%
Education sector		
Higher Education	63.5%	48.9%
VET[a]	8.9%	23.4%
ELICOS[b]	17.0%	17.8%
Non-award	4.5%	5.8%
Schools	6.2%	4.2%

[a]**VET** – Vocational Education and Training.
[b]English Language for Intensive Courses for Overseas Students.
Source: Australian Government, International Student Numbers 2012. Australian Education International, Canberra (2013).

The primary purpose of the trip to Australia by respondents was education. Consistent with those from other source countries, Chinese students like to incorporate travel as part of their Australian experience. Table 20.3 summarizes the travel behaviors of Chinese students. It is worth noting that the vast majority of international students traveled during the course of their studies in Australia (84.9%). The study found that 60.5% of Chinese students planned to travel within Australia during their upcoming summer vacation (at the time of the study). While the respondents may have acquired extensive experience of domestic travel in China, most are new to international travel. On this basis, they tend to travel to capital cities and tourist destinations close to their place of study. The study found that most international students had visited a capital city (77.1%) in the state where they were studying and/or a capital city in another state (61.2%). Travel to regional Australia is perceived as less desired because Chinese students are unfamiliar with regional travel and assume that regional Australia resembles its equivalent in regional China. This implies that most of the land is allocated to farming and agriculture. As a result, Chinese students have low awareness of the opportunities or experiences offered by regional Australia. The following illustrates the respondent's views of regional China:

"In China, outside the city in the countryside, it just means it's undeveloped, so people just are relatively so poor, they just do farm work."

A few focus group participants had visited Australian regional tourism destinations, mostly on trips that were organized by friends living in the relevant region. Respondents had low expectations and were pleasantly surprised by the extent of the tourism offerings of the relevant destinations. For example, one Chinese focus group participant had visited a friend in Melbourne and her friend had taken her on a day-trip tour to a winery and art trail in the surrounding regional area. She said,

TABLE 20.3 Summary of Study Findings on Chinese Student Travelers

Variable	Percentage of Chinese Students
Duration (most recent trip)	
Day visit	29.4
1–3 nights	34.4
4–6 nights	24.7
>1 weeks	11.5
Expenditure (most recent trip)	
>US$100	20.1
US$101–US$300	20.9
US$301–US500	21.4
US$501–US$1,000	22.6
>US$1,000	15.0
Preferred accommodation	
Hotels	48.4
Apartments	28.0
Backpacker hostel	16.9
Camping	3.7
Other	3.0
Mode of transport (most recent trip)	
Car	26.3
Bus	23.1
Boat	17.4
Other	33.2

"I just found out about the tour by talking to my local friends and they introduced me to everything. If they hadn't introduced me to other attractions within the local area, I would have only stayed in the city."

Travel by Chinese students mostly occurs during vacation periods at their respective institution, particularly during inter-semester vacations. Trips typically last for less than a week. This study found that most Chinese students took a day trip (29.4%) or short break of 1–3 nights (34.4%) or 4–6 nights (24.7%) on their most recent trip. Almost three-quarters of respondents reported spending up to US$1,000 on their most recent trip, with 22.6% spending US$501–US$1,000, 21.4% spending US$301–US$500, and 20.1% spending less than US$100. There was also a small portion of high spenders with 14.6% of respondents spending over US$1,000 on their most recent trip.

Parents of Chinese students often encouraged their children to travel as part of their Australian experience but did not want this to jeopardize their studies. One focus group participant described her parents' stance toward travel as part of her international education experience. They wanted her to:

"Go out and travel, go and see the world, go and see Australia. This is going to be great as part of your life experience."

Another student reported having received parental encouragement to travel as part of her study experience and, importantly, to share that travel experience by sending photographs home during her travels:

"I say I want to travel [to] my parents. [They] say oh it's ok because I always stay at home [in China]. So, my parents [are] like you should always travel. Please don't stay at home, send us pictures of you doing things."

Chinese students are also mindful that they are typically the only child within their family and are cautious travelers concerned about safety. One respondent commented that,

"I am the only children in my family, so I won't take the risk. . . If something happened, oh, it will be for your parents' trouble. They can't accept [it]."

Another respondent stated that,

"People really care about the safety. Yeah, they don't want to try the thing a little [too] dangerous, adventure."

Another respondent recognized the thirst for new experiences and adventure among Chinese youth and stated that,

"More and more young people in China really want to try some exciting things in recent years, because their life [is] a little bit boring, so we want to try adventure."

To assure their parents that they are safe and self-reliant, Chinese students communicate regularly about their life in Australia and any travel plans. They often consult with their parents when planning to travel and communicate regularly during and after their trips. They attach considerable importance to the reactions of their parents toward their travel-related activities. Parents often provide them with guidance. At one extreme, a student stated that, "I just follow my parents' ideas". Other respondents made travel decisions independently of their parents. In avoiding to inform parents about their travel plans in case they raised objections, one respondent noted that,

"I am [from] a traditional family, you know. My parents always worry about my safety if I go somewhere. My parents always tell me, 'There are some bad guys outside, and you have to be careful' <laughs>. You know, I've got a job. I can earn by myself, so . . . I have money, if I want to go there, I just go."

Chinese parents generally want to know that their son or daughter is safe, particularly in light of the "one child policy," which makes each a focus of attention. However, many parents also recognize the contributions of acquiring experiences outside study to personal development. They may encourage and fund their children to travel as part of international education on this basis. Furthermore, the photographs that their children send home of studying and traveling in Australia provide social capital when these artifacts and experiences are shared with friends and family. International student travel can have a flow-on marketing effect through this medium, generating positive word-of-mouth and providing destinations with aspirational appeal. For these reasons, it may be argued that international student travel should be encouraged as a means of stimulating overall travel activity. Communication activities should also aim to allay the safety concerns prevailing amongst students and their parents.

The high proportion of respondents who undertook their most recent trip with friends (74.5%) suggest that this provides Chinese students with a sense of security when traveling. Most of the friends were other Chinese students (63.4%). In comparison, only 41.9% of all international students (including the Chinese respondents) traveled with their compatriot students. Chinese students were more likely to travel via public transport than international students, with 23.1% taking the bus and 17.4% boat on their most recent trip (the comparable figures for all international students were 12.0% and 5.8%, respectively.) Relative to all international students (34.7%), they were also less likely to use a car (26.3%). This suggests that Chinese students prefer the security of friends and group transport when traveling. Socializing is also a preferred recreational pursuit, with 32.1% reporting that the main activity during their recreation and personal time is meeting friends.

The travel that they undertook while studying was the first time that many respondents had been away without their parents and their first independent trip. As cautious travelers, they expressed concerns about safety. Unlike their Western counterparts, generally and backpackers in particular, the trip itself represents risk and adventure, rather than the activities undertaken while away. The present study discovered that respondents prefer more mainstream accommodation options, such as hotels (48.4%) and apartments (28.0%), over backpacker hostels (16.9%) and camping (3.7%). In comparison, 22% of all international students preferred backpacker hostels and 5.5% expressed a preference for camping. The research identified that international students, including Chinese, visited natural attractions (75.6%), sightseeing (73.3%), and shopping for pleasure (51.3%) when traveling. Meeting with friends (32.1%), relaxation (31.5%), shopping (15.4%), and travel and tourism activities (13.6%) were also popular during their recreational and personal time.

Chinese students attract a large number of visiting friends and relatives. The present research shows that most (93.4%) encourage their friends and family to visit Australia, with 21.3% reporting that at least 2 friends or family intended to visit them during their stay, 22.4% reported 3–4 friends, 17% reported 5–10 friends, and 12.2% reported more than 10 friends or family intended to visit during their stay. Graduation appears to be the most likely time for parents to visit, with approximately80% of Chinese respondents indicating that their parents would attend their celebrations. Most of the friends and family who were visiting Chinese students were likely to stay between 1 and 2 weeks (39.7%), 22.3% for 2–4 weeks, and 22% were likely to stay more than 1 month. These students play the role of tour guide, accompanying their parents, family, and friends on their travels around Australia. For instance, one Chinese student stated that,

"When my parents [visit]. They're coming in June/July . . . I'm planning a trip with them . . . when family is visiting, the students here they can travel as well."

20.5 CONCLUSIONS

The present chapter has shown the significant role that education has played in the China outbound tourism phenomenon to Australia. Though international students are sometimes viewed as occupying a distinct category of mobility that contrasts with group tourists, it is they who have evidently been laying important groundwork for future generations of independent travelers. Their presence and relative mobility within Australia have familiarized service providers with visitors from Chinese backgrounds. By acquiring an understanding of Chinese students' behaviors, destination authorities and tourism operators can become progressively accustomed to handling more independent travelers from China. However, they should also consider how the travel preferences and behavior of Chinese students differ from those of traditional youth travelers, who are mostly Western backpackers. This represents a challenge for tourism destinations, like Australia, that have developed adventure and youth-oriented experiences, based on servicing backpackers. The emerging Asian and, particularly Chinese, independent travel market are prompting the industry to rethink product offerings and marketing strategies. There have been several recent industry conferences in Australia to discuss this issue, with special sessions at the 2013 Australian Tourism Export Council's Australian Youth Tourism Conference and the 2013 Queensland Government's Destination Q Forum on the rise of the Asian independent travel. It is also evident that repeat visitation by Chinese students, sometimes as a part of alumni networks, represents a significant opportunity for destination-based operators. Students acting as tour guides for their Chinese visiting friends and relatives may also extend the potential impact of this group, further accelerating the shift from group to independent travel among Chinese outbound travelers. An offer of combined work and holiday experience could promote Chinese independent travel, especially if supplemented by the prospect of studying a short course (such as English language). The popularity of the WHM scheme among youth travelers has been evident in other Asian source markets, such as Taiwan and Korea, as well as in traditional Western markets. Destinations are also increasingly aware of their capacity to shape attitudes at home with China, thereby providing a basis for building bilateral relations. International students are evidently an important part of China's "New Wave" of outbound travel and are a potentially important catalyst for change.

REFERENCES

Australian Education International. *Research Snapshot: International Student Numbers 2012*; Australian Education International: Canberra, 2013.

Australian Government. *2020 Summary of Tourism Australia's China 2020 Strategic Plan*: Tourism Australia, Canberra, 2011.

Australian Government, *China - First and Return Visitation: Snapshot 2012*: *Tourism Research Australia* Canberra, 2012.

Australian Government. http://www.austrade.gov.au/Tourism/Tourism-and-business/ADS (accessed 1 September 2014) 2014.

Australian Government. https://www.immi.gov.au/visitors/working-holiday/ (accessed 14 September 2014) 2014.

Australian Government. *International Student Numbers 2012*; Australian International Education: Canberra, 2013a.

Australian Government. Working Holiday Maker Visa Program Report: Canberra, 2013b.

Brown, L. The transformative power of the international sojourn: An ethnographic study of the international student experience. *Ann. Tourism Res.* 2009, 36(3), 502–521.

Chirkov, V.; Vansteenkiste, M.; Tao, R.; Lynch, M. The role of self-determined motivation and goals for study abroad in the adaptation of international students. *Int. J. Intercult. Relations*, 2007, 31(2), 199–222.

Falk, J.; Ballantyne, R.; Packer, J.; Benckendorff, P. Travel and learning: a neglected tourism research area. *Ann. Tourism Res.* 2012, 39(2), 908–927.

Featherstone, M. *Undoing Culture: Globalization, Postmodernism and Identity*; Sage, London, 1995; Vol. 39.

Gardiner, S.; King, B.; Wilkins, H. The travel behaviours of international students nationality-based constraints and opportunities. *J. Vacation Mark.* 2013, 19(4), 287–299.

Hugo, G. *A New Paradigm of International Migration: Implications for Migration Policy and Planning in Australia*; Department of the Parliamentary Library, Canberra, 2004.

Inglis, C. Chinese in Australia. *Int. Migr Rev.* 1972, 6(3), 266–281.

Inkson, K.; Myers, B. "The big OE": self-directed travel and career development. *Career Dev. Int.* 2003, 8(4), 170–181.

Jenkins, J.; Galloway, F. The adjustment problems faced by international and overseas Chinese students studying in Taiwan universities: a comparison of student and faculty/staff perceptions. *Asia Pac. Educ. Rev.* 2009, 10(2), 159–168.

Kim, Y. Y. *Becoming Intercultural: An Integrative Theory of Communication and Cross-Cultural Adaptation*; Sage, Thousand Oaks, CA, 2001.

Milstein, T. Transformation abroad: sojourning and the perceived enhancement of self-efficacy. *Int. J. Intercult. Relations*, 2005, 29(2), 217–238.

Paige, R. M. International students: cross-cultural psychological perspectives. *Appl. Cross Cult. Psychol.* 1990, 14, 367–382.

Pyke, J.; King, B.; Whitelaw, P. The visitor experiences of international students in Australia: a cross-cultural comparison. In CAUTHE Conference 2014: Brisbane, 2014.

Smith, R.; Khawaja, N. A review of the acculturation experiences of international students. *Int. J. Intercult. Relations*, 2011, 35(6), 699–713.

Tourism Australia. *Understanding the Chinese Consumer*: Department of Education & Training, Canberra, 2013.

UNESCO. http://www.uis.unesco.org/Education/Pages/international-student-flow-viz.aspx (accessed May 29) 2015.

Vertovec, S. Cheap calls: the social glue of migrant transnationalism. *Global Netw.* 2004, 4(2), 219–224.

Wang, X.; Walker, G. The effect of face concerns on university students' leisure travel: a cross-cultural comparison. *J. Leisure Res.* 2011, 43(1), 133–147.

Ward, C.; Bochner, S.; Furnham, A. *The Psychology of Culture Shock*; Routledge, Hove, East Sussex, 2001.

Williams, C.; Johnson, L. Why can't we be friends?: multicultural attitudes and friendships with international students. *Int. J. Intercult. Relations*. 2011, 35(1), 41–48.

Xiao, H. Dynamics of China tourism and challenges for destination marketing and management. *JDMM*. 2013, 2(1), 1–3.

Zhang, J.; Goodson, P. Predictors of international students' psychosocial adjustment to life in the United States: a systematic review. *Int. J. Intercult. Relations*, 2011, 35(2), 139–162.

SECTION IV

REFLECTIONS AND FORECASTING

CHAPTER 21

NEW PERSPECTIVE FOR RESEARCH METHODOLOGY IN THE ERA OF CHINA OUTBOUND TOURISM 2.0: A PRACTITIONER'S OBSERVATION

STANLEY CHAN

CONTENTS

21.1 INTRODUCTION

Based on the statistics issued by China National Tourism Administration, the total number of Chinese outbound tourists is actually 5 times of the figure recorded 1 decade ago.[1]

This "outbreak" of Chinese outbound travel business forces each destination to adjust their current travel resources to absorb the huge number of Chinese visitors. The typical example of such phenomenon is the incorporation of more and more "Chinese" elements into the interface with the Chinese visitors. For example, in Germany, Chinese newspapers and boiled water are always provided in the major hotels so as to cater for life habits of the Chinese.[2]

Another more recent example is that Chinese maps with recommended sightseeing venues and attractions have been provided by the tourism bureau in Boston, and it is considered as a key marketing strategy to attract the Chinese tourists.[3]

All in all, destinations around the world have realized that the influx of Chinese tourists has become a sustainable revenue stream, which must be maintained and further explored by a more tailor-made marketing approach. Actually, this discussion has been started within the industry workers in China (please see related discussion in *Green Book of China Tourism* issued by Chinese Academy of Social Science, 2014).

This derives the rise of the "Era of Chinese Outbound Tourism 2.0." Based on my over 10-year experience in conducting tourism research in China, let me take a short review of this new era and the corresponding changes in the market landscape in the next section. After this, we would focus on how the market researchers should react with this new era and market changes.

21.2 DEFINING FEATURES OF CHINA OUTBOUND TOURISM 2.0

21.2.1 SUPPLY SIDE: A SHIFT OF PARADIGM IN MARKETING

In "China Tourism 1.0," the key marketing perspective across most destinations is to define all the Chinese tourists as "Homogeneous Mass." That is, Chinese tourists can be portrayed in "one face." For example, all Chinese tourists are crazy about luxury brands or most Chinese are reluctant to participate in outdoor activities (e.g., skiing or canoeing).

---

[1]The information was extracted from the news report made by the website, www.022net.com, 7, Oct., 2012.

[2]See footnote 1.

[3]The information was extracted from the news report made by the website, www.travel.china.com.cn, 23, April, 2013.

Such thinking has governed the destinations' marketing planning for many years, and the action implication is that a standard packaged tour is good enough to accommodate the outbreak of Chinese travelers. The only thing a destination needs to do is to just reallocate their current resources a little bit via the incorporation of more Chinese elements (e.g., more Chinese servicing staffs to be employed, Chinese guidelines shown in the attractions, etc.) so as to absorb the expanding pool of Chinese visitors.

Obviously, the above marketing approach is no longer workable in the recent times because of the following reasons:

- More and more Chinese tourists give up the travel mode of packaged tour but employ free independent traveler (FIT) approach to hope for enjoying a more in-depth experience of each destination.
- The rise of middle class in China leads a change of the outbound Chinese tourists' demographics. In the past, tours of retired elderly and incentive tours were the major sources of Chinese tourists for the Western destinations. Nowadays, more and more young people are able to afford outbound trips, and they expect to experience the "uniqueness" of each destination. To a certain extent, these young customers want more tailor-made products and get in-depth experience in the exotic culture.
- The rapid development of social media (e.g., Weibo and WeChat) in China is also a dramatic force to help shape the market trend. The key is that most Chinese tourists obtain "consumer education" from such social media. The sharing of past travel experience via these media is recognized as the most trustworthy information source of judging the "value" of a certain destination. In one recent focus group we conducted in Beijing, one young professional told me as follows:

  Before making a travel decision, I always read over 10–15 articles posted by the past travelers on Weibo – these people may be my friends or the friends of my friends.

The implication is that most young Chinese tourists are no longer the passive audience to be influenced only by advertising campaigns or travel agent's persuasion; instead, they may be some "informed public" who have already obtained voluminous information and may have quite abundant "virtual" experience of visiting a destination.

To address the above emerging trends, the travel suppliers must re-define the Chinese tourists as a "heterogeneous customer pool" in which different segments would need different tailor-made products. Moreover, these customers would become more and more pickier, as well as demanding, because they have already formed certain expectations of each destination before making the visit.

21.2.2 DEMAND SIDE: GOING SOPHISTICATED AND EXTENSIVE

In "China Tourism 1.0," most outbound Chinese tourists were urban residents in key cities (e.g., Shanghai, Beijing, or Guangzhou). They required standard packaged travel products to let them establish the first impression of the destination. They were very eager to experience the different aspects of the Western society and then use this experience to help reinforce their social status in their social network. In short, it is a kind of "conspicuous consumption," and outbound travel is more or less taken as a cultural capital for showing off a person's class identity.

In the new era, "China Tourism 2.0," the Chinese tourists display quite different facets:

- Gateway cities in China are no longer the only source of Chinese tourists. Destinations have started to realize that more and more fresh Chinese tourists who visit their destinations are actually those who come from the second-tier and third-tier cities. In the *Annual Report of China Outbound Tourism Development 2013*, China Tourism Academy indicates that the future tourist momentum would certainly be driven by the second and third-tier cities in the middle and Western region. However, up to this moment, no destination owns good knowledge about these people/regions.
- On the other hand, the emerging middle-class in the first-tier cities have already taken outbound travel as a life habit. A housewife in Guangzhou shared with us as follows:

  Outbound travel is a must for my family because my husband and I strongly expect that such travel experience is able to let our child widen his horizon. Also, we would go to Austria again in the coming summer – it will be our third time to visit Vienna which is "the City of Music." We hope our child to develop his future career in music-related areas so that we trust the social atmosphere of Vienna must be beneficial for him.

In other words, the first-tier cities would supply more and more repeated tourists to different destinations. These tourists would carry quite different expectation and motivation to visit a destination on a repeated basis. The above-mentioned example indicates that a child's future could determine a family's interest in a destination. Such travel motivation cannot be found in "Tourism 1.0."

Anyway, these sophisticated tourists would form a big puzzle in front of different destinations. That is, we have to study these sophisticated repeated tourists in a more comprehensive manner and find out the exact triggers to drive their travel interest.

- One more obvious trend to be found in the demand side of "China Tourism 2.0" is that the consumers are hungry for reliable information of each destination. The growing popularity of FIT mode for travel forces potential tourists to chase information via sources other than travel agents. They dislike packaged tours and need something unique from the general information received by other ordinary tourists. On the other hand, scandals about the misbehaviors of certain travel agents (e.g., forced shopping) lower the public trust in the local travel suppliers. All in all, the consumers tend to trust their personal network and the information shared from social media. This emerging pattern of information gathering drives the necessity of the destinations to review the effectiveness of their current promotional campaign and choice of advertising channels.

21.3 CHALLENGES OF AND IMPLICATIONS FOR MARKET RESEARCH IN CHINA OUTBOUND TOURISM 2.0

Based on the above description of the macroenvironment of the latest market situation, all destinations have to study the changing Chinese tourists in a more thorough and continuous way. Certain major emerging requirements for market research in the era of China Outbound Tourism 2.0 would be the following:

- ➢ As the Chinese tourists have formed a sustainable revenue stream to every destination, it is a must for the destinations to fully grasp the tourists' satisfaction with their travel experience.

In the last decade, many traveler satisfaction studies were carried out by different destinations. However, such studies are always in the nature of one-off cross-sectional survey. Such studies always require the tourists to recall their travel experience which would have occurred 1 year ago or even longer. This does not mean that this measurement is wrong, but we certainly need some other indicators, which allow us to tap tourists' immediate experience after visiting a destination.

This is quite important because, as mentioned, the Chinese tourists' characteristics and expectations are ever-changing. Therefore, research data about them must be "fresh" enough so as to enable the research users to review and fine-tune their strategies or allocation of resources in the most reliable, timely, and relevant way.

In addition, these indications of satisfaction should be able to reflect whether a tourist's pre-travel expectation can be met or not. This pre- and post-comparison would offer a fairer way to evaluate a destination's tourism service.

Therefore, the challenge for market researchers is to invent a precise research design and measurement tools to help visualize whether and how recent travel experience satisfies a Chinese tourists using the tourist's pre-trip expectation as a benchmark.

- ➢ In terms of target audience, destinations should be aware that focus must be put on the young middle class in the first-tier cities and those up-market consumers in the second- as well as third-tier cities. Destinations should establish benchmarking statistics about these different groups of target audience, including their demographics, psychographics, as well as travel behavior.

We need to keep tracking the expectation and imagery perception of each destination among this large pool of emerging potential tourists over the time.

- ➢ When destinations realize the need to tailor-make more different products for different tiers of Chinese tourists, product tests would become a need for all marketing teams of the destinations. Such product tests are expected to offer reliable findings about the marketability of a certain travel product based on the Chinese tourists' real experience. Then, how we can use an experimental design in this regard would become a core challenge.
- ➢ In addition, we must keep studying the effectiveness of all different marketing channels in China, especially about how the social media is influencing a Chinese tourists' choice. This implies the need to use certain diary method or similar approaches to better record how a social media affects a potential tourist.
- ➢ Finally, all destinations must keep "nosing" the upcoming policies and issues, as well as trends, which would influence the market; that is, a need to keep talking with the key opinion leaders or industry stakeholders becomes a must.

21.4 LONG-STANDING OBSTACLES IN CONDUCTING RESEARCHES ON CHINESE OUTBOUND TOURISTS

Other than the above-mentioned emerging challenges, we should also pay attention to the following long-standing obstacles in conducting research on the Chinese tourists. These obstacles would anyway exist in the new era and cause difficulties to researchers, especially to Western researchers:

- ➢ Unavailability of comprehensive and reliable secondary statistics about the profile of Chinese outbound tourists – undeniably, we can get the total numbers of annual Chinese visitors to the United States or any other destinations from the national statistics bureau; however, more details, for example, a demographic breakdown or a differentiation between "packaged tour tourists" and "FIT tourists," are still quite difficult for us to obtain. In addition, even for some other national statistics, some crucial indicators look quite "unrealistic," for example, the annual income of the urban people. You can see this data of each city in the respective statistical yearbook, but the figure is always too low in comparison to your perception gained from each city. Obviously, Chinese people are used to underreport their income level.
- ➢ For example, in certain Western tourism researches or corporate/brand imagery studies, one key indicator which is always used for measuring the destination's/ corporate's/brand's performance is the likelihood of a customer to recommend the relevant destinations/brands, etc. The question's format is usually defined as: *"How likely would you recommend this destination/*

brand/corporation¾please use a 10-point scale to rate: 1 – very unlikely, 10 – very likely?"

This question may be quite understandable for the Western respondents, but it is very difficult for a Chinese person to offer answer to this question. Based on some previous qualitative studies I did among Chinese consumers, "to recommend something" is not an easy thing within Chinese culture. Chinese people would not "recommend" anything if they don't know "who is to take this recommendation" or "they have gained very in-depth understanding about what is to be recommended." "To recommend" is something related to the establishment of "Guangxi" (social relationship). Chinese people need to consider many aspects before making such a recommendation. The implication is that, in a survey, this question of "likelihood to recommend" may always gain an answer of "not sure" and "don't know."

What I want to pinpoint is not about how to improve a questionnaire's wording. My concern is that researchers should not apply the global measurement tools directly in research in China just via a literal translation of the scales. Instead, we have to "make sense" of a measurement tool or survey finding via different qualitative studies which should be done before and after a quantitative survey.

➢ Low response rates of random method. First of all, random door-to-door household visit is no longer a feasible method in China because increasingly more urban Chinese are residing at apartments or communities in which there are very strict security control to avoid any stranger's visit.

  Second, random telephone interview is also facing greater and greater difficulty in getting the Chinese' consent for taking up the interview. The major reason may be due to the fact that a lot of Chinese telemarketers are currently using random cold calls to "hook" customers, but most of the Chinese do not like such calls. The outcome is that the Chinese tend to prefer hanging up when receiving any cold call.

  On the other hand, we have found that a telephone interview over 20 minutes is too long for a Chinese and a respondent's fatigue would lead to unreliable response, especially for open-end questions.

  Therefore, many researchers have decided to rely upon the online panels to get representative sample. However, most of the existing panels in China show no sufficient reliability and representativeness. For example, one key issue is that most panels fail to offer sufficient responses from middle-to-old-aged people. Therefore, in the future, market researchers must try their best to find out and/or explore a new way of collection of representative data.

21.5 RESEARCH PROGRAM IN CHINA TOURISM 2.0: HOW TO FIND OUT THE MISSING PIECES OF A JIGSAW PUZZLE?

To address all the above-mentioned issues of conducting tourism research in era 2.0, I don't think there is an orthodoxical approach that is able to offer a "best practice" or a panacea. Instead, as a frontline research practitioner, what I can do should be to outline different possibilities of methodology, which can help find out the missing pieces of our jigsaw puzzle:

21.5.1 USE OF QUALITATIVE STUDY TO RE-INTERPRET THE MOST IMPORTANT EMERGING MARKET SEGMENTS

As mentioned above, the 2 market segments among the Chinese consumers would become the most important sources of outbound tourists is shown in Fig. 21.1.

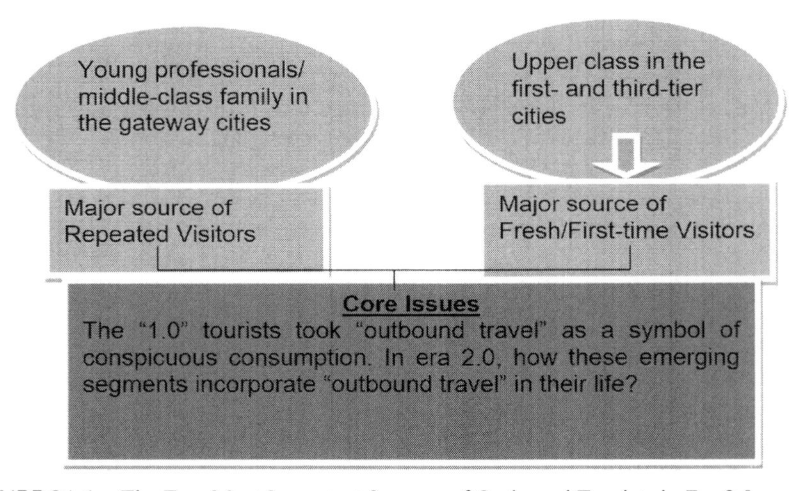

FIGURE 21.1 The Two Most Important Sources of Outbound Tourists in Era 2.0.

Most destinations should have carried out many different cross-sectional surveys to understand the whole market; therefore, researchers should be able to sort out the demographics and psychographics of these 2 emerging segments in an easy way. Based on this, the upcoming core issue is anyway not about whether we have sufficient quantitative data to help "profile" these segments in a static way. On the contrary, the most difficult task or the missing piece of our jigsaw puzzle is how these specific consumers make sense of "outbound travel" and position it in their life. What a researcher needs to do is to accurately portray how these consumers formulate their decision-making process in determining their product choice. This understanding would also cover how new social media and traditional promotional channels affect the consumers' thinking. All in all, we need a more "dynamic" methodology, personally speaking. I would highly recommend the 2 possible approaches shown in Fig. 21.2.

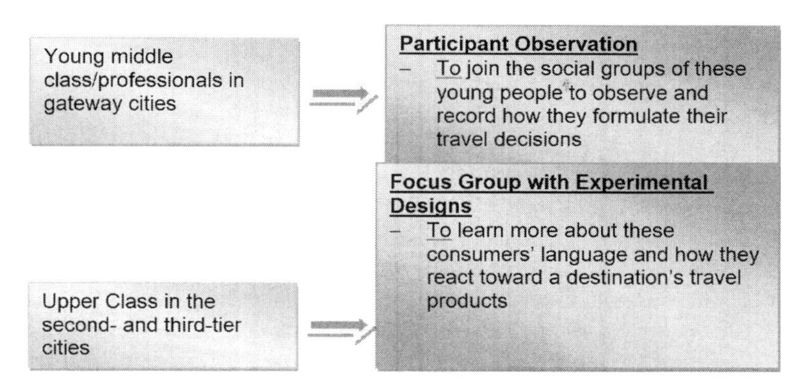

FIGURE 21.2 Recommended Research Approaches To Be Used for Studying the 2 Major Segments of Tourists.

Regarding young middle class and professionals in gateway cities, their most mysterious facet is what their purchase decision-making social processes are. Researchers should go into their real-life environment and processes to observe, record, as well as explore how consumers react with different destinations' imageries, promotions, and word-of-mouth commentaries. I believe that such participant observation which is focusing on dynamic process is certainly able to help add great values to our traditional quantitative survey. That is, young Chinese middle class/professionals can be visualized in terms of not only numeric figures but also their real interactions with the destinations, as well as all other possible influences in driving or preventing their decision-making.

For the study on the fresh visitors in the second- and third-tier cities, these people should not be sophisticated as their counterparts in the gateway cities. What the destinations need is to check out their true preference with reference to different proposed products. Therefore, I recommend using focus groups to help make sense of their consumer language, as well as get them exposed to different test stimuli (e.g., video clips, concept boards, or so on). It must be more fruitful if we can incorporate more elements related to an experimental design into this qualitative approach to enhance the findings' reliability. For example, we may need to set up control groups to benchmark the stimuli's performance.

21.5.2 INDIVIDUAL-BASED PRE-TRIP AND POST-TRIP SURVEY

Currently, most destinations like to conduct annual tracking study to monitor the change of Chinese tourists' preference and perception over the time. That is, in each year, a fresh sample of past visitors would be drawn and interviewed. It follows that the data collected in this year would be compared with that of last year. Such annual tracking survey is undeniably very important for the destination's reference in reviewing its efforts paid in the year for attracting Chinese tourists, especially monitoring the change in terms of overall satisfaction index or similar indicators. However, such comparison between 2 independent samples (sample of "this year" and sample of "last year") would become a bit unfair if the sampling error is great. Please remember what I had mentioned before, an ever-lasting issue in China, that is, the low response rate in random method which would further undermine the validity and reliability of such data comparison. To address the issue, I recommend to conduct an individual-based comparison to measure the tourists' satisfaction with the destinations via the comparison between their expectation formulated before the trip and the actual satisfaction gained immediately after the trip. That is, we need to interview the same tourists twice: once before the trip and once after the trip. This comparison score would offer fairer evaluation of a destination's performance because in comparison with the annual tracking study, this pre/post-trip measurement is more capable in guaranteeing that the respondent's perception change is only due to the trip's experience.

21.6 CONCLUSION

In this chapter, I have given my recommendations on how to enrich the current research practice so as to help researchers get a precise picture of "China Tourism 2.0."

In any case, we need to be more open and creative in constructing more locally relevant research methods to help establish our understanding of the Chinese tourists because China keeps undergoing its kaleidoscopic transformation, but the Chinese people's wealth has been commonly recognized as the major income source for many different industries across the world.

REFERENCES

China Tourism Academy. *Annual Report of China Outbound Tourism Development – 2013*; Tourism Educations Press, Beijing, June 1, 2013.

Chinese Academy of Social Sciences. *Green Book of China Tourism*; No. 12, Social Sciences Academic Press, Beijing, January, 2014.

CHAPTER 22

CHINESE OUTBOUND TOURISM: A CRITICAL INQUIRY

RICH HARRILL, XIANG (ROBERT) LI, and HONGGEN XIAO

CONTENTS

22.1 INTRODUCTION

Driven by relaxed government policies and a growing middle class, the emerging Chinese outbound tourism market is understood by many academics as one of the most significant in the history of tourism (Li et al., 2009). The United Nations World Tourism Organization estimates that China will generate 100 million outbound tourists by 2020 – making China the fourth largest source market.

Arlt (2006) has described China's Approved Destination Status as a carefully conceived method for ensuring that its outbound tourism conforms to the country's social and economic policies. With little doubt, the Chinese outbound market will exercise profound influence on social, political, and economic conditions inside and well beyond China's borders. Despite recent growth of empirical work on the topic, there has been comparatively little theoretical or philosophical reflection of this emerging market in tourism studies.

In his article, "Tourism: A Critical Business," John Tribe (2008) identifies the rise of China as a tourism-generating and -receiving country as a research topic worthy of critical examination. He describes *critical tourism* as based on numerous sociological and philosophical traditions, including postmodernism and the Frankfort School's Critical Theory. Defined by Bernstein (1976),

> *Critical theory aspires to bring the subjects themselves to full self-consciousness of the contradictions implicit in their material existence, to penetrate the ideological mystifications and forms of false consciousness that distort the meaning of existing social conditions (p. 182).*

Taken together, these traditions offer a poststructuralist alternative to tourism research as a narrowly positivist, technical, and comprehensively rational endeavor (Aitchison, 2006, 2001a, 2001b, 1996; Ateljevic et al., 2007; Ateljevic and Doone, 2002; Bianchi, 2010; Bramwell and Lane, 2011; Franklin and Crang, 2001; Fullagar, 2002; Gale and Botterill, 2006; Hollinshead, 2006, 1992; Jamal and Everett, 2004; Pretes, 1995; Pritchard et al., 2011; Mellinger, 1994; Tribe, 2005). Comparatively Tribe (2007, 2008, p.251) describes critical tourism as a politically engaging, interpretive, and strategically incremental activity with the goal of "extending, supplementing, and challenging the dominant discourses that pervade the management and governance of tourism."

Thus, a critical interpretation may afford researchers an enhanced understanding of tourism as a productive and empowering activity for Chinese outbound visitors, while simultaneously revealing tourism as an ideological tool used by China, the United States, and other "tourism states" (Hollinshead, 2010) to suppress, distort, or exclude other worldviews encountered by visitors (Hollinshead, 1999).

The image of the abyss provides a helpful metaphor for researchers seeking explanations of tourism beyond social-psychological motives extending to the international social, cultural, and political implications (Harper and Stein, 1995). Tourists, on arriving at their destinations, may experience social and psychological dissonance between the conceptualizations of self and the new worldviews they encounter. While Chinese tourists use tourism for the purpose of empowerment (Hollinshead, 2010; Swain, 2009), they may also encounter a kind of "postmodern abyss" defined as a disconcerting lack of community, including a perceived absence of morality and values, in places they visit. For researchers, the abyss may describe the disjunctive relationship between postmodernist research and its relevance to practitioners and policymakers working under modernist assumptions. As noted by Beauregard (1991) in the context of urban planning,

> *The postmodern perspective in its various guises thus undermines the intellectual base of the modernist planning project. The modernist planning project is thereby suspended between a modernism whose validity is decaying and reconfiguring, and, a postmodernism whose arguments are convincing yet discomforting. As planning theorists we have failed to formulate a response and failed to work with practitioners to move the planning project from its ambivalent position. (p. 193)*

The purpose of this chapter is to provide a deeper and richer understanding of interpretation of sought-after activities and memorable travel experiences of Chinese outbound tourists to the United States. Theoretically drawing upon Habermas and Foucault, this chapter explores a critical approach to analyzing the Chinese outbound market and identifies the influences of ideology on Chinese outbound travel experience.

22.2 A CRITICAL INTERPRETATION OF CHINESE OUTBOUND TOURISM

22.2.1 HABERMAS' THEORY OF COMMUNICATIVE ACTION

German philosopher Jurgen Habermas' Theory of Communicative Action (1981) is frequently cited in critical tourism studies (Spracklen, 2009; Tribe, 2007, 2008). Communicative action differs from social construction, in that dialogues or "speech situations" among participants– whether tourists, residents, or researchers – can be construed as social and political activity with a direct influence on community development and tourism management. According to planning theorist John Forester (1993), "Communicative action is always interaction between persons, thus political in a very broad sense, reproducing, whether maintaining or altering, social and political relations" (p.24). Methodologically, this theory allows researchers to appreciate the social and political character through which any management plan or policy is achieved (Forester, 1989). Open communications about tourism among multilateral "targets" and "agents" (Cheong and Miller, 2000) rather than unilateral "hosts" and "guests" can be used to enhance what Habermas would term "deliberative democracy," underlying the notion that societies open to tourism are probably, but not necessarily, democratic for residents.

As a methodological approach, communicative action calls for researchers to analyze how tourism studies and management plans are shaped through actions such as "challenging, criticizing, announcing, exposing, threatening, predicting, promising, encouraging, explaining, insulting, forgiving, presenting, recommending, and warning, among others" (Forester, 1993, pp. 24–25). Through interpretive research techniques such as unstructured interviews, focus groups, participant observation, case studies, hermeneutics, and literary criticism (Tribe, 2008), communicative actions on the part of residents, tourists, and policymakers demonstrate how language can conceal and support the discourses of power and domination.

In the contexts of an ongoing global transformation and the still-unfolding East–West contention over values, the new look of international tourism, for example, the emergence of Chinese outbound tourism, serves as a good instance of cross-cultural interactions to observe such communicative actions. Inevitably, the language used by researchers about themselves, residents, and tourists is more than just talk – researchers "bring to bear the moral imagination and shape the moral imagination of others" (Forester, 1996, p.220). In sum, this kind of communicative methodology allows researchers to identify the gaps, silences, and misconstructions in research, thus assisting them to "speak the truth *of* power and facilitate the speech of the *powerless*" (Tribe, 2005, p.377).

22.2.2 FOUCAULT'S PANOPTICON

Discursive power was the primary intellectual concern for the late French philosopher Michel Foucault, whose work offers a more substantive, if not bleak, critique of contemporary tourism (Bramwell and Meyer, 2007; Cheong and Miller, 2006; Hollinshead, 2010, 2008, 1999; Jordan and Aitchison, 2008; Mordue, 2005; Wearing et al., 2010; Wearing and McDonald, 2002; Wang, 2006). Throughout his body of work, most notably *Discipline and Punish: The Birth of the Prison* (1975), Foucault believes some types of power emanate from surveillance, often discretely disguised as social and political norms by those who hold the reins of power. Both Hollinshead (1999) and Jordan and Aitchison (2008) have noted the influence of Foucault's work on Urry's (1990) seminal notion of "tourist gaze." According to Jordan and Aitchison (2008),

For Urry, the power inherent in the gaze of the tourist on local sites and people displays parallels with Foucault's disciplining power of surveillance. Where people and places are subjected to touristic surveillance, argues Urry, there is an inherent (and often problematic) power relationship established. It should be noted that in subsequent work, Urry moves beyond a focus purely on the visual gaze (criticized by some as being rather passive or at least one-dimensional conceptualization of the sensory and embodied practice of tourism) to incorporate a more active, embodied consumption of tourism on the part of tourists. (p. 333)

In the context of critical tourism, the notion of gaze is effectively nuanced and multilayered (Hollinshead, 1999), including observation, consumption, and repression or control. Spatially, these power relationships encompass regions, state, and nations (meta-gaze) and are embodied in management plans and policies. Increasingly, these relationships at the national and international levels are managed and controlled by an electronic surveillance (virtual gaze) used to keep residents in and tourists out, or alternatively, to shroud resorts and expatriate enclaves from locals (Morgan and Pritchard, 2005).

Cheong and Miller (2000) take the influence of Foucault on Urry a step further, noting the similarities between surveillance, the concept of gaze, and Foucault's description of Bentham's Panopticon, an architectural model for an 18th-century prison:

> *The Panopticon is designed in such a way that whomever is at the center sees everything, enabling one to watch many. In absence of a chief inspector in the center of the edifice, any agent – for example, a member of the prisoner's family, a friend, a visitor, and even a servant – can operate the system. Hence, anybody can watch and be watched by someone depending on where they are placed. Important to this spatial arrangement is the place of the individual. The agents construct the gaze as they observe the target. (emphasis added, p.377)*

Comparatively, then, Habermas offers a more hopeful vision for society – using communicative action as a methodology to uncover ideological distortions, while Foucault depicts a gloomier view of a panoptic world that inevitably distorts and where there is little tourism researchers can do to free tourism as an empowering and self-actualizing human activity from the grasp of hegemonic political power (Harper and Stein, 1995, p. 242).

22.2.3 THE POSTMODERN ABYSS

For some travelers, home countries such as China may represent a repressive social and political order from which to escape (Panopticon) to a "free," individualistic society fraught with ethical and moral dilemmas and ambiguities (the abyss). Such desires for diversionary modes of experiences (Cohen, 1979) or to "seek and escape" (Iso-Ahola, 1982; Mannell and Iso-Ahola, 1987) are well-known among researchers, including, but not limited to, strangeness versus familiarity (Cohen, 1972), being away to versus being away from (Wapner et al., 1976), freedom to and freedom from (Neulinger, 1974), anomie versus normlessness (Dann, 1977), consistency versus complexity (Mayo and Jarvis, 1981), and approaching versus avoidance (Mayo and Jarvis, 1981).

However, many of these theories and approaches, borrowed from the positivist social psychology literature, fail to capture the emotional and even emancipatory motivations for escaping the watchful eye of the Panopticon. Jordan and Aitchison (2008) integrate both positivist and critical interpretations: "When taking a holiday, the assumption is that a person is transported physically and emotionally away from his or her home and from the regulatory effects that social norms act to determine and constrain 'normal' behavior" (p.334). When viewed in this context, the value of tourism does not lie entirely in open communication or exchange of ideas, but also in a deeper-seated, healthy curiosity about the outside world, including leisure activities that may deviate from the tourist's social and cultural norms, including cursory encounters with vice and sexuality.

In the abyss, the lack of community can reveal individuals to be self-centered and self-serving and the activities of organizations and corporations to range from obscure to manipulative. As Cheong and Miller (2000) noted, "power is invisible in tourism when it is conceived after the image of rulers and politicians" (p.378). The metaphors of the Panopticon and the abyss can be used as a flexible critical framework for evaluating and questioning the impact of muted discourses and invisible surveillance on tourist exchanges from the local to the international. Following Forester (1993), "we must explore the strategies organizations and their members use to reproduce beliefs, trust, and attention in highly politicized environments. These questions lead immediately to those of the maintenance and vulnerability of hegemonic power" (p.11).

Tribe (2008) alludes to critical tourism studies as reflexive and transformative approaches to interpreting the meaning of tourism and delving into the tourist self. Using critical inquiry to understand the experience of Chinese outbound tourists to the United States, this chapter explores and reveals much about how rising hegemonic power of China is remolded in the travel discourse and how tourists' yearning for community, justice, and sustainability is likely to reshape a West that embodies the vacuous, postmodern abyss. Future critical studies should seek a balance between inquirer positions and participant perspectives, the analysis and interpretation of focus groups, for example. These studies should also seek a cross-cultural instance that aims to balance methodology and critique (Berry, 1979) forming a bridge between the hopefulness of Habermas' communicative action and the despair of Foucault's power relations. Notably, this should also be acknowledged as a limitation of the critical inquiry.

22.2.4 OUTBOUND TOURIST NARRATIVES

We theorize that Chinese tourists frequently employ gaze as a means of empowerment, express curiosity about alternative worldviews, and explore the boundaries of communal norms and values. In these future research situations, researchers will ask focus groups to explain not only how they perceived tourism destinations, but also what they had experienced or wanted to experience at a destination. As an example of what behaviors might be central to this critical inquiry might be participants' reflections about gambling, sexuality, and gun culture (or ownership of handguns) from their perceptions of, and/or recollections from, outbound travel.

GAMBLING

Gambling is pursued by many Chinese outbound tourists, and Chinese society in general has experienced this activity at specific points in time and history. According to Lam (2005), the Yuan (1271–1386), Ming (1386–1644), and Qing (1636–1911) dynasties implemented nationwide gambling bans without much success. The punishment for offenders then included caning, exile, and even execution. Toward the end of the Qing dynasty, wealthy merchants and investors were regular patrons of gambling dens. Lam (2005) noted that gambling control was most successfully implemented from the early years of the People's Republic of China to the Cultural Revolution of the 1960s and 1970s. During this time, gambling activities among Mainland Chinese practically ceased to exist. However, since Deng's reforms in 1978, gambling activities have re-emerged and as a leisure activity have become an important attraction for some Chinese outbound tourists. Still, many Chinese are fascinated by the ubiquity of gambling opportunities in destinations like Las Vegas. Further, for some travelers, gambling is often associated with something to be confirmed, compared, and even contrasted in their outbound travel experience, comparing Las Vegas, Macao, and Hong Kong.

Other travelers define their role as an observer rather than an active gambler. Here, tourists enjoy the role of the voyeur and clearly regarded their own ability to gaze as empowering (Jordan and Aitchison, 2008). For many Chinese, visiting Las Vegas and comparing that destination with gaming facilities in Macao provided an opportunity to reconcile worldviews by comparing China with the United States. However, other Chinese tourists who actively participate in gambling may be attracted to experience risk or to compensate for highly controlled social, political, and even physical environments back at home.

SEXUALITY

Sexuality is a basic human trait and as such is the basis of Foucault's 3-volume *History of Sexuality* (1992, 1984, and 1976). In this work, Foucault makes explicit connections between surveillance and discouraged or repressed sexual behavior. Although sex tourism is often regarded as one of the most socially repugnant forms of tourism, sexuality in this critical context may be associated with freedom of expression and empowerment for both agents and targets (Ryan and Martin, 2000; Ryan and Kinder, 1996). While beach visits are sometimes associated with surprises for Chinese outbound tourists, it could be very intruding for local (typically Western) beachgoers to have photographic sightseers around. On the other hand, some tourists indicate that nudity was desirable in some beach destinations. Finally, outbound tourists' curiosities about the tolerance or acceptance, and even legality of different sexual orientations, are notable, for example, same-sex marriage, strip clubs, sex stores, and sex museums.

As opposed to being personally engaged in commercial sex, Chinese outbound tourists seem to have a genuine interest in having exposure or access to such "views or sceneries." Like nudity beach, such experiences are often associated with "freedom or an unconstrained state" available and affordable through outbound tourism. Nonetheless, tourists have also expressed concerns about the danger or risks associated with the abyss, for example in their views toward prostitution and sexual violence or abuses.

GUN CULTURE

China has some of the most severe gun control laws in the world and has long imposed a ban on gun ownership. Since 1966, the government has prohibited private production, sales, transport, possession, and import or export of bullets and guns. Possessions or sales of firearms will result in punishments ranging from a minimum of 3-year imprisonment to death penalty (Moxley, 2010). Severe prohibitions as such tend to trigger curiosity about the gun culture in the West. For some tourists, gun shops can be a point of interest for Chinese outbound tourists at US destinations.

Fascination with gun culture may be ascribed to the glorification and hyper-reality of violence in American books and movies now distributed widely in China and readily available on the Internet. Perceived positively or negatively, gun ownership may be seen as an expression of autonomy and self-determination. Some outbound Chinese tourists see problems with the United States' lack of gun control (as echoed in repeated campus shooting tragedies). The concerns about gun ownership have also been linked to visitors' shopping safety.

22.2.5 CRITICAL INTERPRETATION

Notwithstanding the fascination and reservation displayed by some tourists, many Chinese potential outbound tourists consider the United States as their dream destination (Burnett et al., 2008) because they view it as a futuristic projection of China, where it is believed that citizens could enjoy a full range of individual and social liberty and yet still manage to exercise tremendous political and economic power on the world stage. Multiculturalism is associated with diverse economic skills and abilities rather than social and cultural fragmentation. However, those who have actually visited the United States often report back to friends and families that US citizens are not insulated from the negative externalities caused by these liberties. Despite its oppressive and at time seductive nature, the Panopticon provides some degree of a safe haven and buffer against the chaos and disorder of the abyss. Iconic destinations such as New York and Los Angeles can seem especially disorderly to first-time Chinese tourists. For example, Mike Davis (1990) in "Fortress Los Angeles: The Militarization of Urban Space" provides an account of what tourists might find in Los Angeles' inhospitable postmodern architecture:

One of the simplest but most mean-spirited of these deterrents is the Rapid Transit District's new barrel-shaped bus bench, which offers a minimal surface for uncomfortable sitting while making sleeping impossible. Such "bumproof" benches are being widely introduced on the periphery of Skid Row. Another invention is the aggressive deployment of outdoor sprinklers. Several years ago, the city opened a Skid Row Park; to ensure that the park could not be used for overnight camping, overhead sprinklers were programmed to drench unsuspecting sleepers at random times during the night. The system was immediately copied by local merchants to drive the homeless away from (public) storefront sidewalks (p.161).

Like tourists from all over the world, Chinese come to the United States to sample high-quality consumer goods and temporarily participate in America's wasteful economy, without staying long enough to succumb to the country's social ills such as fiscal bondage and consumer debt. The same Confucian frugality that compels Chinese tourists to shop for bargains also makes them somewhat uncomfortable with a culture of consumer excess. Many incoming visitors make the connection between China's rise as a lender nation and the decreasing economic fortunes of borrower nations such as the United States. In sum, most Chinese tourists realize that with "big freedoms" come "big problems" associated with loss of self and cultural identity. Others, however, hurl themselves headlong into the abyss, enjoying America as a dreamlike, hedonistic playground – a hyper-real, postmodern, yet imperfect projection of a futuristic China.

It is notable that many of these desired activities were mentioned in proximity to references about freedom or iconic symbols of freedom, such as the Statue of Liberty. These interests, for example, gambling, have been studied by many leisure and tourism researchers under the rubric of purple leisure, dark leisure, or taboo leisure. Much of this literature is derived from the positivist traditions that treat these behaviors as a need to return to psychological equilibrium or one to achieve a higher level of satisfaction.

However, when taken too far, self-empowering activities and behaviors such as gambling, sexuality, and gun culture can also challenge state authority, hence the rather severe punishments for these acts in China. In the context of critical tourism, these experiences can become empowering acts, expanding the boundaries of self until they test the limits of social control. When practiced on a wider scale, these activities might serve to unravel social and political order that supports the economic and institutional foundations of power. Rarely, if ever, are these activities discussed in the research literature as having greater cultural or political implications outside the immediate leisure context. In the absence of critical discourse, to dismiss these activities as merely contributing to the abyss and requiring surveillance only serves the self-reifying logic of the powerful against the powerless.

On a simpler plane, many Chinese are simply curious about these activities that it is merely fun to experience them briefly as part of a guided tour. For example, many focus group participants noted that although they do not gamble, they would like to experience Las Vegas. These visitors are interested in Las Vegas as a highly permissive leisure environment to explore personal boundaries, challenging and expanding their own consciousness and worldviews. At its most basic, tourism is discussed in the popular literature as a highly personalized, transformative experience for tourists with few opportunities for bridging thoughts, desires, and actions beyond the individual or community realm.

Some visitors, on the other hand, would like to participate in these activities to the extent that they become the first step toward self-awareness and self-actualization. As conceivably interpreted by Schopenhauer (1966) and Nietzsche (2012), these acts serve as transcendent expressions of imagination and will. However, to attain greater significance, these actors must seek a collective voice and social expression (Unger, 2007). Indeed, this tentative first step may eventually serve as a route to individual and collective autonomy and emancipation. Finally, for many Chinese tourists, these activities may also lead to a better appreciation of their own culture and social order and even perhaps an inflated sense of nationhood or national pride.

22.2.6 BRIDGING THE ABYSS: CRITICAL PRAGMATISM

The abyss need not leave tourism researchers suspended and directionless. Because tourism is an applied field where experience is coveted, philosophical pragmatism has much in common with the industry's more positive aspects. In comparison with critical theory's emphasis on language and postmodern historical and cultural interpretations, pragmatism has been called the theory of experience – or "learning by doing" – an expression coined by pragmatic philosopher John Dewey (Dewey, 1925). However, because pragmatism in its purest form may be construed as crudely instrumentalist and overly technocratic, pragmatism informed by critical theory – *critical pragmatism* – may be helpful in bridging the postmodern abyss. For Forester (1993), pragmatism informed by critical theory offers a compromise to the vagaries of Habermas' ideal speech situations and Foucault's *noir* theoretical landscapes. According to Forester (1993),

The glass of critical theory is surely half-empty: We are still missing a good deal of guidance about the most fruitful ways to carry out empirical, historically situated, phenomenologically cogent, normatively insightful analyses. Here surely the balance must now shift from necessarily abstract cases, concrete attempts to work out the implications of a *critical communicative* framework in particular cases, specific analytical experiences on the basis of which our collective research abilities and judgments will develop. But the glass is also half-full: the *theory of communicative action* represents certainly the most systematic rethinking of action theory in the context of theoretical rationalization that is available today. (emphases added, pp. 13–14)

What Forester offers tourism researchers in critical pragmatism is a methodological bridge between tourism's positivistic, empirical paradigm and the interpretive paradigm represented by critical theory. As a case study in critical pragmatism, Chinese outbound tourism offers insight into how the specific and concrete voices, experiences, and actions of individual tourists – many leaving their country for the first time – in turn influence and shape the power of destination countries and conditions. This research suggests that Chinese outbound tourists exercise power abroad and at home: Their collective buying power has international economic impact and at the same time shapes material expectations and subsequently the goods, services, and activities available to them at home. Here, critical pragmatism is shown as just as concerned with power relations as Habermas' critical theory or Foucault's postmodern critiques, focusing on the power of individuals to exert institutional pressure and engage in institutional experimentation and reform. Exploration of Chinese outbound tourism may be *critical,* in that it is enlightening and emancipatory for researchers and participants, and *pragmatic,* in that the sheer numbers of Chinese institutions entering the marketplace have the capacity to reshape old and new global institutions for decades to come.

As a historically unique situation, Chinese outbound tourism market may be thought of as the first instigators, experimentalists, and ambassadors of global economic democracy. In the meantime, the emergence of Chinese outbound tourism, plus other rising source markets from Asia (e.g., India and South Korea), constitutes a challenge to the conventional/post-colonial frames of representation in marketing, researching, and interpreting international tourism via binary perspectives, in which Western perceptions of non-Western destinations dominate (Britton, 1979; Echtner and Prasad, 2003; Santos, 2004; Yan and Santos, 2009).

Recent pragmatic thought might be interesting to tourism researchers seeking new directions from critical pragmatism. For example, the inheritors of Dewey's legacy such as Cornel West and Roberto Mangeberia Unger have in recent years created a more politically informed version of pragmatism. In particular, West's (1989) brand of neopragmatism is instructive for conceiving tourism as a context for social and political action. At once critical and pragmatic,

> *This form of neopragmatism explodes the preoccupation with transient vocabularies and discourses. Instead it shuns any linguistic, dialogical, communicative, or conversational models and replaces them with a focus on the multileveled operations of power. This focus indeed takes seriously the power-laden character of language—the ideological weight of certain rhetorics and the political gravity of certain discourses. Yet it refuses to posit language and its distinctive features as a model for understanding other, nondiscursive operations of power such as modes of production, state apparatuses, and bureaucratic institutions. (p.209)*

Not only does West (1989) appear to warn social scientists not to abandon potent political economy critiques, but his own focus on the marginalized and disenfranchised may be instructive for researchers to explore the social and political impact of the differences between the new international tourist class and the classes (stratified by gender and race) that serve them. Indeed, as can be inferred from the above analysis of cross-cultural encounters of the outbound Chinese tourists, due to contention over values and power relationships (e.g., between East and West), emerging outbound tourism in Asia will inevitably involve a reverse and occasionally resistant gaze of the "other," which calls for new pragmatically critical approaches in future tourism studies. Any inadequacy in the understanding of the rhetoric of (new) international tourism will hamper the roles and goals of such target-agent relationships.

Social and political theorist, law professor, and later Brazilian government economic minister, Roberto Mangeberia Unger provides a neopragmatic bridge between the Northern and Southern hemispheres. In his 3-volume *Politics* (1987a, condensed as *Politics: The Central Texts* in 1997), Unger developed a social theory of human empowerment and social experimentation that avoids the ironclad laws of positivism, capitalism, Marxism, and other "deep logic" theories. According to Unger, these social theories reproduce *context-preserving* routines that hinder imaginative, *context-transforming* institutional experimentation. Interestingly, the author claims that developing countries have been the most successful in creating new institutions and alternative forms of government.

Unger (1987a, p.113) argues that non-Western countries have long begun to combine Western-style technology and non-Western varieties of work organizations in different ways of social organization. This chapter on Chinese outbound tourism suggests that in addition to work, some leisure activities favored by tourists may facilitate context-transforming institutional development. Further, researchers drawing on Unger might focus on experimentalist cooperation in hospitality and tourism found in "informal" economies composed of many different fluid and flexible service activities at diverse levels of society (Unger, 2007). In this way, "modes of production" are reintroduced and re-examined in a critical, pragmatic context without succumbing to the flawed logic of capitalist or Marxist meta-narratives.

Unger's (2007) neopragmatic critique may also be used to understand tourism's role in the global economy where "goods and capital are free to roam the world, yet imprison labor within the nations or within communities of relatively homogeneous nation-states" (p.179). Here, tourism research may play a role in exploring how open exchanges of social, cultural, and intellectual capital can create innovative and technologically sophisticated societies that are prosperous both materialistically and spiritually. With the assistance of analytical tools such as geographic information systems, a neopragmatic critique might focus spatially on the social, political, and economic implications of East-West tourism migrations (Chinese outbound tourism to the West) and China's internal urban-rural tourism in terms of how assets and attractions are developed, marketed, and branded for Western tourists.

22.3 CONCLUSION

In conclusion, this chapter focuses primarily on Chinese outbound tourists' expected perceptions, preferred activities, and overseas travel experiences. However, if critical inquiry is used to interpret these desires and preferences rather than explain how these desires and preferences are formed in the participants – an approach that would call for the use of both positivist and interpretive paradigms in a single study as conveyed by critical tourism research. The authors believe that such a syncretic approach would prove extremely valuable to understanding both the behavioral and philosophical underpinnings of Chinese outbound tourism. Indeed, spanning positivist and interpretive paradigms, critical pragmatism may provide an appropriate theoretical basis for mixed-methods research due to its emphasis on criticality of social construction that best produces desired outcomes within critical contexts (Pansiri, 2006). Following Bernstein (1976, p. 235), an adequate social and political theory must be *empirical, interpretive,* and *critical.*

Critical studies of Chinese outbound tourism represent a means of interpreting China within a myriad of contexts, attaching personal voices, experiences, and histories to specific situations and encounters. However, for all its promise, critical tourism is vulnerable to criticism as yet another static worldview of meta-narrative trap. Still, critical studies can raise the level of theoretical debate in tourism as it did in planning and public policy some 30 years ago. From these social sciences, critical studies advanced theory while addressing the widening gap between theory and practice. Further, this approach provided substance to numerous social and political communities liberated by the structuralism and functionalism of the 1960s and 1970s. It remains to be seen if these same communities – now multiplied and fragmented by race, gender, income, and even technology – can be collected, providing the participation and solidarity necessary for implementing tourism as a sustainable management strategy.

Critical tourism studies can assist in creating discursive ground between the iconic China of the Great Wall and the reality of China as a multifaceted and modernized society. However, tourism's potential may not be reached in critical tourism studies without appropriate methods, vocabularies, and symbols. Although only a simplification of complex phenomena, the Panopticon and the abyss image is a potentially useful metaphor for understanding tourism as a transnational and transactional phenomenon. For example, the image may be appropriate to interpret travel from the country to the city – from China's less-developed inland to the more commercialized and Westernized eastern coast (Li et al., 2009). Fundamentally, however, the metaphor gives voice to individual experiences in relation to tourism states. For some, China means economic opportunity. For others, the rise of China represents a shift in world geopolitics. For many, however, China's inbound and outbound tourism represents the opportunity to observe and participate in a unique and fascinating culture with a long and influential history in the human journey.

REFERENCES

Arlt, G. *China's Outbound Tourism*; Routledge: New York, 2006.

Aitchison, C. The critical and the cultural: explaining the divergent paths of leisure studies and tourism studies. *Leisure Stud.* 2006, 25, 417-422.

Aitchison, C. Gender and leisure research: the "codification of knowledge." *Leisure Sci.* 2001a, 23, 1-19.

Aitchison, C. Theorizing other discourses of tourism, gender, and culture: can the subaltern speak (in tourism)? *Tourist Stud.* 2001b, 1(2), 133-147.

Aitchison, C. Patriarchal paradigms and the politics of pedagogy: a framework of feminist analysis of leisure and tourism studies. *World Leisure Recreat.* 1996, 38(4), 38-40.

Ateljevic, I.; Pritchard, A.; Morgan, N. *The Critical Turn in Tourism Studies*; Elsevier: London, 2007.

Ateljevic, I.; Doone, S. Representing New Zealand: tourism imagery and ideology. *Ann. Tourism Res.* 2002, 29, 648-667.

Beauregard, R. Without a net: modernist planning and the postmodern abyss. *J. Plann. Educ. Res.* 1991, 10(3), 189-194.

Bernstein, R. *The Restructuring of Social and Political Theory*; University of Pennsylvania Press: Philadelphia, 1976.

Berry, J. Research in multicultural societies: implications of cross-cultural methods. *J. Cross Cult. Psychol.* 1979, 10(4), 415-434.

Bianchi, R. The "critical turn" in tourism studies: a radical critique. *Tourism Geogr.* 2010, 11, 484-504.

Bramwell, B.; Lane, B. Editorial: crises, sustainable tourism and achieving critical understanding. *J. Sustain. Tourism*, 2011, 19, 1-3.

Bramwell, B.; Meyer, D. Power and tourism policy relations in transition. *Ann. Tourism Res.* 2007, 34, 766-788.

Britton, R. The image of the third world in tourism marketing. *Ann. Tourism Res.* 1979, 6, 318-329.

Burnett, T.; Cook, S.; Li, X. *Emerging International Travel Markets, 2007 ed.*; Travel Industry Association: Washington, DC, 2008.

Cheong and Martin. Power and Tourism: A Foucauldian Observation. *Ann Tourism Res.*, 2000, 27(2), 371-390.

Cheong, S.; Miller, M. Power and tourism: a Foucauldian observation. *Ann. Tourism Res.* 2000, 27, 371-390.

Cohen, E. Toward a sociology of international tourism. *Soc. Res.* 1972, 39, 164-182.

Cohen, E. A phenomenology of tourist experiences. *Sociology*, 1979, 13, 179-201.

Dann, G. Anomie, ego-enhancement and tourism. *Ann. Tourism Res.* 1977, 4, 187-219.

Davis, M. *Fortress Los Angeles: The Militarization of Urban Space*. In *City of Quartz*; Vintage: New York, 1990; p 155-180.

Dewey, J. *Experience and Nature*; Open Court: Chicago, 1925.

Echtner, C.; Prasad, P. The context of third world tourism marketing. *Ann. Tourism Res.* 2003, 30, 660-682.

Forester, J. *The Rationality of Listening, Emotional Sensitivity, and Moral Vision.* In *Explorations in Planning Theory*; Mandelbaum, S., Mazza, L., Burchell, R., Ed.; Center for Policy Research, Rutgers University: New Brunswick, NJ, 1996; p 204-224.

Forester, J. *Critical Theory, Public Policy, and Planning Practice: Towards a Critical Pragmatism*; State University of New York Press: Albany, 1993.

Forester, J. *Planning in the Face of Power*; University of California Press: Berkeley, 1989.

Foucault, M. *The History of Sexuality, Vol. 2: The Use of Pleasure*; Penguin: London, 1992.

Foucault, M. *The History of Sexuality, Vol. 3: The Care of Self*; Penguin: London, 1984.

Foucault, M. *The History of Sexuality, Vol. 1: The Will to Knowledge*; Penguin: London, 1976.

Foucault, M. *Discipline and Punish: The Birth of the Prison*; Gallimard: Paris, 1975.

Franklin, A.; Crang, M. The trouble with tourism and travel theory. *Tourist Stud.* 2001, 1(1), 5-22.

Fullagar, S. Narratives of travel: desire and the movement of feminine, subjectivity. *Leisure Stud.* 2002, 21, 57-74.

Gale, T.; Botterill, D. A realist agenda for tourist studies, or why destination areas really rise and fall in popularity. *Tourism Stud.* 2006, 5(2), 151-174.

Habermas, J. *The Theory of Communicative Action, Vol. 1: Reason and the Rationalization of Society*. Boston: Beacon, 1981.

Harper, T.; Stein, S. Out of the postmodern abyss: preserving the rationale for liberal planning. *J. Plann. Educ. Res.* 1995, 14, 233-244.

Hollinshead, K. "Tourism state" cultural reproduction: the re-making of Nova Scotia. *Tourism Geogr.* 2010, 11, 526-545.

Hollinshead, K. Policing the World Through Tourism: Foucault and you, the Tourism-Judge, Proceedings of the 18th Council for Australian University Tourism and Hospitality Education Conference, Gold Coast (Queensland), Australia, February 11-14, 2008.

Hollinshead, K. The shift to constructivism in social inquiry: some pointers for tourism studies. *Tourism Recreat. Res.* 2006, 31(2), 43-58.

Hollinshead, K. Surveillance of the worlds of tourism: foucault and the eye-of-power. *Tourism Manag.* 1999, 20, 7-23.

Hollinshead, K. "White" gaze, "red" people – shadow visions: the disidentification of "Indians" in cultural tourism. *Leisure Stud.* 1992, 11, 43-64.

Iso-Ahola, S. Toward a social psychological theory of tourism motivation: a rejoinder. *Ann. Tourism Res.* 1982, 9, 1982, 256-262.

Jamal, T.; Evertt, J. Resisting rationalization in the natural and academic life-world: critical tourism research or hermeneutic charity. *Curr. Issues Tourism*, 2004, 7(1), 1-19.

Jordan, F.; Aitchison, C. Tourism and the sexualization of the gaze: solo female tourists' experiences of gendered power, surveillance, and embodiment. *Leisure Stud.* 2008, 27, 329-349.

Lam, D. *The Brief Chinese History of Gambling*; University of Macau: Macau, 2005 http://www.umac.mo/iscg/Publications/InternalPublications.htm (Retrieved January 15, 2011).

Li, X.; Harrill, R.; Uysal, M.; Burnett, T.; Zhan, X. Estimating the size of the Chinese outbound travel market: a demand-side approach. *Tourism Manag.* 2009, 31, 250-259.

Mannell, R.; Iso-Ahola, S. Psychological nature of leisure and tourism experience. *Ann. Tourism Res.* 1987, 14, 314-331.

Mayo, E.; Jarvis, L. *The Psychology of Leisure Travel*. CABI: Boston, 1981.

Mellinger, W. Toward a critical analysis of tourism representations. *Ann. Tourism Res.* 1994, 21, 756-779.

Mordue, T. Tourism, performance and social exclusion in "Old York." *Ann. Tourism Res.* 2005, 32, 179-198.

Morgan, N.; Pritchard, A. Security and social "sorting": traversing the surveillance-tourism dialectic. *Tourist Stud.* 2005, 5(2), 115-132.

Moxley, M. China's Gun Culture Grows. Asia Times [Online] 2010 http://chinaelectionsblog.net/?p=7309 (Retrieved January 11, 2013).

Neulinger, J. *The Psychology of Leisure*; Charles C. Thomas: Springfield IL, 1974.

Nietzsche, F. *Thus Spoke Zarathustra*; Simon & Simon: New York, 2012.

Pansiri, J. Doing tourism research using the pragmatism paradigm: an empirical example. *Tourism Hospit. Plann. Dev.* 2006, 3(3), 223-240.

Pretes, M. Postmodern tourism: the Santa Claus industry. *Ann. Tourism Res.* 1995, 22, 1-15.

Pritchard, A.; Morgan, N.; Ateljevic, I. Hopeful tourism: a new transformative perspective. *Ann. Tourism Res.* 2011, 38, 941–963.

Ryan, C.; Kinder, R. Sex, tourism and sex tourism: fulfilling similar needs? *Tourism Manag.* 1996, 17, 507-518.

Ryan, C.; Martin, A. Tourists and strippers: liminal theater. *Ann. Tourism Res.* 2000, 28, 140-163.

Santos, C. Framing Portugal: representational dynamics. *Ann. Tourism Res.* 2004, 31, 122-138.

Schopenhauer, A. *The World as Will and Representation*; Dover: New York, 1966.

Spracklen, K. *The Meaning and Purpose of Leisure: Habermas and Leisure at the End of Modernity*; Macmillan: Hampshire, UK, 2009.

Swain, M. The cosmopolitan hope of tourism: critical action and worldmaking vistas. *Tourism Geogr.* 2009, 11, 505-525.

Tribe, J. Tourism: a critical business. *J. Travel Res.* 2008, 46, 245-255.

Tribe, J. *Critical Tourism: Rules and Resistance*. In *The Critical Turn in Tourism Studies: Innovative Research Methodologies*; Ateljevic, I., Morgan, N., Pritchard, A., Ed; Elsevier: Amsterdam, 2007; p 29-39.

Tribe, J. The truth about tourism. *Ann. Tourism Res.* 2005, 33, 360-381.

Unger, R. *The Self Awakened: Pragmatism Unbound*; Harvard University Press: Cambridge MA, 2007.

Unger, R. *Social Theory, its Situation and its Task: A Critical Introduction to Politics – A Work in Reconstructive Social Theory*; Cambridge University Press: Cambridge, 1987a.

Unger, R. *False Necessity: Anti-Necessitation Social Theory in Service of Radical Democracy: Part I of Politics – A Work in Reconstructive Social Theory*; Cambridge University Press: Cambridge, 1987b.

Unger, R. *Plasticity Into Power: Comparative-Historical Studies on the International Conditions of Economic and Military Success: Variations on Themes of Politics – A Work in Reconstructive Social Theory*; Cambridge University Press: Cambridge, 1987c.

Unger, R. *Politics: The Central Texts*; Verso: New York, 1997.

Urry, J. *The Tourist Gaze*; Sage: London, 1990.

Wang, X. Travel and cultural understanding: comparing victorian and Chinese literati travel writing. *Tourism Geogr*. 2006, 8, 213-232.

Wapner, S.; Cohen, S.; Kaplan, B. *Experiencing the Environment*; Plenum: New York, 1976.

Wearing, S.; McDonald, M. The development of community-based tourism: re-thinking the relationship between intermediaries and rural and isolated area communities. *J. Sustain. Tourism*, 2002, 10(2), 21-35.

Wearing, S.; Wearing, M.; McDonald, M. Understanding local power and interactional processes in sustainable tourism: exploring village-tour operator relations in Kokoda Track, Papua New Guinea. *J. Sustain. Tourism*, 2010, 18(1), 61-76.

West, C. *The American Evasion of Philosophy: A Genealogy of Pragmatism*; The University of Wisconsin Press: Madison WI, 1989.

Yan, G.; Santos, C. "China, Forever": tourism discourse and self-orientalism. *Ann. Tourism Res.*2009, 36, 295-315.

CHAPTER 23

MARKET TRENDS AND FORECAST OF CHINESE OUTBOUND TOURISM

LI JASON CHEN, GANG LI, LINGYUN ZHANG, and RUIJUAN HU

CONTENTS

23.1 EVOLUTION OF THE CHINESE OUTBOUND TOURISM

Over the past 2 decades, the global tourism destinations have witnessed a drastic growth in Chinese visitors. As the largest outbound tourism market since 2012, China remained on the top of the global spenders' list with a tourism expenditure of US$129 billion in 2013, an increase of almost tenfold since 2000, when it ranked 7th. In fact, China has surpassed the second and third largest spenders the United States and Germany with a gap stretched to over US$42 billion in 2013 (UNWTO, 2014).

The initial development of Chinese outbound tourism was largely driven by the easing restrictions on foreign travel. Prior to 1983, Chinese citizens were only allowed to travel overseas for official, education, and business purposes. As a milestone, visiting family members in Hong Kong and Macao was allowed in 1983. By 1990, traveling on package tours organized by the Chinese Travel Service to Thailand, Singapore, and Malaysia has become possible (Arita et al., 2012; China National Tourism Administration). In 1997, the Approved Destination Status (ADS) scheme was officially recognized, allowing Chinese leisure travelers to join approved group tours to visit destinations where a bilateral agreement between the governments has been established (Arlt, 2006). By 1999, there were 9 nations on the ADS list, which has rapidly expanded to over 140 destinations by 2014. In addition to the ADS scheme, Chinese citizens have been also able to apply for individual visas subject to a string of conditions depending on the destinations. Although visa applications are still complicated, especially for residents in smaller cities and rural areas, the gradual easing of restrictions has paved the way for the surge of the Chinese outbound tourism.

Beginning in the new millennium, the mobility of Chinese people has been further bolstered by the soaring economy and a more pragmatic policy on the outbound tourism (Arlt, 2006). Since China joined the United Nations World Tourism Organization (UNWTO) in 2001, a series of regulation relaxations liberalized the tightly controlled outbound tourism market. In 2002, the quota system controlling the travel to Hong Kong and Macao was abolished. After the outbreak of severe acute respiratory syndrome in 2003, the Individual Visit Scheme was launched to allow individual tourists to visit Hong Kong and Macao. Contributed by the simplification of passport applications in 2005, the number of Chinese passport holders reached 38 million in 2012, when the e-passport was introduced (China.org.cn). Together with other simplifying procedures in obtaining foreign currencies and visas applications, the demand for overseas travel has been further stimulated.

23.2 MARKET TRENDS OF CHINESE OUTBOUND VISITORS

Beyond the dramatic surge in the visitor numbers, there have been substantial changes in the Second Wave of Chinese outbound tourism (Arlt, 2013). In comparison with the first batch, who have been usually stereotyped as tourists only visiting iconic attractions on overly scheduled package tours, the new Chinese tourists are more experienced and keen to explore unfamiliar destinations with diversified needs.

23.2.1 MORE EXPERIENCED TRAVELERS

As the First Wave of Chinese outbound travelers is becoming more seasoned, the market has gradually segmented into inexperienced and experienced travelers (Arlt, 2013). The inexperienced segment is currently predominant in the market, contributed by millions of first-time travelers every year (Mintel, 2013). As a result, most of outbound tourists still prefer traveling on package tours visiting mainstream destinations. On the other hand, the experienced travelers are inclined to accumulate unique experiences to share with peers, rather than glancing through "must-see" attractions on a packed multicountry itinerary (Arlt, 2013). The experienced segment is currently outstripped, but it is estimated to expand to one-third of all international trips and 40% of overall Chinese outbound expenditure by 2020 (Mintel, 2013).

23.2.2 IMMINENT GROWTH IN THE FOREIGN INDEPENDENT TRAVEL MARKET

Underlined by the significant growth of the experienced travelers, the foreign independent travel (FIT) market has shown its potential in the future Chinese outbound tourism. In the survey conducted by Hotels.com (2014), 67% of respondents indicate their preference for making their own travel arrangements. It is arguable that there might be a gap between respondents' preference and their actual behavior. The sampling bias in the online survey could also exaggerate the size of this market. However, the continuous increase of this percentage over the past 3 years suggests a clear trend of the FIT travel (from about 50% in 2012 to 62 in 2013 and 67 in 2014) (Hotels.com, 2012, 2013, and 2014). The increasing number of FIT travelers tend to be wealthier and traveling longer than their tour group counterparts (Hotels.com, 2014), which will divert the demand from short-haul destinations to medium and long-haul destinations and spreading further to the off-the-beaten-track destinations (Arlt, 2013).

23.2.3 DIVERSIFIED NEEDS

The coexistence of the demand for group tours and FIT travel in the foreseeable future creates different needs in the outbound travel market. While group tour goers still crave for iconic places, mainstream destinations are no longer on the wish lists of FIT tourists. Independent travelers' shopping preference has also been shifting away from mainstream brands to style (Arlt, 2013). Even within the FIT market, the independent trips are not always completely self-organized by travelers, due to obstacles such as language barriers and time cost. Parts of the arrangements such as itinerary planning, visa applications, and ticket and hotel booking are still often arranged by travel agents on a case-by-case basis. The ever-evolving and diversified needs require tourism to provide customized itinerary and personalized services.

23.2.4 USE OF TECHNOLOGY

The increasingly important role of technology in tourism, such as the Internet, mobile devices, GPS, social media, augmented reality, and online review platforms, defines the era of "smart tourism." The "connected" Chinese tourists are able to travel more confidently and independently. Although the traditional sources, including travel guide books, friends, and travel websites, still play the most important role in travel decision-making, other sources such as online review sites and social media have witnessed a steady rise over the time. Hotels.com (2014) found that about 53% of Chinese tourists are booking their accommodation either on the web or via mobile apps. The usage of mobile apps is particularly higher for young travelers aged below 35, and people prefer independent travel. On the other hand, the percentage of booking through travel agents has been falling.

23.2.5 DEMAND FOR ACCESSIBLE/BARRIER-FREE TOURISM

Many Chinese tourists have received good education; some have studied abroad. Speaking simple English is becoming less of problem. However, many tourists still encounter the language barrier when traveling abroad, especially in non-English-speaking countries. According to Hotel.com (2014,2013), more than 50% of travelers reported that in-house Mandarin-speaking staff and the ability to accept Chinese payment methods, such as China Union Pay and Alipay, are the most important offering from international hotels. Providing Chinese-specific services has also been perceived as a symbol of showing respect to the Chinese culture (Arlt, 2006), which is often expected from the tourists (Li et al., 2011). It therefore has been a phenomenon that in-house Mandarin-speaking staff are hired to engage with tourists from China. More tourism practitioners across all the tourism sectors have realized the importance of providing Chinese-specific services throughout different stages of tourist experiences, such as the offer of translated booking websites, travel guides, welcome materials, kettles, Chinese tea, Chinese TV programs, and engaging the translated online review sites and Chinese social media.

With the development of the FIT market, visa restrictions imposed by both the Chinese and foreign governments have become one of the major barriers preventing Chinese passport holders from traveling abroad. Many destinations have been keen on simplifying the procedures in visa applications. By June 2014, 49 countries and territories have agreed to provide visa-free or visa on arrival access to Chinese visitors (Ministry of Public Security of China). Destinations such as South Korea, the United States, the United Kingdom, and the Schengen region have been continuously simplifying visa applications for Chinese visitors.

23.3 FORECASTING THE DEMAND FOR CHINESE OUTBOUND TOURISM

Riding the tide of the new wave of Chinese outbound travelers, it is vital for the tourism industry to gauge the trend of visitor arrivals for tourism destinations around the world to cope with the Chinese demand in the future. Therefore, this section aims to forecast the total Chinese outbound travelers by 2020, as well as the arrivals to each of the international destinations.

23.3.1 METHODOLOGY

DATA

The data of visitor arrivals origin from China to 143 international destinations during the period of 1995 and 2012 were gathered from UNWTO (UNWTO, 2014). In line with the data source, the measures of visitors may vary across destinations, including both visitor arrivals, tourist arrivals, and their variations. The total numbers of Chinese outbound visitors during 1994 and 2013 were collected from CEIC (2014). To ensure the observations for each destination are sufficient for the modeling and forecasting purposes, when the available time series for a country or region is shorter than 8 years, the destination is dropped from the analysis. This results in 111 destinations remaining for the analysis.

TREATMENT OF MISSING VALUES

Among the 111 destinations, the observations of visitor arrivals are not always available throughout the time span between 1995 and 2012. The data for 15 destinations contain missing values in the middle of the time series. In order to use the available information as much as possible, the multiple imputation (MI) method is used to deal with the missing values. Although the MI approach was initially developed to impute the non-responses in surveys, Honaker and King (2010) extend the method to cope with time series and cross-sectional data. In contrast with single imputation methods such as mean replacement, the MI method does not "make up" information which was originally missing. Instead, it provides measurable uncertainties caused by the missing information in the form of variance of quantities.

The R package "Amelia II" (Honaker et al., 2013) is used to perform the MI procedure for each of the destinations with missing values. Following the suggestion by King et al. (2001), 5 complete data sets are generated, based on which model estimation and forecasting are carried out. The produced 5 sets of forecasts are then combined using the pooling method described by Rubin (1987) and Schafer (1997). Essentially, the pooled point forecasts are the means of results from the 5 imputations. And the combined prediction intervals account for the uncertainties from both the forecasting and the missing data in terms of the variance within each imputation and the variance between MIs.

MODEL SPECIFICATION

Exponential smoothing methods have been widely used since the 1950s. However, a big limitation of these methods is the inability to produce prediction intervals. As a remedy, scholars such as Gardner (1985) and Hyndman et al. (2002) extend exponential smoothing methods to various state space models. Based on the developments in exponential smoothing methods, Hyndman et al. (2002, 2008) propose an innovations state space framework, which performs particularity well in about 6-period-ahead forecasts. Therefore, the exponential smoothing approach under the state space framework by Hyndman et al. is employed to forecast the total Chinese outbound travelers, as well as the visitor arrivals to selected 111 destinations by 2020.

According to Hyndman et al. (2008), a linear innovations state space model can be written as Equations (1) and (2):

$$y_t = w' x_{t-1} + \varepsilon_t \tag{1}$$

$$x_t = F x_{t-1} + g \varepsilon_t , \tag{2}$$

where y_t is the observation of outbound visitors at time t; x_t is a state vector containing the unobserved components of level, trend, and seasonality; t, w, F, g are coefficients; and ε_t is a white noise series. Equation (1) is the measurement equation, describing the relationship between the unobserved states x_{t-1} and the observed variable y_t. Equation (2) is the transition equation, specifying the evolution of the states over time. The use of identical errors (or innovations) in these 2 equations makes it an innovations state space model.

Depending on the combination of different specifications of the trend component (5 versions including none, additive, additive damped, multiplicative, and multiplicative damped trends), the seasonal component (3 versions including none, additive, and multiplicative), and the error term (either additive or multiplicative errors), 30 different models can be specified.

To select the best-performing trend component, seasonal component, and error term, Hyndman et al. (2002) provide a model selection procedure based on the Akaike's Information Criterion (AIC). However, as the number of observations for each destination is relatively small, the AIC-based procedures may not have sufficient data to choose the best performing model. Therefore, Hyndman et al. (2008) recommends using the AIC to choose among the linear nonseasonal models for annual time series. Following this model selection procedure, the nonseasonal model with additive trend and additive errors (known as Holt's linear model with additive errors) is selected for most destinations. The R package "forecast" (Hyndman and Khandakar, 2008) is used to estimate the models and generate point forecasts and prediction intervals by 2020. For destinations treated with the MI method, the combined forecasts are constructed as the final results.

23.3.2 RESULTS AND DISCUSSIONS

FORECASTS OF TOTAL CHINESE OUTBOUND VISITORS

As shown in Fig. 23.1, the point forecasts of total Chinese outbound travelers are to reach 199.14 million by 2020, doubled the number in 2013. If the growth pattern remains unchanged as observed in the historical data, the number of travelers could have a 95% of chance falling into the predictive intervals between 136.36 million and 261.91 million by 2020.

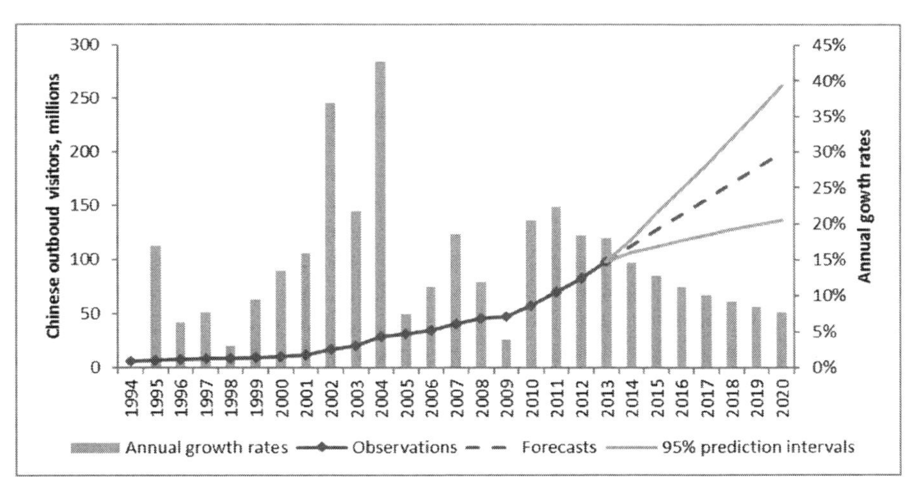

FIGURE 23.1 Actual Trend and Forecasts of Total Chinese Outbound Visitors (1994–2020).

The increase in China's gross domestic product has been gradually slowing down since 2008, yet the annual growth rate still stands well above 7.7% over the past 2 years (World Bank, 2014a). The mounting disposable income, the long-term trend of appreciation in Chinese Yuan, the expansion of transportation networks, and the easing of restrictions on overseas travel will continue to underpin the development of China as an international tourism source market.

During the forecasting period, the gross outbound departure rate (that is the number of trips made by every 100 people) over the total population is expected to essentially double from 2013 to 2020 based on the population projection by World Bank (2014b). Comparing with other tourism source markets, such as the United Kingdom and Canada, the outbound departure rate is still relatively low even by 2020, which implies a promising potential of the Chinese outbound tourism.

The growing needs of international travel have been met by easing restrictions and simplifying procedures. The ADS agreements with certain destinations, such as Australia, New Zealand, Japan and the United States, were firstly launched as a pilot program in first-tier cities before gradually spreading out to other regions in China. It thus can be expected that the outbound tourism demand from smaller cities will be stimulated by the progress of the ADS scheme. Individual visa applications are becoming easier and more straightforward. By June 2014, 49 countries and territories have agreed to provide visa-free or visa on arrival access to Chinese visitors (Ministry of Public Security of China).

Meanwhile, the glooming future of advanced economies (IMF, 2014) highlights the importance for destinations in developed countries to attract tourists, especially from the emerging markets like China. Policies such as simplifying visa applications and tailored services for Chinese visitors have been rolled out in various destinations to harvest a share of the mushrooming Chinese outbound travel market.

From 2015 to 2020, the Chinese outbound travel is predicted to increase at an average annual rate of 9.41%. Due to the specification of the linear trend in the modeling, the annual growth rates of the point forecasts gradually decline over the forecasting period. The growth trend in Chinese outbound visitors may face uncertainties and challenges from factors such as the renminbi exchange rates, inflation, airline fuel surcharge, competitions from domestic tourism market, and one-off events. In the first quarter of 2014, the Chinese Yuan has witnessed the biggest decline since 2005 when the new currency regime was introduced (BBC). As the People's Bank of China, the Chinese central bank, aims to liberalize the currency within the decade, such fluctuations are expected to occur more frequently. In the meantime, the introduction of Hainan Pilot Tax Rebate Program in 2011 and the launch of Shanghai Pilot Free-Trade Zone in 2013 may partially divert tourists' shopping expenditures to the domestic market (China Tourism Academy, 2014). Hong Kong will no longer be the only home to Disneyland attraction in the Greater China region after the launch of Disney Resort in Shanghai. The unpredictable one-off events, such as political unrests, abductions of tourists, and health hazards, will also have a significant impact on specific markets in the short run. For instance, the Beijing Olympics in 2008 partially diverted or postponed the outbound demand. The outbreak of the H1N1 virus in 2009 caused a drop of trips to a number of destinations. As the uncertainties and risks in the future may break the trend captured from the historical data, the forecasting results need to be interpreted with cautions.

FORECASTS OF CHINESE OUTBOUND VISITORS BY DESTINATION

The point forecasts and 90% and 95% prediction intervals of Chinese visitors received by 111 international destinations are projected by 2020. Given the limited space, only the top 20 international destinations are shown in Table 23.1 based on the point forecasts in 2020. The trends and forecasts of Chinese visitor arrivals to these destinations are shown in Fig. 23.2.

Sharing a similar pattern with the global outbound tourism, which sees about 80% of international travel, takes place within travelers' own regions, with about 4 out of 5 worldwide arrivals originating from the same region. Among the top 10 international destinations for Chinese travelers, most are within Asia except the United States and France, while more European destinations can be found from the top 20 list. Hong Kong, Thailand, and Macao are predicted to be the top 3 most visited destinations by 2020. The majority of the top 20 destinations in 2012 would remain attractive to Chinese travelers by 2020. As the only exceptions, Philippines and Italy, which were ranked in 19 and 20 positions, respectively, in 2012, would drop out of the 2020 list and be replaced by Nigeria and Maldives. Taiwan was ranked 5th among the global destinations in 2012, but it is not included in this study since the observations for Taiwan are insufficient for the forecasting. As a destination newly opened to independent Mainland Chinese tourists, Taiwan has seen its tourist arrivals skyrocketed from 0.86 million in 2009 to 2.46 million in 2012, at an average growth rate of 41.89% (UNWTO, 2014). Since residents of more Chinese cities will be allowed to visit Taiwan, Taiwan is very likely to see itself as one of the major destinations by 2020.

TABLE 23.1 Top 20 International Destinations of Chinese Outbound Visitors in 2020

Destination	Point Forecast (000s)	Lower Limit of 95% Prediction Intervals (000s)	Upper Limit of 95% Prediction Intervals (000s)	Average Growth Rate (2015-2020)	Rank in 2012[a]
Hong Kong	22,276	18,346	26,206	4.59%	1
Thailand	11,079	4,113	18,046	13.59%	4
Macao	10,193	6,628	13,759	7.30%	2

South Korea	7,385	4,288	10,482	10.32%	3
The US	4,532	2,405	6,659	11.64%	7
Singapore	3,023	1,799	4,248	4.68%	5
Malaysia	2,419	1,381	3,457	5.39%	6
France	2,142	1,504	2,781	5.05%	10
Vietnam	1,978	1,022	2,933	3.88%	8
Japan	1,913	813	3,014	3.89%	9
Indonesia	1,575	1,006	2,144	9.18%	13
Switzerland	1,559	381	2,736	10.55%	15
Australia	1,117	734	1,500	6.63%	14
Austria	1,112	270	1,955	11.73%	16
Russia	1,068	927	1,209	1.99%	11
Cambodia	995	228	1,762	11.33%	17
Germany	965	679	1,251	2.93%	12
Nigeria	742	332	1,152	12.56%	23
Maldives	578	179	977	9.92%	21
Canada	567	310	825	7.62%	18
Total outbound	199,139	136,364	261,914	9.41%	–

[a]The ranks in 2012 exclude Taiwan. In line with UNWTO (UNWTO, 2014), the measures of travelers may vary across destinations (i.e., tourist arrivals at national borders for Hong Kong, Thailand, the United States, Malaysia, and France; tourist arrivals in hotels and similar establishments for Macao; visitor arrivals at national borders for South Korea, Singapore, Vietnam, and Japan.) To be consistent, they are all referred as travelers or visitors in this study.

Compared with the ranking in 2012, Thailand is predicted to move up 2 places to overtake Macao's position as the second most popular destination for Mainland Chinese by 2020. Both the United States and France are expected to rise up 2 spots to 5th and 8th, respectively, in the raking by 2020. Russia and Germany may slightly lose their relative popularity to Chinese tourists, whose forecasts are mainly pulled down by the slip in 2008 and 2009. The ranks for the rest of the top destinations are generally consistent with their ranks in 2012.

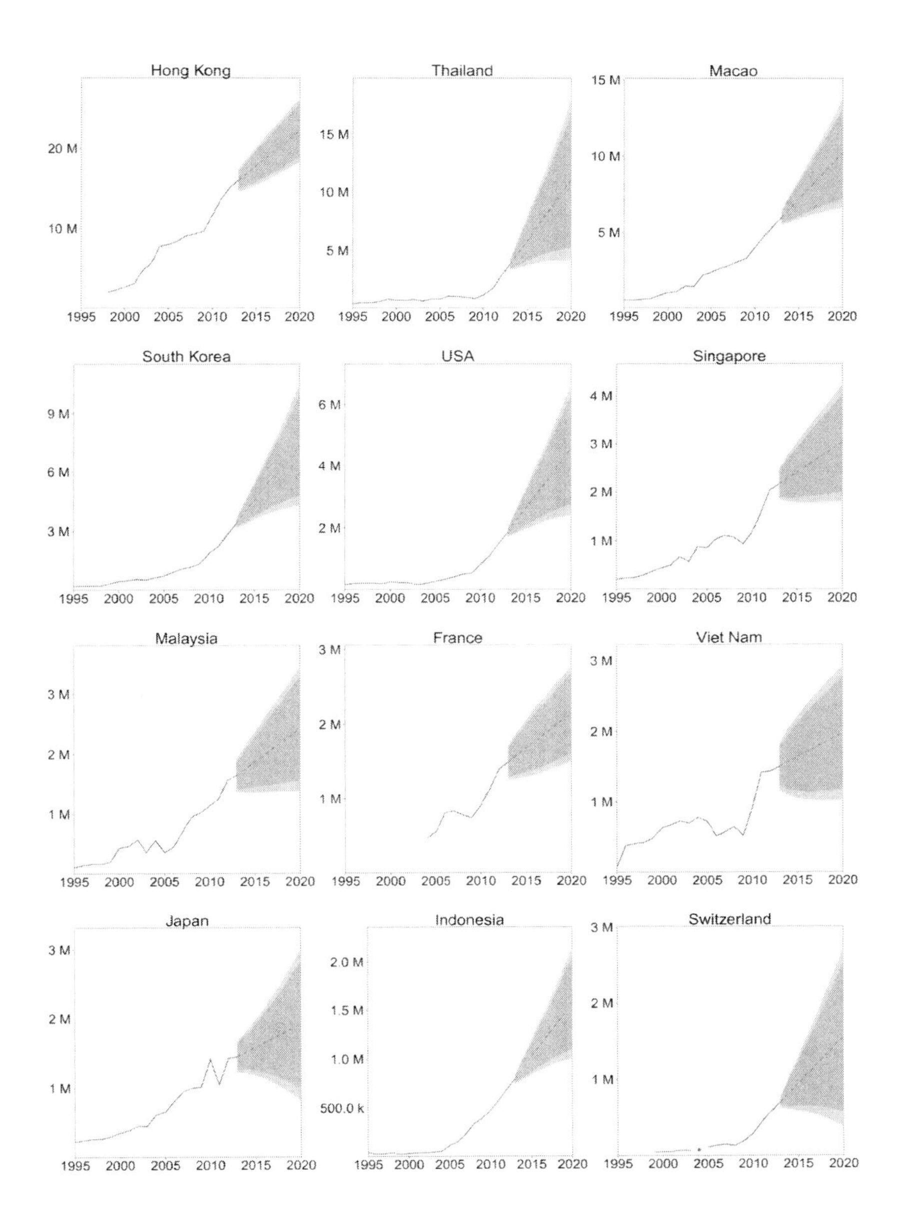
Hong Kong
20 M
10 M
1995 2000 2005 2010 2015 2020
Thailand
15 M
10 M
5 M
1995 2000 2005 2010 2015 2020
Macao
15 M
10 M
5 M
1995 2000 2005 2010 2015 2020
South Korea
9 M
6 M
3 M
1995 2000 2005 2010 2015 2020
USA
6 M
4 M
2 M
1995 2000 2005 2010 2015 2020
Singapore
4 M
3 M
2 M
1 M
1995 2000 2005 2010 2015 2020
Malaysia
3 M
2 M
1 M
1995 2000 2005 2010 2015 2020
France
3 M
2 M
1 M
1995 2000 2005 2010 2015 2020
Viet Nam
3 M
2 M
1 M
1995 2000 2005 2010 2015 2020
Japan
3 M
2 M
1 M
1995 2000 2005 2010 2015 2020
Indonesia
2.0 M
1.5 M
1.0 M
500.0 k
1995 2000 2005 2010 2015 2020
Switzerland
3 M
2 M
1 M
1995 2000 2005 2010 2015 2020

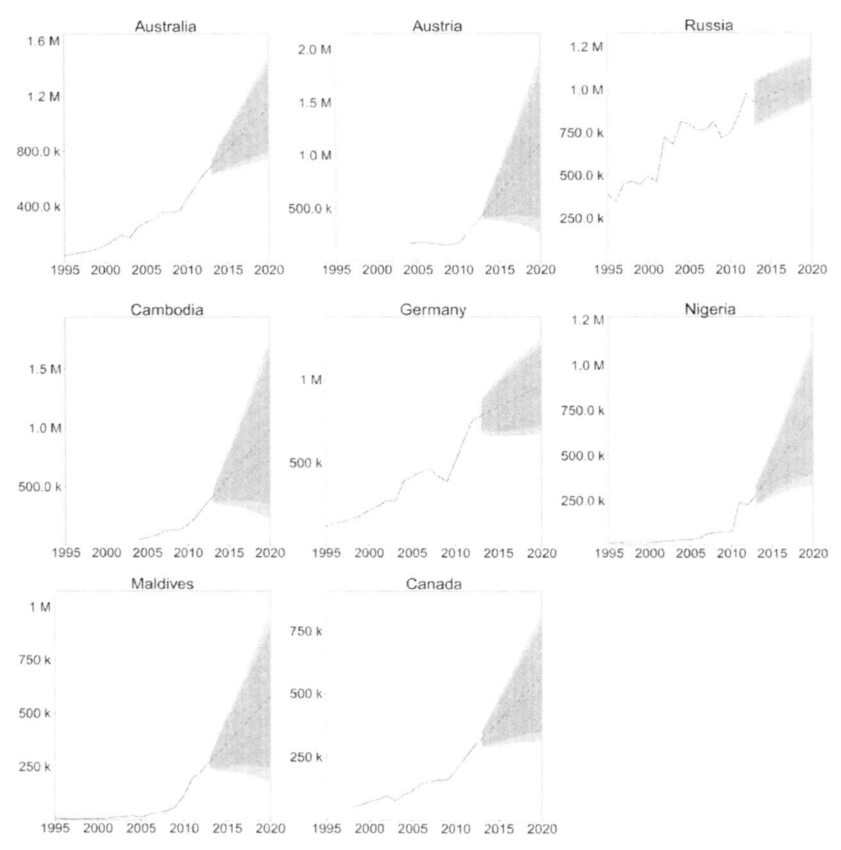

FIGURE 23.2 Top 20 Destinations for Chinese Outbound Travelers – Actual Trends, Point Forecasts, 90% and 95% Prediction Intervals (1995–2020).

ASIA AND THE PACIFIC

Hong Kong is predicted to remain on top of the list of popular destinations by attracting 22.28 million Chinese tourists in 2020, with a compound growth rate of 4.59% annually between 2015 and 2020. The 95% prediction intervals suggest that the number could go from 18.35 million up to 26.21 million. Thailand and Macao are to take the 2nd and 3rd places in Asia, as well as over the world by 2020, with Chinese tourist arrivals of 11.08 million and 10.09 million, respectively. Although it attracted only 439 visitors from China in 2008, Palau is likely to be the fastest growing global destination for Chinese tourists by 2020. Its visitor arrivals have soared tenfold in 4 years to 4,471 in 2012, thanks to the visa-on-arrival arrangement launched in 2009. If the trend continues, it is expected to receive 26.65 thousand Chinese travelers by 2020 with an average growth rate of 15.82% between 2015 and 2020. Swelling annually at a compound rate of 13.59% from 2015 to 2020, Thailand sees itself as the 2nd largest destination, as well as the second fastest growing destination for Chinese visitors in Asia Pacific, followed by Bhutan with a growth rate of 12.62% yearly.

EUROPE

As the top tourism destination in the world, France is estimated to remain in its position as the most visited European destination by Chinese travelers, ranking the 8th among all international destinations by 2020. France is followed by Switzerland and Austria, receiving 1.56 million and 1.11 million Chinese visitors by 2020, respectively. By ranking the growth rates in European destinations, Bosnia and Herzegovina topped the list with an average annual rate of 13% between 2015 and 2020. Other fast-growing destinations in Europe include Slovenia, Portugal, and Austria, with a growth rate of 11.89%, 11.86%, and 11.73%, respectively.

THE UNITED STATES OF AMERICA

Ranked the 5th amongst all international destinations for Chinese visitors, it is predicted that 4.53 million Chinese tourists will make their trips to the United States by 2020, making the United States the top destination among its American counterparts, distantly followed by Canada (0.57 million) and Brazil (0.14 million). In terms of growth rates, Paraguay is expected to be the top American destination attracting Chinese visitors with a compound rate of 14.96% per year. This is followed by Cayman Islands, Bahamas, and the United States, with a growth rate of 13.18%, 12.02%, and 11.64%, respectively.

AFRICA

In the forecast results, no African destinations have yet shown its popularity by reaching the 1-million milestone in the market of Chinese outbound travel. The projections suggest that Nigeria, South Africa, and Angola are likely to be the most visited destination in Africa, receiving 741.82 thousand, 476.32 thousand, and 103.76 thousand Chinese arrivals by 2020, respectively. However, as far as the growth rates are concerned, the potential can be observed from destinations such as Seychelles, South Africa, Nigeria, and Mauritius, which are expected to be growing at a faster rate of 15.10%, 13.01%, 12.56%, and 11.04%, respectively.

MIDDLE EAST

Due to the ongoing unrest in some destinations of the region, the growth of Chinese visitors to Middle East is expected to be relatively modest. It is estimated that the top 3 destinations for Chinese travelers are Egypt, Saudi Arabia, and Bahrain, receiving 87.89 thousand, 48.38 thousand, and 47.06 thousand visitors by 2020 from China, respectively. In terms of average annual growth rates between 2015 and 2020, Bahrain (6.70%), Jordan (4.82%), and Saudi Arabia (4.77%) are predicted to be the top performers in the Middle East region.

One has to be cautious when interpreting the forecasting results due to the long forecasting horizon compared with the relatively short data span, as well as the existence of missing values. The 90% and 95% prediction intervals are relatively robust forecasts, as the forecasting error and the uncertainties from MIs are accounted for.

23.4 CONCLUSIONS

The emerging trends of Chinese outbound visitors impose enormous opportunities and new challenges to the global tourism industry. The potential increase of independent travelers means more tourism businesses in both China and international destinations are getting directly involved with Chinese visitors, who demand tailored services to go farther and experience deeper from their distinctive trips. The emerging FIT market opens up opportunities for off-the-beaten-track destinations to share the benefit from the growing Chinese outbound travel. The era of smart tourism features the wide use of social media and mobile devices, which requires the industry to promptly adapt innovative E-business models from digital marketing and online booking to personalized customer engagement on social media and mobile platforms.

On the other hand, as pointed out by Arlt (2006), group tourism may still remain as a major segment of outbound tourism in a group-orientated society such as Japan and Taiwan, even if all the restrictions and barriers were lifted. The diversifications in market segmentation and tourist needs create challenges for tourism companies to achieve economies of scale, as the one-size-fits-all strategy will no longer work (Mintel, 2013). It requires the industry to develop innovative and tailored tourism products to meet different needs. Entering the new era, more destination management organizations and industry participants will be enticed by higher margins in the FIT market over the packaged tours and will join the battle on the increasingly competitive landscape to capitalize the next wave of FIT and smart tourists from China.

REFERENCES

Arita, S.; Croix, S. L.; Mak, J. How Big? The Impact of Approved Destination Status on Mainland Chinese Travel Abroad; Working Paper 2012–13; University of Hawaii Economic Research Organization, University of Hawaii: Manoa, 2012.

Arlt, W. China's Outbound Tourism; Routledge: London, 2006.

Arlt, W. G. The second wave of Chinese outbound tourism. Tourism Plann. Dev. 2013, 10, 126–133.

BBC. The Curious Case of China's Falling Yuan. [Online]; BBC News, February 28, 2014. http://www.bbc.co.uk/news/business-26385213 (accessed Jun 23, 2014).

CEIC. Chinese Outbound Visitors by Country [Online]; CEIC China economic & industry data database, 2014. https://www.ceicdata.com/en/countries/china (accessed Jun 10, 2014).

China National Tourism Administration. The Official ADS Destinations. [Online]; China National Tourism Administration, May 13, 2009. http://www.cnta.gov.cn/html/2009-5/2009-5-13-10-53-54953.html (accessed Jul 1, 2014).

China Tourism Academy. Annual Report of China Outbound Tourism Development 2014; Tourism Education Press: Beijing, China, 2014.

China.org.cn. China Launches Electronic Passports. [Online]; China.org.cn, May 16, 2012. http://www.china.org.cn/china/2012-05/16/content_25392418.htm (accessed Jul 2, 2014).

Gardner, E. S. Exponential smoothing: The state of the art. J. Forecast. 1985, 4, 1–28.

Honaker, J.; King, G. What to do about missing values in time-series cross-section data. Am. J. Pol. Sci. 2010, 54, 561–581.

Honaker, J.; King, G.; Blackwell, M. Amelia II: A program for missing data, 2013. Journal of Statistical Software 2011, 45(7), 1–47.

Hotels.com. Chinese International Travel Monitor 2012; [Online]; Hotels.com, 2012. http://press.hotels.com/en-gb/files/2012/07/Hotels.com-Chinese-International-Traveller-MapCITM.pdf (accessed Jul 24, 2014).

Hotels.com. Chinese International Travel Monitor 2013; [Online]; Hotels.com, 2013. http://press.hotels.com/citmcn/files/2013/08/cs4-1-9a.pdf (accessed Jul 24, 2014)

Hotels.com. Chinese International Travel Monitor 2014 [Online]; Hotels.com, 2014.http://press.hotels.com/content/themes/CITM/assets/pdf/CITM_UK_PDF_2014.pdf (accessed Jul 24, 2014).

Hyndman, R. J.; Khandakar, Y. Automatic time series forecasting: The forecast package for R. J. Stat. Softw. 2008, 27, 1–22.

Hyndman, R. J.; Koehler, A. B.; Ord, J. K.; Snyder, R. D. Forecasting with Exponential Smoothing the State Space Approach; Springer: Berlin, 2008.

Hyndman, R. J.; Koehler, A. B.; Snyder, R. D.; Grose, S. A state space framework for automatic forecasting using exponential smoothing methods. Int. J. Forecast. 2002, 18, 439–454.

IMF. World Economic Outlook April 2014: Recovery Strengthens, Remains Uneven; World Economic and Financial Surveys; IMF: Washington, DC, 2014.

King, G.; Honaker, J.; Joseph, A.; Scheve, K. Analyzing incomplete political science data: An alternative algorithm for multiple imputation. Am. Pol. Sci. Rev. 2001, 95, 49–69.

Li, X.; Lai, C.; Harrill, R.; Kline, S.; Wang, L. When east meets west: an exploratory study on Chinese outbound tourists' travel expectations. Tourism Manage. 2011, 32, 741–749.

Ministry of Public Security of China. Destinations with Visa-Free and Visa-on-Arrival Access. [Online]; Ministry of Public Security of China, Posted June 27, 2014. http://www.mps.gov.cn/n16/n84147/n84196/4069471.html (accessed Jul 1, 2014).

Mintel. China Outbound – May 2013; [Online]; Mintel Group Ltd., 2013. http://reports.mintel.com/display/718321/ (accessed Jul 21, 2014).

Rubin, D. B. Multiple Imputation for Nonresponse in Surveys; Wiley: New York, 1987.

Schafer, J. L. Analysis of Incomplete Multivariate Data; Chapman & Hall/CRC: Boca Raton, FL, 1997.

UNWTO. Outbound Tourism Data (Calculated on the Basis of Arrivals Data in Destination Countries); Dataset; UNWTO: Madrid, 2014.

UNWTO. UNWTO Tourism Highlights, 2014 ed.; UNWTO: Madrid, 2014.

World Bank. GDP Growth [Online]; World Bank National Accounts Data, 2014a. http://data.worldbank.org (accessed Jul 1, 2014).

World Bank. Population Estimates and Projections [Online]; World Bank Health, Nutrition Population Statistics, 2014b. 2014 http://datatopics.worldbank.org/hnp/Population.aspx (accessed Jul 1, 2014).

INDEX

D

E

F

G

H

I

J

K

L

M

N

O

P

Q

R

S

T

Y

Z